APPROACHES to ECONOMIC DEVELOPMENT

To Lidiya and Teagan from Mom
and to Madeline

APPROACHES to ECONOMIC DEVELOPMENT

Readings from *Economic Development Quarterly*

John P. Blair / Laura A. Reese

Editors

SAGE Publications

International Educational and Professional Publisher

Thousand Oaks London New Delhi

For information:

SAGE Publications, Inc.
2455 Teller Road
Thousand Oaks, California 91320
E-mail: order@sagepub.com

SAGE Publications Ltd.
6 Bonhill Street
London EC2A 4PU
United Kingdom

SAGE Publications India Pvt. Ltd.
M-32 Market
Greater Kailash I
New Delhi 110 048 India

Printed in the United States of America

Library of Congress Cataloging-in-Publication Data

Main entry under title:

Approaches to economic development: Readings from *Economic Development Quarterly* /
edited by John P. Blair and Laura A. Reese.
p. cm.
Includes index.
ISBN 0-7619-1883-3 (acid-free paper)
ISBN 0-7619-1884-1 (pbk. : acid-free paper)
1. United States—Economic conditions—1981- 2. Economic development.
I. Blair, John P., 1947- II. Reese, Laura A. (Laura Ann), 1958-
III. Economic development quarterly.
HC106.8 .A66 1998 98-40126
338.9—ddc21

This book is printed on acid-free paper.

01 02 03 04 05 7 6 5 4 3 2

Acquiring Editor: Catherine Rossbach
Editorial Assistant: Caroline Sherman
Production Editor: Astrid Virding
Editorial Assistant: Nevair Kabakain
Typesetter/Designer: Rose Tylak/Marion Warren
Cover Designer: Candice Harman
Copyeditor: Joyce Kuhn
Indexer: Paul Corrington

Contents

❋❋❋

Acknowledgments

❀❀❀

The quality of this book was greatly enhanced by suggestions from a variety of individuals widely recognized and respected in the field of economic development. The following individuals consulted in the selection of articles and the organization of the book:

Greg Andronovich, University of California, Long Beach
Bob Beauregard, The New School, New York
Dick Bingham, Cleveland State University and Founding Editor, *EDQ*
Virginia Carlson, University of Wisconsin, Milwaukee
Margaret Dewar, University of Michigan
Sabina Dietrick, University of Pittsburgh
Edward Hill, Cleveland State University and Editor, *EDQ*
Mike Peddle, Northern Illinois University

Special thanks are also due Catherine Rossbach, of Sage, for bringing the project to a quick and successful conclusion.

Introduction

❋❋❋

During the 1980s and 1990s, the skills and knowledge expected of economic development (ED) professionals increased significantly. Jerold Thomas listed over 25 skills needed by an economic developer in five different categories: leadership, theoretical, practical, political, and change skills.[1] The greater knowledge needs of ED practitioners and the increased importance of ED to localities have contributed to an emerging partnership between academia and ED professionals. The readings in this book have their origins in that partnership.

The first part of this introduction describes factors that have contributed to the closer relationship between the academic and applied sides of ED. The second part explains why the readings selected for this book are important additions to the knowledge base needed by ED professionals.

The Emerging Bridge

The current demand for ED practitioners (and thus the interest in studying the subject) may be traced to political and economic changes in the 1980s. During that period, many individuals were experiencing dislocations due to deindustrialization, pressures to increase net corporate earnings, and globalization. At the same time, the political climate shifted to the right. The ability of private markets to produce optimal outcomes was trumpeted while the efficiency of federal government involvement in the economy was suspect. To the extent that there was a recognized federal economic responsibility, it was to improve the national economy. Tilting economic policy to favor one region over another was impossible to avoid in practice but was eschewed in conservative political ideology. Instead, the strategy suggested by the slogan, "a rising tide lifts all boats," dominated policy making. Local ED activities that were once influenced by significant federal involvement "devolved" to states and localities. However, the shift in emphasis was not a radical change because localities have historically played a major role in the development of their local communities dating at least to the great canal and railroad building period starting in the early 1800s.

State and local ED activities are also a form of government involvement in the economy and thus violate the spirit of laissez-faire. Nevertheless, the perceived rewards to localities from channeling economic activity were too large to be deflected by strict free market ideology. Politicians of all parties felt a need to strengthen their local economies. The need for local development efforts was particularly strong in areas with high unemployment or underemployment, particularly in urban areas. When asked what was being done about important economic issues, office seekers needed an answer. Consequently, states and localities expanded ED efforts, regardless of political party dominance. Currently, the bulk of these efforts continues to be at the state and local level in the United States.

The local emphasis on ED contributed to a growing demand for ED specialists and encouraged many communities to require that other officials such as city managers, finance directors, and zoning officials have some knowledge concerning ED. Often, the ED person serves in other official capacities, particularly in small communities.

As ED has become an increasingly important, accepted, and complicated function of local governments, the knowledge that practitioners require has increased, forming the foundation for the bridge between the ED profession and academia. Local ED efforts were once overwhelmingly a combination of public relations and industrial recruitment. Even today, boosterism and smokestack chasing remain important activities. However, more is currently expected of ED practitioners. ED practice has evolved more sophisticated strategies and techniques, stimulated partly by sharper intergovernmental competition and the greater use of business incentives. Knowledge of legal, economic, social, environmental, and political factors must be combined for an effective ED program to be formed.

The complexities of planning have further tied ED to academic analysis. ED is currently recognized as an *ongoing process* that even prosperous communities need. It is necessary not only to stimulate new activity but to assist in the regeneration of jobs. As part of this ongoing process, encouragement of an atmosphere that meets the needs of existing local firms and allows new entrepreneurial businesses to germinate has become as important as direct subsidies but is more subtle in design and implementation. Accordingly, the preparation of comprehensive ED strategies for particular communities has emerged as a key part of the development process.

Planning has led officials to think more rigorously about what they want from ED. Job creation and fiscal improvement remain the most important ED goals, but they now share the stage with a variety of other objectives. Currently, ED officials must consider various quality-of-life goals, special needs of special groups, neighborhood revitalization, equity concerns, property values, open citizen participation, and so forth. The wider scope of ED has also contributed to the greater educational needs of practitioners.

Recognizing the greater responsibilities and knowledge requirements of ED practitioners, the public and private agencies that hire economic developers seek individuals with appropriate college education and professional designations. To accommodate educational requirements, universities have initiated course work and formal concentrations in ED. Working with universities to accommodate needs of experienced professionals, two professional organizations have been created: the American Economic Development Council and the National Council of Urban Economic Development.

On the academic side, three factors stimulated increased university interest in ED. First, knowledge of ED has been an asset to students seeking jobs. Public officials are increasingly required to be knowledgeable in ED even if their primary job is not ED. Second, many universities have embraced community outreach as part of their missions, and ED has been useful in those efforts. Many states have provided universities additional funds for institutes and centers that provide assistance to community ED efforts. In an era of tight budgets, universities have been anxious to attract additional money. These centers have contributed to the development of applied scholarship in ED and have helped provide practitioners with better conceptual and quantitative tools. Third, practitioners have improved their capacity to take advantage of academic analysis, increasing the market for academic research in the field. Currently, about 90% of practitioners hold a college degree, and nearly half have advanced degrees.[2] Not only do practitioners constitute an audience for academic writings, but they are writing about their experiences and informing academics about real-world issues. Thus, the bridge goes two ways and both academics and ED officials have benefited.

The response of universities to the educational and research needs of the ED profession has been positive but slow. One reason for the delayed response was that until the past couple of decades the practice of local ED was less sophisticated. There was only a small audience of practitioners who cared to read academic perspectives on local ED or take advanced course work. But as the profession has evolved, the practitioner-academic dialogue has strengthened.

Another reason why universities have only in the past 20 years been addressing concerns of state and local practitioners is that ED was considered an anomalous problem in the United States. ED concerns were focused on the poor, Third World countries. The United States was seen as a prosperous country with "pockets of poverty," such as inner cities, and isolated rural areas, such as Appalachia. During the 1960s and early 1970s, local ED was seen as a concern limited to *The Other America.*[3]

The interdisciplinary nature of local ED was a third obstacle to its growth in academia. Universities, like many bureaucracies, sometimes find it difficult to launch new ventures that disturb established turf and have no clear administrative home. Universities change slowly. That may be good because it reduces resources devoted to the study of trendy, but ultimately less meritorious, subjects. However, this same caution about course expansion and subjects that crossed disciplinary lines operated to slow the acceptance of ED in universities.

The Plan of This Volume

We envisioned this book, first and foremost, for advanced undergraduate and graduate courses that either include ED as a major portion of the class or focus entirely on ED. Teachers, practitioners, and researchers who wish to stay abreast of the field will also find this volume useful. The articles presented in this book, all selected from the journal *Economic Development Quarterly* (*EDQ*), illustrate some of the best writings that have bridged the academic/application gap and continue to be relevant to ED practice today.

EDQ, founded with the goal of encouraging exchange between practice and research in ED, was established in 1987 with Richard D. Bingham, Gail Garfield Schwartz, and Sammis B. White as editors. Bingham played a leading role and was later designated Founding Editor. The tradition of bridging academic research and practice has continued under the leadership of the current editor, Edward W. Hill. The contents of the journal are evidence of its success in bridging the academic/practitioner gap. Both academics and ED professionals write for and read *EDQ*. It is recognized by scholars, practitioners, and teachers as the premier journal in the field of ED.

A necessary condition for inclusion of an article in this book is classroom suitability. The articles may not represent "the best" of *EDQ* from an academic or technical standpoint, but they do in a pedagogical sense. With this as the overriding goal, there were three additional article selection criteria:

1. They have influenced the way ED issues are conceptualized, taught, and practiced. In short, they have made a significant contribution to the field of ED and deal with topics that are essential to learning about and practicing in the field.
2. They are readable and understandable to undergraduate students as well as to in-service graduate students. Avoided were articles that assume knowledge of methods that cannot be explained by an instructor within the course of a 50-minute lecture or that approach an issue that cannot be placed in context in the natural flow of a course.
3. They address "real-world" issues useful in practice and so aid students in placing the course work in context: in-service students in particular will be more responsive to and engaged by such pieces.

A word about the process used for selecting articles for this volume is also useful. Based on the three criteria above and the overriding goal of providing a set of readings that serves the pedagogical goals of teaching ED, we individually reviewed all past issues of *EDQ*. Two separate lists of about 20 articles each were compiled and compared. Those appearing on both lists were immediately included in the volume, and discussions and negotiations ensued over the remaining articles. The final list of articles was then distributed to a number of other scholars in the ED field for review. The extent of consensus was notable, and we revised the final volume to reflect reviewers' comments.

The interdisciplinary nature of ED practice is apparent from the variety of articles selected. We as editors are also an interdisciplinary team and contributed equally to this volume. One of us is an economist (Blair) and the other a political scientist working in a department of planning and geography (Reese).

The overall plan or design of the reader flows from the selection process described above. Essentially, there are two ways to approach such a task: deductively, operating from themes in ED that should be presented, and inductively, selecting articles and than arranging them in a way that makes sense. Obviously, the latter approach was employed here for two reasons. First, we wanted the substantive and pedagogical quality of the articles to drive the process, and second, selecting articles to address particular predetermined themes proved problematic. For example, reviewers of the volume correctly noted that we do not include such important issues as globalization and race/gender in ED as central themes. The fact that the volume does not explicitly focus on these issues, in the form of a category of articles, is in no way meant to imply that these are not critical to understanding ED. Rather, across the history of *EDQ,* only a relatively small number of articles have focused on such topics, a clear indication of where future research is needed.

The articles presented in the form of chapters in this volume reflect five major themes from the *EDQ* literature: (a) strategic paradigms that frame thinking in the field; (b) descriptions of, and recommendations for, actual policies and programs; (c) consideration of neighborhoods and social equity; (d) state level and regional issues; and (e) planning for and evaluating ED policies and the political environment in which all the former are embedded.

Part I on strategic paradigms contains articles that have influenced the parameters of thought about ED policy. They suggest general designs for policies and provide a context for more specialized articles that follow. Although these articles do not exhaust the possibilities, the spirit of the analysis may encourage readers to see the possibilities that apply to their own communities.

Each of the articles in Part I encourages ED planners to sharpen the target of policy intervention. Thompson says that traditional industrial categories employed to target business are too broad. There are many types of economic activity within an industry. For example, both textiles and advanced imaging industries have research components. Therefore, strategies should better target the kinds of economic activities within an industry a community seeks to attract. Bartik elaborates on the traditional economic approach to ED. He implies that government programs should correct market failures and then allow the market to operate. Unlike Thompson, Bartik does not suggest targeting specific activities. Kasarda is concerned with the mismatch between the job skill requirements and the skills of many urban residents, particularly the underclass. He suggests strategies to overcome the structural unemployment problem. As the pace of change quickens, it is likely that structural unemployment will become an increasing problem. Porter is also concerned with inner-city development, making what has become a rather controversial case that market forces within the inner city are sufficient to sustain development. Thus, he argues that, although government and community assistance and development programs are useful, long-range development will depend on the investment of business within the inner city to take advantage of "unrecognized and untapped" demand and economic resources.

The policies and programs described in Part II are less abstract than strategic paradigms, and so the articles presented are more specific in potential application and employ real-world examples to make the policy approaches more concrete. Each policy may be a useful tool that contributes to more than one of the strategic paradigms described above.

Ledebur and Woodward as well as Ihlanfeldt suggest improvements in the use of financial incentives (subsidies) that have been used to attract and retain business. Ledebur and Woodward are concerned that communities might not receive the jobs, taxes, and other benefits that they were promised when they agreed to provide assistance to businesses, whereas Ihlanfeldt is concerned about potential inefficiencies and unfairness that ED subsidies may create. Ranney and Betancur describe an alternative approach to labor force development, using brief case studies to illustrate programs in action. Their concern with labor force development dovetails well with Karsarda's analysis in Part I. Doeringer and Terkla suggest that economic developers need to understand better the ties that cause some businesses to cluster in an area. They focus on why activities tend to locate together and suggest specific steps to improve the identification and understanding of such clusters.

ED decisions frequently center on the plight of neighborhoods, particularly low- and moderate-income neighborhoods, and this is the focus of Part III. The concern with neighborhoods is useful not only in assisting individuals in those areas where help is most needed but in controlling and reversing physical deterioration that could spark a vicious downward cycle.

Wiewel, Brown, and Morris describe mechanisms, including the role of markets, investment, corporate decisions, and politics, that link ED in neighborhoods to larger regional events. They also provide an excellent review of important theories of neighborhood change. Hill and Bier, in a pathbreaking work, also address neighborhood change by tracing the impact of national economic events on neighborhood change. Their analysis helps explain why neighborhood development plans may be disrupted by forces over which the practitioner has little control. In a similar vein, Nowak notes the limitations of community development efforts posed by the regional economy and suggests job creation and workforce strategies that address the reality that neighborhoods are intricately tied to the larger region of which they are a part. Thus, appropriate linkages are more important than "self-containment" or "self-sufficiency" of neighborhoods. Blair and Endres describe the importance of the underground economy in distressed neighborhoods. The article points to problems with over- reliance on official statistics by local officials and shows that many unrecognized assets may be found in poor neighborhoods. Planners may utilize these hidden assets as part of a larger neighborhood revitalization effort.

The operation of ED programs in a decentralized governmental system is the topic of Part IV. Federal, state and local governments all participate in local ED efforts. Together the articles provide a picture of state-level activity, intercity competition, and opportunities for cooperation. They describe complexities that local ED policies encounter when applied in a decentralized governmental system.

Clark and Gaile describe the initiatives taken by local governments during the 1980s as a result of federal cutbacks and suggest that American cities have developed a high-risk, "entrepreneurial" strategy. Although most of the focus of Part IV is on local governments, states play a large role in the development process. They enable local governments, and they have significant programs of their own. Eisinger shows that state ED programs are in flux. They are reducing efforts in some areas and possibly returning to an industrial recruitment strategy. He concludes that political factors rather than program efficiency explain the changes. Goetz and Kayser explore intergovernmental competition at the subregional level. This competition can affect local governments, causing them to offer excessive incentives to attract firms because cities fear the firm may locate in another nearby jurisdiction. The authors suggest that cooperation among governments in some spheres may improve the welfare of the region but that elimination of intergovernmental competition is unlikely. Nelson, Drummond, and Swicki examine the issue of exurban industrialization, the process of industry moving to the outer edge of the urban fringe. In some instances, this process encourages inefficient use of infrastructure and poses a challenge for intergovernmental cooperation.

Part V provides a sense of the political environment in which ED policies are framed, considered, planned, implemented, and evaluated. It then moves to more explicitly address issues in choosing and planning policies and evaluating those policies once in place.

Wolman and Spitzley provide a meta-analysis of the extant literature on the "politics" of ED, and through the description of many transmission mechanisms, they show how politics permeates the ED process. Yet they also illustrate how our knowledge of such processes remains limited. Rubin provides a more specific focus on the environment in which ED professionals operate and how this affects policy and implementation decisions. Moving to planning and evaluation, Reese and Fasenfest discuss the challenges to, and the "politics" of, evaluating ED policies and argue that academic and practitioner evaluations should include a broader conception of the "goals" of development policies. Finally, in a more applied sense, Marvel and Shkurti present an evaluation of the economic impact of Honda Motor Company on the state of Ohio, asking whether the benefits justified the costs of the subsidies. They provide a useful application of cost/benefit analysis that is transferable to other projects and situations.

As a group, the articles provide the student of ED a sense of the breadth and depth of existing knowledge and research in the field of primarily local ED. They are interdisciplinary and speak both to academics and practitioners. Again, although they are not meant to represent "the best" of all articles published in *EDQ,* they provide an essential framework for learning about and understanding the field and provide the base from which critical classroom discussion and elaboration can emanate. Finally, they also show where our knowledge is limited and where future research, by both the academic and practitioner communities, can fill important gaps. This issue is revisited at the end of the reader.

Notes

1. Jerold R. Thomas, "Skills Needed by the Economic Developer," *Economic Development Review* 13 (1995): 9-13, at p. 9.

2. Richard D. Bingham and Robert Mier, *Dilemmas of Urban Economic Development: Issues in Theory and Practice* (Thousand Oaks, CA: Sage, 1997), p. xi.

3. Michael Harrington, *The Other America: Poverty in the United States* (New York: Macmillan, 1962).

PART I

Strategic Paradigms

Chapter 1

Policy-Based Analysis for Local Economic Development

Wilbur R. Thompson

❋❋❋

Local economies, especially, need to distinguish carefully between what they make and what they do. Dominated by the ideas of economists, the emphasis in local economic development has been on the local *industry-mix*—what a place makes. But growing concern over "sunset occupations" (e.g., machine operators and assemblers) and the readily available census data cross-classifying industries and occupations should have helped lead scholars and practitioners to pay more attention to the local *occupation-mix*—what a place does. Seeing clearly the occupational-functional role of a place—research, innovation, decision making, production—opens up a perspective that suggests more policy ideas and provides more program handles than the industry approach, especially to state and local governments, more involved in education, human relations, and city planning.

The Conventional Industry Approach

I argued, decades ago, that "a city is a bundle of industries in space—tell me your industries and I will tell your fortune." The local industry-mix is still a good place to begin local analysis for development planning. But if local analysts and strategists are to get beyond chasing raw growth and progress into the greater sophistication of planning for *economic development,* then they must see beyond "growth industries" and look into the many less dramatic characteristics of the many and diverse industries, both the ones they now have and alternative industries in the national pool.

In another place, the growth rates, cycle patterns, wage rates, concentration ratios foreign trade position, fixed assets, female-to-male employment ratios, energy and water use, and the occupation-mix of 150-odd manufacturing industries have been carefully profiled.[1] Space prevents a full recapitulation, but that data will be drawn on here, selectively and in context. Local industry analysis for targeting can be much more than it has been.

But even better industry analysis is not enough. It takes two cross hairs to hit a target. Local competitive (comparative) advantage is the other major line of sight needed. Develop-

Reprinted from *Economic Development Quarterly,* Vol. 1, No. 3, 1987.

ment strategists must compare and trade off between the industries they need most and the work they can do best—better than competing places. The easiest and perhaps the best way to cross over from targeting industries to evaluating local competitive strengths (weaknesses) is through a cross-classification of industries and occupations, assuming as we do here that human resources are the key determinant of local economic development.

From Industries to Occupations

Each industry assembles a distinctive occupation-mix that is the single most powerful determinant of the current level of local income. But, just as important, the local profile of professional and skilled workers largely determines the current level of education in that labor market and the combination of education and work experience heavily influences the local propensity to "entrepreneur" some new set of industries.

For example, within manufacturing alone, the percentage employed as engineers varies widely from an extreme of 25% in guided missiles and space vehicles and 13% in aircraft to below 1% in most food, apparel, and wood processing industries. An industry-mix that generates a high proportion of scientists and engineers may almost ensure timely and recurring reindustrialization of the local economy. The local occupation-mix is seen here as the key to a new, complementary approach.

The Occupational-Functional Approach

Just as local development analysts do well to master hundreds of industries, so too they must avoid being distracted and confused by the data on hundreds of occupations. We have found it both convenient and incisive to identify five broad "occupational-functional" classes that define five local economic roles or development paths:

(1) Entrepreneurship: emphasizing innovation and risk-taking;
(2) Central administration: serving as a headquarters city;
(3) Research and development: a center of science and technology;
(4) Precision operations: a place of skilled work; and
(5) Routine operations: a pool of low-skilled labor.

The necessary disaggregation will come in time as we gain new insights from the hands-on experience. Precision work is already being subdivided in our data base into blue-collar skills (tool and die workers), white-collar occupations (auditing and advertising), or new "white coat" laboratory technologies. Nor can even these subdivisions be so clean-cut; consider new academic programs offering a joint degree in engineering and business as preparation for *research management* or, in a variation, entrepreneurship in high-tech industries.

State and local policymakers will choose, explicitly or implicitly, to travel one or more of these five paths, so they, through their staffs, will need to learn more about the many ways in which these paths are substitutive or complementary or sequential. Those who target industries founded on higher economic functions should come to appreciate that higher skills command higher incomes that raise land rents, and that affluent households demand better public services that raise tax rates and other business costs, and this acts in turn to repel routine work and dislocate lower-skilled workers.

Again, development analysts should come to appreciate how local success in nurturing entrepreneurship could lead to large organizations and impose heavy responsibilities of central administration, which may in turn constrain further venturing, perhaps until complacency creates a crisis, and challenge leads to response, expressed in a new round of entrepreneurship. We find strong evidence of long waves in entrepreneurship and oligopoly, most notably today in the industrial Midwest.[2] Local development managers have also become intellectually lazy and have come to depend too much on adroitly marketing a poorly crafted product.

Toward Targeting Only Part of an Industry

Conventional industry targeting falls short to the degree that the strategist acts as if it were necessary to target the whole industry, from laboratory research through the headquarters, down to the assembly line. A given local economy may have the hub location from which to carry on the headquarters function, another may be blessed by a strong university that attracts corporate research facilities, despite its out-of-the-way location. But that high-cost headquarters hub-location, and perhaps the college town too, may have to accept the flight of the routine operations of that company (and others) to smaller, low-wage, low-tax, low-rent places. The occupational-functional approach in effect disassembles a potential target industry and matches the special locational needs of a given operation to the functional strengths of the home locality.

The increasingly complex interrelationships between the location of the headquarters, research laboratories, and production facilities of a company are well illustrated by the recent interest of prestigious hospitals in experimenting with heart surgery in branch clinics at considerable distance (e.g., the Mayo Clinic in Scottsdale, Arizona, and Cleveland Clinic in Fort Lauderdale). Rochester and Cleveland may experience reduced pressure to reindustrialize—may remain medical centers, despite population shifts—through functional specialization. By trading on the headquarters name and reputation, *central administration* in the medical industry is retained, supported by their state-of-the-art heart *research*—high-tech fonts to which practicing surgeons would retreat periodically for continuing education—even though *precision production* (the practice of surgery) is spun-off to places closer to the final market.

This dispersion of industry functions has led, in the best recent practice, to the reclassification of cities as "command and control centers" and production centers.[3] Note, there are functionally specialized places (e.g., headquarters cities and "research triangles") only because businesses do disperse functions, and that this classification is distinct from industrial specialization (e.g., manufacturing areas and financial centers), where the city may or may not have very diversified functions. Compare *functionally diversified* Detroit to Flint, which is both industrially and functionally specialized. But, mostly, we mix and confuse what a place makes and does.

Where no man has yet feared to tread is into global comparisons. What does it mean to rank El Paso high as a low-cost site for routine operations, with Ciudad Juarez across the river? Moreover, it is hard to resist speculating on the functional roles of Japan (precision production and aggressive management?), Taiwan (routine operations and entrepreneurship?), and Korea (evolving from routine to precision production?). While the data requirements are chastening, the hard reality is that international specialization and trade must, more and more, be seen in terms of *functional* comparative advantage.

The Political Geography of Local Economic Development

Local economic analysis most appropriately focuses on metropolitan areas, our best approximation of the local economy—the job commuting radius that bounds the local labor market. The policy problem is that there is simply no metropolitan government that coincides; there is no comparable political economy. Even the seeming exceptions to the rule—city-county consolidations—typically lag events. By the time consolidation takes effect, most of the new suburban growth is in contiguous counties, well within commuting distance, but beyond the reach of the new "metropolitan area government."

If then that neat local economic analysis typically falls squarely between the state and municipal-county policymaking and public management stools, to whom does one hand the completed study? The urban-regional, political economist has a hard choice: either carry the analysis to the state level by extending its scope to all or most of the subeconomies of the state, to show clearly the spatial trade-offs in state resource allocation, or break out the separate *intra*metropolitan area fortunes and options in a way that forces local public officials and managers to coordinate their development strategies. Or, better, do both.

In current research at Cleveland State University, we have prepared a preliminary economic-topographic map of Ohio, from a general data base equally applicable to other states.[4] We seek to sketch, even if only roughly at first, the industrial and occupational functional contours of a state by seeing it as a *federation* of local economies. This practice contrasts sharply with those highly aggregated, statewide business climate indexes that gloss over the great differences between local economies and then concentrate too narrowly on routine production.[5]

The very considerable social gain that comes with the drafting of an economic topographic map that encompasses the whole state system of local subeconomies is that the analyst captures the attention of state officials, if for no other reason than the fact that interregional trade-offs become so very explicit. The funding of graduate programs in science and technology is seen almost too clearly as a zero-sum game in the setting of the research and development pattern for the state.

Again, state policy, planning, and public works often create new urban centers and always affect the competitive position and prospects of existing ones that act, in turn, to assign headquarters roles to some cities rather than others. Those who allocate state funds for higher education and transportation have more control over the population settlement pattern of the state for decades to come than they know or perhaps want to know.

So even if we reject the state as a true regional economy and see it as a federation of local economies, when we reach the policy stage, we must often return there because only the state has both the governing powers and the spatial reach needed to frame and carry out a comprehensive *local* economic development strategy. (And statewide business climate studies do recover a modicum of respectability as the relevant spatial entities for comparing unemployment and workmen's compensation packages and state tax rates.)

In a recent retreat with Ohio local economic development "managers," I became frustrated when all my good ideas about strategy were successively dismissed as being beyond their powers. Finally, in exasperation, I struck back by accusing them of posing as managers when they were at best traders. But they could be born again if they became better analysts of the local position, prospects, and potential, and helped their principals trade better in the arenas where the power does reside, for example, in the corridors of the state capitol. "We will support your bid for the new airport, if you support our case for the new state-endowed chair in material sciences." But this assumes both better regional analyses and that fractious local officials can come to agree on the appropriate metropolitan area development path.

In reflection, 25 years ago, metropolitan area government was argued on the grounds of efficiency in the production of local public services (e.g., scale economies in public transit, utilities, and libraries) or on equity (e.g., tax base sharing between high- and low-income municipalities). The case for ("two-tier") regional *governance* is extended here to local economic development policy, planning, and practice.

From Analysis to Policy

The challenge to state and local public officials will be to assemble a series of mutually consistent, and, if possible, mutually reinforcing, policies from the various responsibilities and powers of state and local government. The challenge to local development analysts is to assemble the data base needed to make informed decisions. We turn now to just six selected policy foci, but this is challenge enough, especially if we also accept responsibility to draw out some of the more important interconnections between what should not be separate policies and programs.

Local Incomes Policy: From Any Job to Good Jobs

Most observers would agree that the level of income is the single best measure of economic well-being, and that its rate of change is the best measure of economic progress—economic development. I used to dramatize income analysis by arguing that there are three ways that a community can grow richer: through skill, power, and luck. A good local data base will assemble information and construct indexes that pinpoint, or surround, the comparative local position in all three dimensions.

Growth Rates as the Luck of the Draw

If either local economic development analysts or strategists were granted the power of a single insight into the future, surely they would choose to foresee the coming rates of growth of various industries. There is no quarrel here, in theory, with the heavy emphasis on "sunrise industries" as preferred targets. Growth, almost any kind, tightens the local labor market and leads to overtime, second earners in the household, and rising wage rates. But empirical work falters here because most of the new industries are, by definition, not in established industry classes, but rather are hidden in catchall classes, such as "miscellaneous chemicals" and "instruments, n.e.c. (not elsewhere classified)." This is a game for venture capitalists and stock market speculators, especially if targeting is extended to growth companies for a much sharper focus. Practice tends to fall short.

The Ambiguity of Average Hourly Earnings

If the high local wage rate reflects the higher skills and productivity of the local labor force, local strategists need industry wage data to target industries that will avoid *underemployment.* For example, aircraft manufacturing reported 1985 average hourly earnings of $14.65 (Class 2 of our ten classes), solidly based on high skills, with engineers accounting for 12.5% of total employment (Class 1) and only 1.4% as unskilled handlers, cleaners, and helpers (Class 10). In sharp contrast, the motor vehicle industry reported even higher 1985 average hourly earnings of $16.53 (Class 1), insecurely founded on only 3.8% employed as engineers and nearly 40% as semiskilled machine operators and assemblers, more than twice that in aircraft. Development planners in auto towns feel forced to shop for replacement industries, in which high wages have been wrung from a combination of oligopoly price power and union wage power, to satisfy the (too) great expectations of the local labor force.

Market Concentration Is Only a Home Remedy

"Good jobs" is too often synonymous with overpaid work, and explains much of the resistance to change found in places long dependent on wage and product price power. Market power leaves a complex local legacy: present power enriches the host locality but *fading* power hobbles that locality in its search for new industries. Moreover, domestic monopoly power is just not what it used to be. Our postwar experience suggests that development strategists should not look separately at some national concentration ratio or at some foreign trade index of a given industry, taken alone, but should bring the two indexes together, with their projected trends, to assess the long-run market power of the industry, especially those facing growing foreign competition.

Labor Market Balance: Males and Females in Joint Supply

Half of the households cannot today buy a new home without two members in the labor force, and so a local income policy must extend to identifying and closing gaps in the local demand for labor. For example, in compiling data on the ratio of female-to-male employees, one becomes more aware of the very wide variation between industries. The steel industry, for example, employs nearly ten times as many men as women, and the apparel industry employs four times as many women as men, making it very difficult in either steel or apparel towns to keep two persons employed.

But, again, the complementary occupational perspective adds deeper insight. Male and female jobs in appropriate balance are especially important in managerial and professional occupations, to attract and hold these key talents. That 4 to 1 ratio of females to males in apparel reverses to 1 to 2 in favor of males in managerial occupations, but this is still better than the all-manufacturing figure of 1 to 5. If, increasingly, good jobs for highly educated females are necessary to attract and hold males as well as females, achieving labor market balance in the higher functions (occupations) becomes a critical strategic matter. Occupational balance may be more important, and attainable, than industrial diversification, and may be the most underappreciated development problem of small areas.

Finally, the objective should not be to attract only (mostly) high skill jobs. A spectrum of skills is important to providing the "occupational ladders" that are indispensable to personal development, the *sine qua non* of community development. The charge then is to bring together comparative data on all of these dimensions to pinpoint holes in the local labor market-local income policy as local labor market analysis. Mr. Mayor, do you know where your labor market specialist is? It is later than you think.

Local Foreign Policy

Heavy engagement in foreign trade opens up both local opportunity and vulnerability, for either an industry or a locality. Global markets offer the best chance of winning big and losing big. Local development analysts need to prepare their policymakers and strategists with foreign trade data on each promising industry and not just to pick winners. Exports as a proportion of industry shipments and imports as a proportion of domestic consumption are proxies for community risk-taking.

In some industries, domestic companies dominate their foreign competition, with export-import ratios of 4 to 1 or more (e.g., oil field machinery, aircraft, and biological products). But in many more industries, heavy shipments flow both ways. For example, in farm machinery and paper-making machinery, about one-quarter of all shipments of U.S. producers reexported and about one-quarter of domestic consumption is imported. These are games for high rollers. Given that more than one-half of all manufacturing industries report foreign trade of less than 10%, that is, are domestic industries, a locality can also choose to escape that fierce (low-wage) foreign competition by not targeting those heavy, balanced flow industries.

Local policymakers should choose industry targets only after paying close attention to levels and trends in industry experts and imports, weighed against the community's willingness to take the risk of competing in world markets. But, viewed from the national side, our great need for exports demands that the federal government find ways to dissuade localities from turning inward toward domestic (transportation-cost-protected) markets or lobbying for tariff protection, and to induce them to take the greater risks of competing in global markets.

This is not to argue for beggar-thy-neighbor export subsidies almost sure to invite retaliation, nor for supporting local industries that will never become competitive. The "infant industry" argument returned in force with protection for the "new" small car industry, and a decade later the outlook is just as bleak. But the national interest in exports may support the case for continuing payments to workers dislocated by foreign competition, not just on equity grounds as a form of unemployment compensation, but as a device to induce localities to take foreign trade risks in the national interest. But, as always, we will need to tie payments to the timely development of new skills and sometimes to a willingness to migrate to localities where the risks of foreign trade have paid off.

Again, a disassembling of an industry data by separable functions adds insight into local foreign trade analysis by identifying the local "balance of trade in occupations." Local development analysis should bring together data on industry shipments with information on local functional specialization in that industry, to infer what a place *does* (not *makes*) in foreign trade. Conventional wisdom that has us exporting high skills and importing low skill work is as obsolete as many of the products we make.

It is chastening to reflect on the fact that Marysville, Ohio, has recaptured routine operations in automobile production (Honda of Japan) but the states of Ohio and Michigan and the nation are, more and more, importing research and engineering and central management in motor vehicles. How do we measure the balance of trade in functions and occupations—from *industry* trade data? Integrating information (data) on industries and occupations would be a good beginning.

Localities can do more than react alertly but passively to global trends. They can, again, see themselves more as traders than pure managers, this time in Washington, and lobby in a more balanced way for their interests. State and local lobbyists will roll better logs only with a more sophisticated appreciation of the interests of local consumers and exporters, to balance their long-standing concern with the welfare of import-impacted constituents. Their principal lobbyists in the capitol would, of course, be their representatives and. in broader coalition with other localities across the state, their senators. Do state and local development staffs have the requisite specialists in international trade, the sharp analysts of *long-run* effects who can do more than woo foreign visitors and close a deal?

Local Stabilization Policy

Local development groups operate with remarkably primitive knowledge about interindustry differences in cycles. We find wide variation within a sample of 150 (SIC 4-digit) manufacturing industries, ranging from many that were stable or grew through both the 1973-1976 and the 1978-1983 recessions (e.g., pulp mills and surgical instruments) to truck trailer manufacture that lost 39% and 48% of employment, respectively. One should expect localities to contemplate achieving stability through industrial diversification, but not without some reasonable appreciation of the trade-off between that goal and the need to achieve *specialized* expertise to compete in world markets.

Stabilization of the local economy can perhaps be attained in ways consistent with heavy engagement in foreign trade, through occupation adaptability. A community could build a base in general education strong enough to support both the specialization needed to compete and the adaptability needed to adapt to rapid change. It would not be enough just to have more engineers and managers, unless these high specialists are both versatile and adaptable. To exaggerate only a little, almost everyone and every place, with high aspirations, must have a touch of both scholarship and entrepreneurship—a love of learning and a lack of fear of change.

State Higher Education and Research Policy as Implicit Regional Planning

The state legislative and executive branches have made more implicit than explicit policy in arranging the settlement pattern of the state in their budgeting of funds for higher education. The state of Ohio has, for example, arranged graduate education so that slightly over one-half (50.5%) of its 1975-1980 state crop of science and engineering doctorates were harvested at Ohio State University. With only 11.5% of the state population residing in the Columbus metropolitan area, this one locality has 4.4 times its *pro rata* share of those doctoral graduates and a similar advantage in supporting faculty and technical facilities at this highest level of academic science and technology.

Ohio does not exhibit an extreme form of spatial concentration in doctoral education. North Carolina graduated 92.5% of its 1975-1980 doctorates in science and engineering from universities in the Raleigh-Durham-Chapel Hill metropolitan area, ten times its 9% of the 1980 state population. In interesting contrast, the University of Minnesota accounted for almost all (98.3%) of the 1975-1980 doctorates in science and engineering in that state, but given the 51.9% of the state population residing in the Minneapolis-St. Paul metropolitan area, the spatial concentration is the lowest of these three places (1.9 times its population share). Finally, the Detroit-Ann Arbor consolidated metropolitan area exhibits an even lower spatial concentration, with only 1.3 times its share.

Neither the high-cost local economies of Flint and Youngstown nor the isolated apparel and textile towns of the Piedmont have so many development options that they can afford to write off every high-tech by-path, even though they start the race hobbled by the chance location of the state's major graduate programs beyond reasonable commuting distance. The importance of having at least one strong state university graduate program in science or engineering in each of the major metropolitan areas is becoming increasingly apparent.

The state does not have to choose between extremes: giving every locality its share of every program in science and technology, or concentrating most of these graduate programs on a single campus. Population share grants would, of course, deny economies of scale—critical mass—in graduate education and research. But extreme centralization of doctoral work on one campus presumes a synergism between disparate fields that probably does not exist—a gross exaggeration of cross-fertilization between fields. How often are ideas exchanged between the chemistry and mechanical engineering faculties, compared to professional contacts between academics and their practicing counterparts in an associated industry town? We should look harder for the dotted lines along which we could tear apart fields of study and research that are only distantly related.

Careful decentralization of graduate programs in science and engineering would strengthen the hand of second-level, high-tech places that have few other development options. Progress in communication technology offers new substitutes for personal transportation and reduces the need to cluster in just one large "science city." This is not to argue that the nation does not benefit from North Carolina's experiment in the centralization of research, but the federal government would do well to encourage experimentation with other patterns.

Politicizing state funds for graduate education does open the way to pork barrel responses, as seen in a recent Congressional resolution earmarking defense funds for research in ten specified universities.[6] But one should not underestimate the vested interests and the political momentum biasing our present high-concentrated systems of graduate education. Bureaucrats prefer to cluster close to the centers of power, alumni (old school) ties in state legislatures cannot be taken lightly, and the faculty have hidden agendas. Still, the have-nots do have the votes and few areas can afford to waive any potential they might have in research and development, especially those saddled with sunset industries and occupations and hobbled by the higher costs of fading market power.

There are also diseconomies of scale in research and the dislocation costs of places left behind deserve inclusion in the state accounting. Again, almost every city needs a touch of science and technology to nurture a sophisticated citizenry and electorate. Perhaps some dispersion of that heavy concentration of science in Huntsville and Oak Ridge would have moderated the recent anti-intellectual movements in Alabama and Tennessee. Town and gown writ large across the state.

Careful study may well reveal that clustering programs and faculty in science and technology tends to tip the balance toward basic research because academic values dominate the reward system. Conversely, decentralizing science and engineering programs to places with the associated industries would turn the academic work more toward commercial applications and speed technological transfer. Not only would increased contact between faculty and business research people change peer pressure more toward applied fields but more profit-making opportunities would come to be seen and attract talent. The argument here is not which is better but rather that we seem, through state higher education policy, to be making implicit choices between pure and applied research, and lamenting that there are not many jobs in those Nobel

Prizes—nor much patent protection either today. What do we want?

Local Fiscal Policy and the Location of Industry

One of the most intriguing indexes stumbled upon in our empirical work was interindustry differences in fixed assets per worker. Surely, fiscal managers could manage better with this rough fiscal benefit/cost index for a given industry. The numerator of this ratio measures the industry's *direct* contribution to the property tax base and the denominator estimates the employees' household demand for public services.

This fiscal index was formed from readily available data on fixed assets per worker contained in the (never repeated) *1976 Annual Survey of Manufacturers: Industrial Profiles.* This direct measure was supplemented and updated with a second estimate of fixed assets from the *Census of Manufacturers:* cumulative capital expenditures for the 11 years, 1972 to 1982. We find an arresting range from pulp mills with \$156,000 in 1976 fixed assets per worker and \$193,000 cumulative capital expenditures per worker, down to figures below \$10,000 in both indexes for household furniture and below \$5,000 in both apparel and footwear.

Combining fixed assets per worker with average hourly earnings data rounds out our industry fiscal impact index, bringing in fiscal surplus and deficit households of the workers. Compare, again, pulp mills with 1985 average hourly earnings of \$16.29, with furniture at \$6.42 and most apparel manufacturing paying about \$5.00. To those that have shall be given. The staff analyst should stand ready with figures such as these to guide decision makers in passing out the limited supply of gratuities used to attract businesses.

Making Sense of Taxes and Location

The many statistical studies on state and local taxes and industry location were inconclusive because the question was posed much too broadly—taxes and aggregated activity. Disaggregating the component industries and asking about taxes and textiles and taxes and machinery helps a little in separating the tax-sensitive from insensitive activities. (Remember the wry classroom illustration in the public finance course: the apocryphal case of the owner of the stone quarry who threatened to move his business out of town if taxes were raised again.) But major progress here comes only by distinguishing between the various functions within any given industry.

If taxes are negatively correlated with (repel) some *functions* and positively correlated with others (attracting them through the better public services financed by those taxes), small wonder that scholars kept finding a near-zero correlation coefficient between taxes and all activity. But the occupational-functional approach lights the way through the maze. Routine operations (stamping metal, molding plastic, or back office paper work) generally seek the cheapest place to do business, leaning toward low-tax locations, bearing easily the *inconvenience* of mediocre schools, libraries, and recreation.

But the research work of those same industries will accept, even press for, higher state and local taxes for excellence in local schools and universities. Again, headquarters decision makers can readily accept the higher public service (and tax) costs of the higher density that comes with larger city size, to get better airline schedules and business services, and to gain richer cultural offerings. High taxes discourage some and encourage other pursuits. What economic role does the community aspire to play? If development analysts could evaluate better the local potential on the different development paths, policymakers could make better tax policy.

Urban Policy: Tracing Regional Development Paths into Intraurban Space

Our work is not finished until we trace each of these Five Paths across intraurban space. Can municipal officials plan land use patterns and guide transportation investments in ways that will strongly support the preferred mix of regional development paths? Hub radial-rail patterns may be more consistent with local specialization in nonmanufacturing (e.g., financial) headquarters, while cars and buses on expressways probably fit better the multinodal spatial patterns of manufacturing areas and the transportation preferences of factory workers.

For example, the 1980 census commuting data indicate that over 12% of the executives and managers but less than 4% of the "handlers, cleaners and helpers" commute to the central business district. This suggests that a local economic development strategy designed to support a central administration (headquarters) path should seriously consider hub radial, rapid transit *and* charge fares high enough to cover most of the cost of a high-quality ride.

A generation ago, I argued that the federal authorities should encourage experimentation by making more highway grants in some places (e.g., Detroit and Los Angeles) and more mass transit grants in others (e.g., New York and San Francisco) "to encourage innovation in urban form."[7] The orientation then was more toward urban efficiency and choice—adapting internally to the different economic roles of cities. A similar strategy today might lean more toward urban policy designed to change or enhance the economic role of a city.

A simplistic "worst-first" strategy in the replacement of local infrastructure, unmindful of changing technology or a new local role, is today not good enough to compete. We have begun the work of coordinating municipal and county with metropolitan development strategies by comparing central city Cleveland to its suburbs, compared to other central city-suburban relationships, in three intraurban roles: as a metropolitan area export district, as a regional trade and service center, and as a place to live.[8] In general, an intrametropolitan model opens the way to extending asset accounting, tax sharing, user charges, and other tools of local public management beyond intraurban efficiency and equity into area development.

While, in metropolitan area development strategy, one fork in the road to implementation leads to the statehouse, the other fork leads toward the offices of mayors, city managers, county commissioners, and local school superintendents. The attention of these trustees for legally separate parts of the local economy can be captured and held only if the various metropolitan area paths to development are plotted in intrametropolitan space—through or around their turf. Will those new headquarters locate downtown or in a suburban center or out along the beltway? If an "advanced manufacturing" scenario is chosen and unfolds, which subdivisions would fall heir to the new tax base and which would reap the whirlwind of traffic noise and smoke?

Lacking regional government, a strong case can be made for less formal methods of tax sharing to achieve equity between municipalities and thereby smooth the way to cooperation in regional development strategy. In general, local economic development planners need better intra-area spatial models and municipal analyses that can be plugged into the regional industrial and occupational-functional analyses. Only then can they help their principals become better traders, first on the floor of the metropolitan area "exchange" and then later in the halls of the state capitol.

It is useful in coordinating intraurban and regional development policy to distinguish between two broad classes of local market-oriented activities: derivative and developmental local services. The derivative services are, by definition, routine in nature and draw largely on low skills that are performed nearly the same everywhere (e.g., most retail trade) and are not critical to local development. Developmental services are defined as precision operations, such as education and city management, and are performed better in some places than others. The latter are as important to local growth and development as the local export industries, less so in the short run and more so in the long run.

While efficiency is important in both derivative and developmental services—they both affect the purchasing power of local income—the latter are by definition major sources of invention and innovation and can give the local economy a competitive edge in interregional trade.

Only when we have gotten beyond recalibrating local multipliers and sorted out the development role of local services will we be well on the way toward a true urban economics of regional development.

Notes

1. Wilbur R. Thompson, "Analyzing National Industries and Assessing Local Occupational Strengths: The Cross-Hairs of Targeting," *Urban Studies* (forthcoming).

2. Wilbur R. Thompson and Philip R. Thompson, "From Industries to Occupations: Rethinking Local Economic Development," *Commentary,* Fall 1985, pp. 12-18.

3. Thierry Noyelle and Thomas M. Stanback, *Economic Transformation of American Cities* (Totowa, NJ: Allanheld and Rowman, 1983).

4. Philip R. Thompson, "Toward an Economic-Topographic Map of Ohio: The Seven Largest Metropolitan Areas," *Ohio Economic Trends Review 1,* no. 3 (1986), pp. 1-22.

5. Grant Thornton, *The Seventh Annual Study of Manufacturing Climates of the Forty-Eight Contiguous States of America* (Chicago: Author, 1986).

6. Colin Norman, "Congress Approves Deals for Ten Universities," *Science 23* 1, no. 4735 (1986), p. 211.

7. Wilbur R. Thompson, *A Preface to Urban Economics* (Baltimore: Johns Hopkins Press, 1965), p. 198.

8. Wilbur R. and Philip R. Thompson, "Accounting for Municipal Economic Development," *Journal of Urban Affairs* 8, no. 2 (1986), pp. 47-66.

Chapter 2

The Market Failure Approach to Regional Economic Development Policy

Timothy J. Bartik

❀❀❀

This article will present the *market failure* approach to regional economic development policy.[1] This approach clarifies the goals of regional economic development policy, thus providing a framework for evaluating regional economic development programs.

Regional economic development policies are defined here as policies that seek to increase the wealth of a metropolitan area or state by providing direct assistance to business.[2] Examples of such policies include assistance to businesses in finding plant sites; tax abatements and other subsidies to new branch plants; business incubators, small business development centers, entrepreneurial training programs, and other small business assistance; state capital funds and other financial market interventions that increase business capital supply; state support for applied business research programs; industrial extension services to improve business productivity; and export assistance programs.[3]

To distinguish the two currently dominant philosophies of economic development policy, they are labeled the *traditional* approach and the *new wave* approach.[4] The traditional approach emphasizes job growth as the unifying goal for regional economic development policy. Regional job growth is believed by traditional economic developers to be most enhanced by focusing on a region's export base. Attracting export-oriented manufacturing branch plants, typically owned by large corporations, is an important emphasis for traditional economic developers.

The new wave approach is more eclectic but emphasizes various forms of economic innovation as a unifying goal. The new wave approach includes policies encouraging small business start-ups and growth, technology development, and business modernization.

The traditional and new wave approaches provide policy guidance that is too vague. Job

AUTHOR'S NOTE: I appreciate helpful comments on a previous version of this article by Richard Bingham, George Erickcek, and four referees. Reprinted from *Economic Development Quarterly,* Vol. 4, No. 4, November 1990.

Reprinted from *Economic Development Quarterly,* Vol. 4, No. 4, November 1990.

growth and economic innovation may well be good things. But job growth and innovation occur in the private market without government intervention, and policies to increase job growth and innovation have costs. Without some understanding of the likely magnitude of the social benefits from job growth and innovation, it is impossible to determine rationally the appropriate size of the government's economic development budget. Without some understanding of differences in the social benefits of different types of job growth and innovation, it is impossible to decide rationally how to allocate an economic development budget among different projects and programs.

The market failure approach aims regional economic development policies at the goal of correcting private market failures. Market failure is the failure of private markets to achieve *economic efficiency,* a situation in which no change would result in net dollar benefits, summed over all members of society. If benefits exceed costs from some change, gains from trade should cause markets to facilitate the change.[5] Market failure occurs when no such market forms.

Market failure is caused by impediments to the formation or operation of markets. Public goods—defined by economists as goods, such as national defense, which simultaneously affect many people—are difficult to provide without government intervention because it is difficult for private providers to exclude nonpayers from consumption of a public good. Externalities, such as pollution, result in inefficiency because a polluter ignores damage to pollution victims and because no market encourages such consideration. Imperfect information can cause market failure by preventing awareness of market transactions that have net benefits.

Using the market failure approach, regional economic development policies will encourage the expansion of benefits that private markets fail to recognize adequately. Regional economic development policies will be efficient if the value of these nonmarket benefits exceed program costs.

For most of this article, I adopt the perspective of a regional policymaker: relevant benefits and costs are those occurring in the region; positive or negative spillover effects on other regions are ignored.

The next section of this article discusses specific types of market failure and their implications for regional economic development policy. The conclusion highlights the strengths and limitations of the market failure approach.

Discussion of Market Failures Justifying Regional Economic Development Policy

For each market failure I will define its nature, explain how it might justify specific regional economic development policies, and consider how it affects policy evaluation. For simplicity, the discussion of each market failure assumes the existence of only that particular market failure.[6]

Unemployment

Not all unemployment justifies government intervention. Unemployment indicates a market failure if individuals without employment are willing to work at the prevailing wage for jobs for which they are qualified. Under these conditions, we can classify such unemployment as involuntary.[7]

The best recent theory of involuntary unemployment is efficiency-wage theory.[8] According to this theory, firms willingly increase wages above the lowest wage at which qualified unemployed persons could be hired because higher wages lower company costs. An employee will work harder at higher wages because higher wages both increase the loss to the employee if he or she is fired and enhance the worker's sense of receiving fair treatment. For similar reasons, higher wages lower workers' *quit* rates, reducing firms' turnover costs.

Reducing involuntary employment for current residents is a nonmarket benefit that is a possible goal to be maximized by regional economic development policy:[9] the unemployed receive an *employment benefit* when hired that is equal to the difference between the actual wage and the lowest wage for which they would accept the job. Regional economic development policies that reduce unemployment would pass a benefit-cost test if they cost less than this employment benefit.

Research indicates the size of employment benefits that result, on average, from regional economic development policies. Empirical estimates show that in the short run, 40% to 50% of the new jobs in a metropolitan area go to in-migrants, while in the long run 70% to 80% of the new jobs go to in-migrants.[10] Thus, in the long run only one of five jobs created by an economic development program would be expected to lead to employment benefits for current residents.

A number of surveys ask unemployed individuals the lowest *reservation wage* at which they would accept a job.[11] In a typical metropolitan area, with moderate unemployment, reservation wages average nearly 85% of the unemployed person's previous wage.[12] Multiplier effects for new manufacturing plants are about 2; that is, for every new manufacturing job brought to an area, we would expect an additional job to be created among local business suppliers and retailers. Hence, an *average* economic development program attracting an average manufacturing employer to an average metropolitan area might provide long-run employment benefits of about 6% of the wages paid by the plant itself (.06 = .20 proportion of jobs to current residents × .15 employment benefit × multiplier of 2). If the program changed the probability of the plant's choosing the region from .90 to 1, then program costs, up to the equivalent of .6% of the plant's payroll, could pass a benefit-cost test.[13]

This ballpark estimate of average employment benefits can be made more accurate for specific cases that have additional information available. More new jobs will go to current residents, and fewer to in-migrants, for economic growth that is well-matched to the available skills of unemployed residents. Estimating this larger effect on unemployment would require detailed information on the development policy and the population's skill mix. Reservation wages would be lower in metropolitan areas or neighborhoods with high unemployment. Surveys can estimate the reservation wage of particular groups. Finally, the estimation of multipliers is routinely done by many economics consulting firms.

Underemployment

Efficiency-wage theory also implies that private markets suffer from involuntary underemployment: workers in some industries want jobs in other industries that pay more and for which they are qualified. Efficiency-wage premiums vary across industries because of industry differences in the costs of worker turnover, the ease of motivating workers through monitoring, and relative profitability.[14] In industries with hard-to-monitor productivity, costly worker turnover, and high average profitability, firms must pay high wages to be considered fair, elicit hard work, and suffer few quits.

The empirical evidence shows sizable wage differences across industries that cannot be explained by differences in workers' skills.[15] Industries with positive-wage premiums include instruments, chemicals, transportation and communications, professional services, primary and fabricated metals, machinery, transport equipment, paper and mining.[16] These wage differentials are not due primarily to unionization. Industry wage differentials were similar in the 1920s to those today and are also similar in weak union countries such as South Korea.[17]

Shifting a regional economy toward high-wage premium industries provides nonmarket benefits that are a possible goal of regional economic development policy.[18] If the average-wage premium per worker increases because of a changing industrial structure, the average worker then receives an *upgrading benefit* equal to the difference in the wage for a given skill level. Because this wage increase is caused by shifts in industrial structure, rather than an across-the-board increase in wages, it will not necessarily decrease business profitability. Available evidence suggests that industries with above-average-wage premiums have above-average profit rates.[19] Hence, shifts in the industrial structure provide benefits for workers that are not offset by costs to firms and that are not taken into account by firms in making output and employment decisions. These nonmarket benefits may allow development policies that shift the industrial structure to pass a benefit-cost test.

Measuring upgrading benefits from regional economic development policies is straightforward. If new jobs from a regional development program, including the jobs from multiplier effects, have higher-industrial-wage premiums than prevail in the region, the extra-wage premium for current residents should be counted as an upgrading benefit.[20] For example, if the new jobs have an average industrial wage premium that is 10% above the average regional premium, the program then provides an extra 10% of wages benefit for every new job that goes to current residents. In the case of the new manufacturing plant discussed in the previous section, the total employment and upgrading benefits would be 10% of the payroll of the new plant, equal to the multiplier of 2 × the .20 proportion of jobs that go to current residents × the sum of the .15 proportion of wages representing employment benefits from an average job plus the .10 proportion of wages representing upgrading benefits from these above-average jobs (.10 = 2*.20*(.15 + .10)). If the economic development program changed the probability of the plant's choosing the region from .90 to 1, then a program subsidy of up to 1% of the plant's payroll could have positive net benefits.

Fiscal Benefits

A perfectly efficient state and local tax system would set the taxes paid by a business or household to the marginal cost of providing that business or household with public services, that is, it would be a user-fee system. Because tax systems consider goals other than efficiency (e.g., distributional equity), state and local taxes often diverge from marginal public service costs.

Empirical evidence suggests that costs for providing public services to a typical manufacturer are only 30% to 50% of the normal state and local taxes collected.[21] This raises the prospect that regional economic development may increase regional tax revenues faster than public service costs, providing a *fiscal benefit* to the region's residents. But business growth attracts population growth. Households, particularly households with children, frequently pay less in taxes than they require in public service costs.[22] The net fiscal benefit of regional economic development depends upon the particular circumstances of the region and type of growth.

A substantial consulting industry now provides economic development policymakers with fiscal impact analyses fine-tuned to a particular region and project. In general, one would expect greater fiscal benefits if the region has underutilized public infrastructure (roads, schools, and the like) and if growth has small effects on the in-migration of households with children.

Fiscal benefits of development, when they exist, provide an efficiency rationale for regional economic development programs. Subsidy costs up to the fiscal benefit amount might still pass a benefit-cost test.

Agglomeration Economies

Agglomeration economies are cost savings that accrue to firms in a particular industry in larger cities (the *urbanization economy* type of agglomeration economy) or in cities that have more of that industry (the *localization economy* type of agglomeration economy). Two distinct economic forces lie behind agglomeration economies. The first is *thick-market* externalities.[23] A larger market for a good makes it easier for buyers and sellers of specialized types of that good to find each other. Lower transaction costs resulting from a thick market encourage greater specialization, an idea as old as Adam

Smith's dictum that "the division of labor is limited by the extent of the market."[24]

Industries using specialized inputs with high transport costs, such as skilled labor, benefit most from thick-market externalities. The magnitude of this type of thick-market externality will be related to the industry's size in a metropolitan area. In addition, groups of industries purchasing closely related services (e.g., temporary office and clerical help) may benefit from thick-market externalities related to the size of the group. The magnitude of this thick-market externality will depend on the overall population or employment of the metropolitan area.

A second force behind agglomeration economies is *human capital* externalities.[25] In innovative industries, informal idea exchange between workers in different firms may fuel growth. Informal exchange is encouraged by spatial proximity, which allows for social interaction and firms' stealing workers from one another. A firm's innovation and growth will depend on industry human capital in its metropolitan area.

A firm making a location or expansion decision fails to recognize that decision's agglomeration economy benefits for other firms. The goal of increasing nonmarket agglomeration benefits may provide a rationale for regional development assistance to particular firms.

The main practical problem with the agglomeration economy rationale is the lack of good quantitative information on the benefits of a small increase in industry agglomeration. Empirical evidence shows the existence of agglomeration economies,[26] which are particularly important for industries (e.g., transport equipment) concentrated in relatively few cities. But the importance of agglomeration economies in explaining average differences across cities in industry productivity does not show that small increases in a city's industry size will boost industry productivity. For example, suppose that some industry-specialized services require a minimum industry concentration in a city, but once that critical mass is reached, there are no gains to further agglomeration. For most cities industry expansion, therefore, provides no agglomeration benefits. Only cities just below the critical mass would benefit from a program encouraging industry expansion. We do not know how the agglomeration benefits from a small expansion of an industry vary with city and industry size.

Human Capital

Increasing the human capital of individuals clearly has benefits in increased earnings for those individuals, but there is no market failure unless those individuals underinvest in human capital acquisition.

There are four possible reasons for underinvestment in human capital. First, individuals may have difficulty financing training or education because lenders cannot repossess human capital. Second, education may increase social stability by instilling civic virtues or by providing a sense of opportunity for the poor. Third, human capital may have externality benefits, because one worker's ideas enhance the creativity of other workers. Fourth, human capital's value is hard to measure before acquisition.

Market failures impeding human capital acquisition traditionally have provided an efficiency rationale for government assistance to education. These same problems could also rationalize regional economic development programs of entrepreneurial training and small business development. The skills needed to become an employer are just as much human capital as the skills needed to become an employee.

These entrepreneurial training and small business development programs can be evaluated using techniques similar to those used to evaluate ordinary job training programs. There is a vast literature on the evaluation of job training and education programs. This literature focuses on measuring the earning gains from training and education; the more intangible social stability benefits and human capital externality benefits of education are ignored. The key evaluation issue in this literature is the determination of what participants' earnings would have been if they had not signed up for the program. Social experiments, with randomly chosen treatment and control groups, are extremely helpful in answering this question. If regional leaders wish to evaluate entrepreneurial training and small business development programs, these same experimental methods can be used.

Research and Innovation Spillovers

The research, development, and risk undertaken by a business or by an individual to create a new product or production technique may lead to the product or technique's adoption by others. This spillover effect results in market failure because businesses and individuals will not consider such social gains, but rather only their own private gain, thus leading to underinvestment in research, development, and innovation. This argument has traditionally been used to justify government support for basic research, but also holds for more applied research and for innovative activity in general.[27]

The innovation spillover argument may justify economic development programs that subsidize applied research (e.g., Pennsylvania's Ben Franklin Partnership Program) or product development (e.g., the Connecticut Product Development Corporation). Such programs should focus on the types of research and innovation that lead to the greatest spillovers, rather than on research and innovation as such. The regional spillover benefits from R&D subsidies are the gain in profits for the industry in that region.

Evaluating regional R&D subsidies is difficult, primarily because every research project is unique. Statistical analysis of the average effect of industry R&D is unlikely to be applicable. Evaluation of regional R&D subsidy programs probably requires individual case studies of a sample of research projects. Interviews with industry observers may determine both whether a given subsidized innovation was widely adopted within the region and its likely profit impact.

Other Imperfections in Information Markets

Information in general, not just human capital acquisition or research knowledge, may be underprovided by private markets. Private production of information may be inhibited because potential purchasers find it difficult to evaluate information prior to its acquisition. In addition, information producers will not take into account the benefits of information spillovers to nonpurchasers.

These problems with private information provision may rationalize government programs that provide information intended to encourage economic development: industrial extension services that provide information on modernization; export information programs; and marketing programs providing information on potential new branch plant sites. The benefits to assisted businesses may exceed costs for these types of information, yet government provision is necessary because of forces impeding private information provision.

Government information provision programs will be most effective when they complement rather than substitute for privately produced information. Obviously there is much private provision of information to businesses from consultants and other sources. Government information provision should include the basic information that is least likely to be optimally provided by the private sector. For example, before a business can intelligently decide whether it needs a consulting engineer, the business must understand the nature of its production problems. Such information may be better provided by the government because a private consultant has an incentive to artificially expand demand for his or her services. Government information provision also has advantages when the government has better access to information. For example, state governments should have the best information on state taxes, regulations, and public services.

The methodology is straightforward for evaluating benefits from government information provision to private businesses. The benefits are the extra profits for the assisted business. The costs are simply the government's costs of acquiring and disseminating the information. Benefits may be measured by surveys that ask assisted businesses what effects the program had on profits. A more objective evaluation would statistically compare profits of assisted businesses with unassisted businesses.

Imperfect Capital Markets

Private capital markets fail to achieve efficiency if socially profitable loans or investments are not made. There are three possible causes of such market failure. First, financial markets in the U.S. are regulated. Competition among banks is restricted, and this may limit credit availability. Furthermore, risks taken by financial institutions are supposedly restricted by government regulators. This may prevent financial institutions from making risky loans or investments despite an expected good return. One defect in this argument is that supposed stringent regulations have not prevented many U.S. savings and loans from making excessively risky investments.

Second, even without regulation, the absence of complete insurance markets may inefficiently restrict the amount of risk that financial institutions are willing to take. Insurance markets for loans and investments are difficult to create because of *moral hazard* problems: in a world of imperfect information, financial institutions will try to insure the loans and investments that are most likely to have problems, and insurers will find providing insurance for only the bad risks an unprofitable venture.

Finally, private market interest rates may be above the optimal social discount rate. In particular, this may be true for long-term loans and investments. Society may judge that individuals devote too little savings to long-term investments that benefit future generations.[28]

These arguments may provide a rationale for economic development programs that directly provide capital to business, or indirectly encourage greater private capital availability to business (e.g., loan guarantee programs, or subsidies for new types of financial institutions such as Michigan's Business and Industrial Development Corporation [BIDCO] program). If the rationale for these programs is excessive risk aversion by the private market, then government intervention should focus on expanding credit availability to riskier ventures whose expected profitability yields a normal rate of return. If this rationale is valid, the portfolio of government-induced investments should yield a normal rate of return.

On the other hand, if the rationale for government intervention in financial markets is that private interest rates are too high, then government intervention should provide credit to business ventures that yield a return below the private interest rate, but above the social discount rate. The government program should be evaluated on the basis of whether it yields a rate of return above the social discount rate, not on whether it yields a normal profit rate. An extensive literature discusses the social discount rate issue.[29]

Our daily observations of the business world may lead us to exaggerate the extent of private capital market failure. After the fact, it is clear that many successful business ventures should have received greater support from private financial institutions. But the problem lies in identifying such opportunities before the business venture is undertaken. In a world of uncertainty, any society's rational procedure for allocating scare capital will sometimes lead to mistakes. Private markets can be said to have failed only if we can create institutional mechanisms that yield a better selection of investments. It is far from obvious that government intervention in capital markets improves the selection of investments.

Multiple Market Failures

An evaluation of an actual regional economic development program will need to include many of these nonmarket benefits. For example, a program providing information to businesses may simultaneously lower business costs and increase regional job growth. Job growth benefits, such as employment benefits, upgrading benefits, and agglomeration economy benefits, should be included in the analysis.

Program Design Issues

Implicit in the above discussion is that the design of regional economic development programs, not just their evaluation, should be altered by the market failure approach. For a given program cost, programs should be designed to maximize nonmarket benefits. For example, economic development programs should seek to increase employment benefits by placing greater focus on creating jobs that are accessible to residents of high-unemployment neighborhoods.

Conclusion: The Strengths and Limitations of the Market Failure Approach

The market failure approach has two strengths, as compared to the traditional or new wave approaches to regional economic development policy. First, focusing policy on what private markets are unable to do allows a wise use of limited government resources. "If it ain't broke, don't fix it" is a good traditional American philosophy.

Second, the market failure approach leads to goals that are measurable in the common currency of dollar benefits.[30] For a fixed government economic development budget, an agency can make rational decisions under the market failure approach about tradeoffs among different types of nonmarket benefits. In addition, leaders can make rational decisions about whether nonmarket benefits from an expansion of economic development programs are worth the costs.

The market failure approach has three limitations. First, as discussed above, we do not always have precise information about the magnitude of some of these nonmarket benefits. But better benefit measurement is feasible if policymakers will encourage the collection of the needed data. This article provides the conceptual framework that is an essential first step in benefit measurement.

Second, the market failure approach by itself does not consider distributional effects of regional economic development policies. A full benefit-cost analysis of these policies should consider both efficiency and distributional effects. Adding in distributional effects makes policy analysis more difficult. Furthermore, the mobility of capital and labor limit the size of the redistributional effects of regional policy. Hence, public finance theorists have usually concluded that the federal government should bear the prime responsibility for income redistribution, and state and local governments should focus more on efficiency issues.

Third, any regional perspective on policy overlooks the benefits and costs of one region's policies for other regions. One region's economic development policies may have negative spillover effects on other regions; for example, if the policy reduces job growth in other regions, this may reduce employment, upgrading, fiscal, and agglomeration benefits in those other regions. One region's economic development policies may create positive spillover effects for other regions; for example, the spillover effects of R&D may help firms in other regions.

Thus, the *national* efficiency of regional economic development policies would be improved by national policies that limit regional policies with large negative spillovers and that encourage regional policies with large positive spillovers. Such national policies seem unlikely in the U.S. in the immediate future.

In the absence of federal government intervention, however, encouraging regional governments to pursue regionally efficient policies is likely, on average, to increase the efficiency of the national economy. In general, regional economic development is not a zero-sum game. For example, employment, upgrading, fiscal, and agglomeration economy benefits will generally be greater in some regions than in others. Those regions that experience the greatest benefits have the greatest incentive to vigorously pursue development policies. The efficiency benefits for these regions will usually exceed the negative spillover effects for other regions.[31] In addition, national employment will expand, particularly in high-wage premium industries, if many regions pursue employment and upgrading benefits. Finally, regional economic development policies in any region that provide positive spillovers will assist the national economy.

Despite these limitations, the market failure approach provides a useful conceptual framework for economic development policy. A clear target is essential to sound policy design and evaluation.

Notes

1. The market failure approach to policy did not originate with me. All economists are indoctrinated with this approach to analyzing policy. Furthermore, a market failure approach has been used occasionally in the past—without really explaining the assumptions behind this approach—to analyze particular issues in regional economic development policy. For example, see George Borts and Jerome Stein, *Economic Growth in a Free Market* (New York: Columbia University Press, 1964); Roger Vaughan, Robert Pollard, and Barbara Dyer, *The Wealth of States* (Washington, DC: Council of State Planning Agencies, 1985). But I believe this article is original in explicitly identifying this market failure approach, explaining its assumptions, applying it to a wide variety of economic development policies, and explaining its implications for evaluating policies, all in a manner that is accessible to noneconomists.

2. Thus this analysis is not meant to be applied directly to communities that are a small part of a metropolitan area. Many of the nonmarket benefits of economic development that are discussed in this article would occur outside such communities and would probably not provide an adequate rationale for smaller communities pursuing such policies. For example, because a small community within a metropolitan area does not constitute even a quasi-independent labor market, most of the employment benefits of any community economic development program will occur outside the community. The only nonmarket benefit relevant for a small community is the fiscal benefit from economic development.

3. Hence, I am not including analysis of policies that indirectly spur economic development by increasing the quantity or quality of regional labor (e.g., education and training policies, or policies to attract in-migrants by improving amenities) or of policies that provide infrastructure (e.g., roads) that aids both households and businesses. These labor-supply and infrastructure policies are less controversial than regional economic development policies that directly aid business.

4. There are endless schemes for classifying regional economic development policies. But the two-part division suggested here corresponds to other observers' views of regional economic development policy. For example, Peter K. Eisinger, in *The Rise of the Entrepreneurial State* (Madison, WI: University of Wisconsin Press, 1988), divides economic development policies into "supply-side" and "demand-side" types, with this division corresponding to my division into traditional and new wave approaches. I should also point out here that although the new wave approach may receive more media attention, the traditional approach probably receives the bulk of economic development resources in most state and local governments.

5. For example, if individual *1* has a unit of good *x* that he values at \$T, and individual *2* values that unit of the good at \$S > \$T, then an economic-efficient change is to transfer that unit of *x* from individual *1* to individual *2* as benefits of the change (\$S) exceed the costs (\$T). One way of accomplishing this change is a free market in good *x*, with a price between \$S and \$T; individual *1* could then sell the good *x* to individual *2*, and both parties to the transaction would gain.

6. This assumption also simplifies the task of benefit-cost analysis. In general with multiple market failures, virtually every price in the economy will be distorted from a true estimate of social benefits and costs. This distorts our measurement of program costs. For example, with involuntary unemployment, the wage costs to a regional economic development program of hiring clerical help will overstate the true social costs. By ignoring these effects of distorted prices on the social value of program costs in the article's analysis, I am implicitly assuming that these types of distortions in program costs are of secondary importance in doing a good benefit cost analysis of these programs.

7. This article, of course, cannot hope to settle the longstanding debate over how much of regional unemployment is involuntary. However, as mentioned in the text, surveys do indicate that most unemployed individuals state a willingness to work at wages below the current market wage. Furthermore, interregional mobility appears unlikely to solve this problem because of high psychological and financial moving costs. For evidence of high moving costs, see L. F. Dunn, "Measuring the Value of Community," *Journal of Urban Economics* 6 (1979): 371-382.

8. George Akerlof and Janet Yellen, *Efficiency Wage Models of the Labor Market* (New York: Cambridge University Press, 1986).

9. From the perspective of regional policymakers, policy effects on in-migrants are not crucial. In addition, in-migrants do not gain greatly from the policies of their chosen region. From the perspective of in-migrants, this region is just one of a number of similar regions. Economic changes in this one region cannot substantially alter the economic opportunities open to in-migrants. If this one region had not changed its policies, the in-migrants could have moved to another, similar region. On the other hand, financial and psychological moving costs mean that what happens in a person's current region is of uniquely large importance. For more formal elaboration of a model with asymmetry of local policy effects between in-migrants and current residents, see Timothy J. Bartik, "Neighborhood Revitalization's Effects on Tenants and the Benefit-Cost Analysis of Government Neighborhood Programs," *Journal of Urban Economics* 19 (March 1986): 234-248.

10. Michael Greenwood and Gary Hunt, "Migration and Interregional Employment Redistribution in the United States," *American Economic Review* 74 (December 1984): 957-969; Richard F. Muth, "Migration: Chicken or Egg?" *Southern Economic Journal* 37 (January 1971): 295-306; Timothy J. Bartik, *Who Benefits from State and Local Eco-*

nomic Development Policies? (Kalamazoo, MI: W. E. Upjohn Institute for Employment Research, 1991).

11. See Robert J. Gordon, "The Welfare Cost of Higher Unemployment," *Brookings Papers on Economic Activity* 1 (1973): 133-206; Sammis B. White, "Reservation Wages: Your Community May Be Competitive," *Economic Development Quarterly* 1 (1987): 18-29; and Harry J. Holzer, "Black Youth Nonemployment: Duration and Job Search," in *The Black Youth Employment Crisis,* ed. R. B. Freeman and H. Holzer (Chicago: University of Chicago Press, 1986) for estimated reservation wages.

12. In addition, there will be fiscal benefits of hiring the unemployed, equal to the savings in welfare and UI costs plus the extra taxes paid once employed. Fiscal benefits of growth are discussed later in the article.

13. If labor costs for manufacturing are ten times the state and local taxes it pays, then a subsidy of .6% of the payroll is equivalent to a 6% reduction in the firm's state and local taxes.

14. See William Dickens and Lawrence Katz, "Inter-Industry Wage Differences and Industry Characteristics," in *Unemployment and the Structure of Labor Markets,* eds. K. Lang and J. Leonard (Oxford: Basil Blackwell, 1987), p. 49.

15. See Alan Krueger and Lawrence Summers, "Efficiency Wages and the Inter-Industry Wage Structure," *Econometrica* 56 (March 1988): 259-294; William Dickens and Lawrence Katz, "Inter-Industry Wage Differences and Industry Characteristics," in *Unemployment and the Structure of Labor Markets,* ed. K. Lang and J. Leonard (Oxford, England: Basil Blackwell, 1987); and Lawrence Katz and Lawrence Summers, "Industry Rents: Evidence and Implications," *Brookings Papers on Economic Activity* (Microeconomics issue, 1989): 209-290.

16. The list of high-wage premium industries in the text comes from Table II in Krueger and Summers, 1988, which lists industry-wage premiums by two-digit SIC (standard industrial classification) code. Krueger and Summers also calculate industry wage differentials for industries classified according to their three-digit SIC code. While some of these industries with high positive wage premiums are experiencing economic problems in the U.S. at present, this is by no means a list of sunset industries. For example, the computer, scientific and medical instruments, and pharmaceutical industries are all expected to experience rapid growth in the U.S. to the year 2000 (Valerie A. Personick, "Industry Output and Employment: A Slower Trend for the Nineties," *Monthly Labor Review* 112 [November 1989]: 25-41), and all are estimated by Krueger and Summers to have positive-wage premiums. In deciding on industry targets for an economic development strategy, a policymaker would, of course, have to consider the future prospects of the industry, as well as its wage premium.

17. See Alan B. Krueger and Lawrence H. Summers, "Reflections on the Inter-Industry Wage Structure," in *Unemployment and the Structure of Labor Markets,* ed. K. Lang and J. Leonard (London: Basil Blackwell, 1987).

18. A similar argument is made by Katz and Summers, 1989, who encourage federal policymakers to consider the effects of policies on the mix of high-wage premium industries versus other industries.

Note that high-wage premium industries are not the same as high-wage industries. High-wage premium industries pay well, relative to the skills required.

19. Katz and Summers, 1989, pp. 241-247; Dickens and Katz, 1987, pp. 78-79.

20. Note that this extra-wage premium is the change in the level of wages *relative to the skills required,* not the actual change in wages.

21. Timothy J. Bartik, "The Effects of Property Taxes and Other Local Public Policies on the Intrametropolitan Pattern of Business Location," in *Industry Location and Public Policy,* ed. H. Herzog and A. Schlottmann (University of Tennessee Press, 1990); William A. Fischel, "Fiscal and Environmental Considerations in the Location of Firms in Suburban Communities," in *Fiscal Zoning and Land Use Controls,* ed. Edwin Mills and Wallace Oates (Lexington, MA: Lexington Books, 1975), pp. 119-174; Helen F. Ladd, "Local Education Expenditures, Fiscal Capacity, and the Composition of the Property Tax Base," *National Tax Journal* 28 (1975): 145-158; William F. Fox and C. Warren Neel, "Saturn: The Tennessee Lessons," *Forum for Applied Research and Public Policy* 2 (1987): 133-206. Fox and Neel's research indicates that this excess of normal taxes on manufacturing over direct public service costs even applies to *greenfield* construction, that is, new plants in relatively undeveloped areas.

22. Peter K. Eisinger, 1988, p. 47.

23. Peter Diamond, "Aggregate Demand Management in Search Equilibrium," *Journal of Political Economy* 90 (1982): 881-894; Robert Hall, "Invariance Properties of Solow's Productivity Residual," (Working Paper 3034, Cambridge, MA: National Bureau of Economic Research, 1989).

24. Chapter III in Adam Smith's *The Wealth of Nations* (1776).

25. Robert Lucas, "On the Mechanics of Economic Development," *Journal of Monetary Economics* 22 (July, 1988): 3-42; Jane Jacobs, *The Economy of Cities* (New York: Random House, 1969).

26. For example, see J. Vernon Henderson, *Urban Development* (New York: Oxford University Press, 1988).

27. The argument may hold to a lesser degree for applied research than for basic research, assuming that the ratio of spillover benefits to private benefits is less for applied than for basic research. However, some subsidy for applied research, at a lesser level than for basic research, would still be warranted.

28. See Stephen Marglin, "The Social Rate of Discount and the Optimal Rate of Investment," *Quarterly Journal of Economics* 77 (1963a): 95-111.

29. See Stephen Marglin, "The Opportunity Costs of Public Investment," *Quarterly Journal of Economics* 77 (1963): 274-289; Arnold Harberger, "The Opportunity Costs of Public Investment Financed by Borrowing," in *Cost-Benefit Analysis,* ed. R. Layard (New York: Penguin Books, 1972); and Martin Feldstein, "The Inadequacy of Weighted Discount Rates," in *Cost Benefit Analysis,* ed. R. Layard (New York: Penguin Books, 1972). Among other things, this literature focuses on the social opportunity cost of private investment displaced by the project's financing. For state

and local programs most financing comes from taxation, which primarily displaces consumption.

30. The market failure approach can still encourage useful evaluations even if data limitations prevent measurement of the dollar benefits of providing nonmarket benefits. For example, if the profit effect of providing information about exporting cannot be measured, surveys could at least ask firms whether government information played a significant role in helping firms increase exports. (See Robert T. Kudrle and Cynthia M. Kite, "The Evaluation of State Programs for International Business Development," *Economic Development Quarterly* 3 [1989]: 288-300, for an excellent discussion of different types of evaluation of state regional economic development programs.) Such surveys would provide a rough qualitative assessment of the degree to which the regional economic development program is successfully addressing a market failure. The market failure approach would still play a useful role in focusing policy attention on areas where private markets are most inefficient. But, as discussed in the text, dollar measurement of benefits provides a better basis for decisions about program mix and program size.

31. An extreme economic view that has sometimes been taken is that perfectly rational regional governments will everywhere adopt business subsidies that exactly match the marginal benefits of new-business growth (see Michael Kieschnick, "Taxes and Growth," in *State Taxation Policy,* ed. M. Barker [Durham, NC: Duke University Press, 1983] for a critique of such models). In such cases the geographic pattern of subsidies will encourage the optimal redistribution of economic activity. Each region's self-interested decisions about subsidies will maximize national economic efficiency. There are no net negative spillover effects on other regions because the net effect of reduced business growth in other regions is zero. The negative effect of reduced nonmarket benefits in these other regions is exactly matched by these regions' savings in subsidies. While the real world does not have perfect subsidies, economic forces tend to push us toward a pattern of subsidies that improves economic efficiency relative to a zero-subsidy world.

Chapter 3

City Jobs and Residents on a Collision Course

The Urban Underclass Dilemma

John D. Kasarda

❁❁❁

America's cities have always been at the forefront of national economic change. The largest cities of the North spawned our industrial revolution in the late 19th and early 20th centuries, generating massive numbers of blue-collar jobs that served to attract and economically upgrade millions of disadvantaged migrants. More recently, these same cities were instrumental in transforming the U.S. economy from goods processing to basic services (during the 1950s and 1960s) and from a basic service economy to one of information processing and administrative control (during the 1970s and 1980s).

The transformation of major northern cities from centers of goods processing to centers of information processing was accompanied by marked changes in their employment bases. As more efficient transportation and production technologies evolved, manufacturing dispersed to the suburbs, exurbs, nonmetropolitan areas, and abroad. Warehousing activities shifted to more regionally accessible beltways and interstate highways. Retail and consumer service establishments, following their suburbanizing clientele, relocated to peripheral shopping centers and malls.

While most parts of the metropolitan core experienced an economic base hemorrhage, pockets of inner-city employment resurgence emerged. Nowhere is this more visible than with the central business district office-building boom, where jobs in administration, finance, communications, and the professions have mushroomed since 1970. Employment growth in these knowledge-intensive, white-collar industries, however, has not typically compensated either numerically or substantively for employment declines in manufacturing, trade, and blue-collar service industries that had once constituted the economic backbone of these cities and sustained their less-educated residents.

Aggravating inner-city employment declines in these traditional urban industries was the exodus of white middle-income residents and the neighborhood business establishments that

Reprinted from *Economic Development Quarterly,* Vol. 4, No. 4, November 1990.

once served them. This exodus drained the city tax base and further diminished the number of low-skill service jobs, such as domestic workers, gas station attendants, and local delivery personnel. Concurrently, many secondary commercial areas of cities withered as the lower-income levels of minority residential groups that replaced the suburbanizing whites could not economically sustain them.

It is particularly noteworthy that cities that lost the largest number of blue-collar and other jobs with lower educational requisites during the past two decades also added to their working-age populations large numbers of blacks with no education beyond high school. Most of these new labor force entrants were offspring of the millions of southern blacks who had migrated to northern cities during the 25-year period following World War II when employment opportunities requiring only limited education were far more plentiful. These lower-skill jobs have been replaced by more knowledge-intensive jobs that demand education beyond high school and, hence, are not *functionally* accessible to most urban blacks, even though the new jobs are in relatively close proximity to them (that is, they are *spatially* accessible).[1]

The simultaneous, yet conflicting, transformations of the employment and demographic bases of the cities contributed to a number of serious problems, including a widening gap between the skill needs of urban job growth industries and the skill levels of minority resident labor (with correspondingly high rates of localized structural unemployment), greater distances separating inner-city blacks from suburban areas of blue-collar job growth, and increasing levels of urban poverty. Associated with these problems have been a plethora of social dysfunctions, such as high rates of family dissolution, out-of-wedlock births, drug abuse, crime, and dropping out of school that have further aggravated the predicament of the economically displaced. The spatial concentration of poverty, joblessness, and social dysfunctions, in turn, contributed to the formation of a segregated, culturally isolated, immobilized subgroup that has come to be labeled the *urban underclass.*

The Rise of the Urban Underclass

Economic distress created by urban industrial change and white flight is only part of the story of the formation of the urban underclass. With important civil rights gains during the 1960s and 1970s, selective black flight from the ghettos accelerated, resulting in a socioeconomic and spatial bifurcation of urban black communities. As William Julius Wilson points out, prior to the 1960s black, inner city communities were far more heterogeneous in socioeconomic mix and family structure because de facto and de jure segregation bound together blacks of all income levels.[2] The presence of working-class and middle-income blacks within or near the ghettos sustained such essential local institutions as neighborhood clubs, churches, schools, and organized recreational activities for youth. Working-class and middle-income black residents also provided community leadership, mainstream role models for youth, greater familial stability, and sanctions against deviant behavior.

Yet, as Wilson argues, it was these more economically stable blacks who disproportionately benefited from civil rights gains, such as affirmative action and open housing, which removed artificial barriers to job access and facilitated their exodus from ghetto neighborhoods. Left behind in increasingly isolated concentrations were the most disadvantaged with the least to offer in terms of marketable skills, role models, and economic and familial stability. Under such conditions, ghetto problems became magnified, and socialization of the younger residents to mainstream values and positive work ethics atrophied.

Let me add four extensions to Wilson's thesis. First, with the flight of working- and middle-class blacks from the ghettos, not only were mainstream role models, normative guidance, and neighborhood leadership resources lost, but it also became extremely difficult for most small, black-owned stores and shops that served ghetto residents to survive. It was often these locally owned neighborhood establishments that provided ghetto youth with their initial job experience and in so doing also offered visible models of employed teenagers. When these establishments closed, both important functions were lost.

Second, prior to the 1960s, de jure and de facto racial segregation in business and shopping patterns resulted in protected markets, with black earnings being expended primarily in black-owned establishments. Money earned by blacks who worked in white-owned businesses (or for white households) was much more likely to be funneled to a black-owned neighborhood establishment or local black professional than became the case in the 1970s and 1980s. Black income thus was multiplied through black chains of exchange rather than flowing out of the black community, which is more likely to be the case today.[3] As a result, not only was aggregate black community income diminished but, in turn, the number of blacks who could be employed in the neighborhood substantially declined.

Third, it is well documented that affirmative action programs were far more effective in the public sector than in the private sector. My analysis of changes in white-collar employment in major northern cities between 1970 and 1980, using the Bureau of the Census machine-readable Public Use Microdata Sample files shows that upper-echelon, white-collar employment gains by central-city blacks were skewed toward the public sector, whereas such gains by non-Hispanic whites and others were almost exclusively in the private sector. By the mid-1970s administrative growth in the public sector had already begun to slow, especially in the major cities, and it slowed further during the 1980s' era of urban fiscal austerity.[4] Also during the 1970s, a burst of entrepreneurship and small business growth commenced that bypassed blacks. In contrast to dramatic gains by most other racial and ethnic groups, the number of black-owned firms with employees actually declined.[5] It seems plausible that the differential success in affirmative action in the public sector disproportionately attracted better-educated, more-talented blacks from private-sector pursuits, where most upper-income growth opportunities emerged between 1975 and the present. Entering the more secure public sector, I propose, also reduces the prospects of these persons starting their own businesses and thus economically bolstering the black community by providing additional employment opportunities for its residents.

Fourth, since the early 1970s certain federal policies have been guided by the reasonable principle that public assistance should be targeted to areas where the needs are the greatest, as measured by such factors as job loss, poverty rate, and persistence of unemployment.[6] The idea was that the most distressed areas should receive the largest allocation of government funds for subsistence and local support services for the economically displaced and others left behind. While these policies unquestionably helped relieve pressing problems, such as the inability of the unemployed to afford private-sector housing or obtain adequate nutrition and health care, they did nothing to reduce the skills mismatch between the resident labor force and available urban jobs. In fact, spatially concentrated assistance may have inadvertently increased the mismatch and the plight of educationally disadvantaged residents by binding them to inner-city areas of severe blue-collar job decline and in areas that, by program definition, are the most distressed.

For those individuals with some resources and for the fortunate proportion whose efforts to break the bonds of poverty succeed, spatially concentrated public assistance will not impede their mobility. But for many inner-city poor without skills and few economic options, local concentrations of public assistance and community services can be sticking forces. Given their lack of skills, they may see themselves as better off with their marginal, but secure, in-place government assistance than they would be by taking a chance and moving in search of a minimum wage, entry-level job, often in an unknown environment.

America's New Ethnic Immigrants

If spatial confinement and poor education are such handicaps to gainful employment in industrially transforming cities, why is it that America's new urban immigrant groups, especially Asians, have had such high success in carving out employment niches and climbing the socioeconomic ladder? Like blacks, many Asian immigrants arrived with limited education and financial resources and are spatially concentrated in inner-city enclaves. Recent research on ethnic entrepreneurism casts light on the reasons for these immigrants' success and is suggestive for reducing underclass problems.[7] These studies show the critical importance of ethnic solidarity and kinship networks in fostering social mobility in segregated enclaves through self-employment.

Many Asian immigrants and certain Hispanic groups have been able to use ethnic-based methods to (1) assemble capital, (2) establish internal markets, (3) circumvent discrimination, and (4) generate employment in their enclaves that is relatively insulated from swings in the national economy and urban structural transformation. Ethnic businesses are typically family-owned and -operated, often drawing upon unpaid family labor to staff functions during start-up periods of scarce resources and also upon ethnic contacts to obtain credit, advice, and patronage. The businesses are characterized by thriftiness and long hours of intense, hard work, with continuous reinvestment of profits. As they expand, ethnic-enclave establishments display strong hiring preferences for their own members, many of whom would likely face employment discrimination by firms outside their enclave. They also do business with their own. A San Francisco study found that a dollar turns over five or six times in the Chinese business community, while in most black communities dollars leave before they turn over even once.[8]

Kinship and household structures of ethnic immigrants have significantly facilitated their entrepreneurial successes. Among recently arrived Asian immigrants, for instance, other relatives (those beyond the immediate family) constitute a substantial portion of households: 55% among Filipinos and Vietnamese, 49% among Koreans, 46% among the Chinese, and 41% among Asian Indians.[9] In addition to serving as a valuable source of family business labor, these extended-kin members enable immigrant households to function more efficiently as economic units by sharing fixed household costs such as rents or mortgages, furnishing child-care services, and providing economic security against loss of employment by other household members.[10] In short, by capitalizing on ethnic and family solidarity, many new immigrant businesses, ranging from laundries to restaurants to greengroceries, have started and flourished in once downtrodden urban neighborhoods, providing employment and mobility options to group members in what is in other respects an unfavorable economic environment.[11]

Native urban blacks, in contrast, have been burdened by conditions that have impeded their entry and success in enclave employment, including lack of self-help business associations, limited economic solidarity, and family fragmentation. A survey by *Black Enterprise* magazine reported that 70% of self-employed blacks consider lack of community support as one of their most formidable problems.[12] This, together with the documented flight of black-earned income to nonblack establishments, led well-known black journalist Tony Brown to comment, "The Chinese are helping the Chinese, the Koreans help the Koreans, Cubans help Cubans, but blacks are helping everyone else. We have been conducting the most successful business boycott in American history—against ourselves."[13] Apparently racial political unity that has led to significant black electoral successes in major northern cities during the past two decades has not carried over to the economic sphere.

Given the demonstrated importance of family cohesiveness and kinship networks in pooling resources to start businesses, provide day-care assistance, and contribute labor to family business ventures, the black underclass is at a distinct disadvantage. As Wilson found, approximately two-thirds of black families living in Chicago's ghettos are mother-only households.[14] These households are the poorest segment of our society, with female householders earning only one-third as much as married-male householders.[15] My point is that it takes discretionary resources to start a small business and it requires patrons with money to sustain that business, both of which are in limited supply in underclass neighborhoods.

All the factors previously mentioned have converged to depress black self-employment rates during the past two decades.[16] According to 1980 census data, lesser-educated urban blacks are consistently underrepresented in self-employment, when compared to other racial-ethnic groups, especially Asians.[17] Korean immigrant self-employment rates, for example, range from a low of 19% in Chicago to a high of 35% in New York.[18]

Just as striking are Asian-black contrasts from the most recent census survey of minority-owned businesses. Between 1977 and 1982 the number of Asian-American-owned firms with paid employment expanded by 160%. During the same period, the number of black-owned firms with employees actually declined by 3%.[19] With small business formation becoming the backbone of job creation in America's new economy, blacks are falling further behind in this critical arena.

Financial weakness and family fragmentation among the black underclass not only preclude capital mobilization for self-employment, but also create barriers to their children's educational attainment and eventual employment prospects. Living in a mother-only household increases the risk of young blacks dropping out of school by 70%.[20] The link between female headship and welfare dependency in the urban underclass is also well established, leading to legitimate concerns about the intergenerational transfer of joblessness. At the root of this concern is the paucity of employment among welfare mothers and its effect on the attitudes of their children toward work. Of those receiving welfare benefits 85% have no reported source of income other than public assistance.[21] Furthermore, 65% of the recipients of Aid to Families with Dependent Children at any one time are in an interval of dependency that has lasted for at least eight years.[22] One does not require a deep sociological imagination to sense the attitudinal and behavioral consequences of growing up in an impoverished household where there is no activity associated with the world of work and a household that, in turn, is spatially embedded in a commercially abandoned locality where pimps, drug pushers, and unemployed street people have replaced working fathers as the predominant socializing agents.

Fostering Social and Spatial Mobility

It is clear from a substantial amount of research that strengthening the black family and reducing the exceptionally high percentage of impoverished mother-only households must be key focuses of policies to rekindle social mobility among today's urban underclass. These policies should be complemented with programs that improve opportunities for ghetto youths to be reared in household and neighborhood environments where adult work is the norm and to attend public schools that will provide them with necessary skills and social networks for employment in a rapidly transforming economy. In this regard, it has been shown that low-income black youths who moved from inner-city Chicago to predominantly white suburbs as part of a subsidized housing experiment performed remarkably well, both academically and socially.[23] Follow-up research on poor black mothers who relocated from inner-city public housing to these suburbs showed that the women substantially improved their job prospects compared to a control group who moved from public housing to better apartments within the city.[24]

I stressed earlier that, as cities have economically changed from centers of goods processing to centers of information processing, there has been a significant rise in education required for urban employment. If greater portions of disadvantaged black youths do not acquire the formal education to be hired by the white-collar service industries that are beginning to dominate urban employment bases, their jobless rates will remain high. For this reason, and because demographic forces portend potential shortages of educationally qualified resident labor for the white-collar industries expanding in the cities, there have been appropriate calls from the public and private sectors to upgrade city schools, reduce black youth drop-out rates, and increase the proportion of those continuing for higher education.

Such policies, however, are unlikely to alleviate the unemployment problems currently facing large numbers of economically displaced older blacks and yet-to-be-placed younger ones with serious educational deficiencies—those caught in the web of urban change. Their unemployment will persist because the educational qualifications demanded by most of today's urban-growth industries are difficult to impart through short-term, nontraditional programs.[25]

The implausibility of rebuilding urban blue-collar job bases, or of providing sufficient education to large numbers of displaced black labor so that they may be reemployed in expanding white-collar industries, necessitates a renewed look at the traditional means by which Americans have adapted to economic displacement—that is, spatial mobility. Despite the mass loss of lower-skill jobs in many cities during the past decade, there have been substantial increases in these jobs nationwide. For example, between 1975 and 1985 more than 2.1 million nonadministrative jobs were *added* in eating and drinking establishments, which is more than the total number of production jobs that existed in 1985 in America's automobile, primary metals and textile industries combined.[26] Unfortunately, essentially all of the net national growth in jobs with low educational requisites has occurred in the suburbs, exurbs, and nonmetropolitan areas, which are far removed from large concentrations of poorly educated minorities. It seems both an irony and a tragedy that we have such surpluses of unemployed lesser-skilled labor in the inner cities at the same time that suburban businesses are facing serious shortages in lesser-skilled labor.

The inability of lesser-skilled urban blacks to follow decentralizing lesser-skilled jobs increasingly isolates them from shifting loci of employment opportunity and has contributed to their high rates of joblessness. To reduce this isolation and improve their employment prospects, a number of spatial-mobility strategies should be considered, including (1) a computerized job-opportunity network providing up-to-date information on available jobs throughout a particular metropolitan area, the region, and the nation; (2) partial underwriting of more distant job searches by the unemployed; (3) need-based temporary relocation assistance, once a job has been secured; (4) housing vouchers for those whose income levels require such assistance, as opposed to additional spatially fixed public housing complexes; (5) stricter enforcement of existing fair-housing and fair-hiring laws; (6) public-private cooperative efforts to van-pool unemployed inner-city residents to suburban businesses facing labor shortages; and (7) a thorough review of all spatially targeted low-income public assistance programs to ensure that they are not inadvertently anchoring those with limited resources to distressed areas in which there are few opportunities for permanent or meaningful employment.

The aim of these people-to-jobs strategies is not only to bring a better balance between local supplies and demands for labor, but also to facilitate the means by which disadvantaged Americans have historically obtained economic opportunity and a better life. In this regard, it is not fortuitous that the three great symbols of social and economic opportunity of America's disadvantaged all relate to spatial mobility—the Statue of Liberty, the underground railway, and the covered wagon.

Notes

1. Data documenting trends in the urban economic and demographic restructuring upon which much of this article is based may be found in John D. Kasarda, "Urban Change and Minority Opportunities" (pp. 33-67); in *The New Urban Reality*, ed. Paul Peterson (Washington, DC: The Brookings Institution, 1985); idem, "Urban Industrial Transition and the Underclass," in *The Annals of the American Academy of Political and Social Science* 501 (January 1989): 26-47; and idem, "Structural Factors Affecting the Location and Timing of Urban Underclass Growth," in *Urban Geography* 11(3): 234-264.

2. William Julius Wilson, *The Truly Disadvantaged* (Chicago: University of Chicago Press, 1987).

3. Joel Kotkin, "The Reluctant Entrepreneurs," *Inc.* 8(9): 81-86; Rick Wartzman, "St. Louis Blues: A Blighted Inner City Bespeaks the Sad State of Black Commerce," *Wall Street Journal* 10 May 1988, p. 1.

4. Philip I. Moss, "Employment Gains by Minorities, Women in Large City Government, 1976-83," *Monthly Labor Review* (November 1988): 18-24.

5. Robert L. Boyd, "Ethnic Entrepreneurs in the New Economy: Business Enterprise among Asian Americans and

Blacks in a Changing Urban Environment" (Ph.D. Diss., University of North Carolina at Chapel Hill, 1989); U.S. Bureau of the Census, *The 1977 Survey of Minority-Owned Business Enterprises: Black,* MB77-1 (Washington, DC: U.S. Department of Commerce, 1979); U.S. Bureau of the Census, *The 1982 Survey of Minority-Owned Business Enterprises: Black* MD82-1 (Washington, DC: U.S. Department of Commerce, 1985).

6. U.S. Department of Housing and Urban Development, *The President's National Urban Policy Report, 1978* (Washington, DC: U.S. Department of Housing and Urban Development, 1978); U.S. Department of Housing and Urban Development, *The President's National Urban Policy Report, 1980* (Washington, DC: U.S. Department of Housing and Urban Development, 1980); Bert E. Swanson and Ronald K. Vogel, "Rating America's Cities—Credit Risk, Urban Distress and the Quality of Life," *Journal of Urban Affairs* 8(2): 67-84.

7. Ivan Light, *Ethnic Enterprise in America* (Berkeley: University of California Press, 1972); Edna Bonacich and John Modell, *The Economic Basis of Ethnic Solidarity* (Berkeley: University of California Press, 1980); Kenneth L. Wilson and Alejandro Portes, "Immigrant Enclaves: An Analysis of Labor Market Experiences of Cubans in Miami," *American Journal of Sociology,* 86 (1980): 295-319; Roger Waldinger, *Through the Eye of the Needle: Immigrants and Enterprise in New York's Garment Trades* (New York: New York University Press, 1986); Thomas R. Bailey, *Immigrant and Native Workers: Contrasts and Competition* (Boulder, CO: Westview Press, 1987).

8. Kotkin, "The Reluctant Entrepreneurs"; Wartzman, "St. Louis Blues."

9. Peter S. Xenos, Robert W. Gardner, Herbert R. Barringer, and Michael J. Levin, "Asian Americans: Growth and Change in the 1970s" in *Pacific Bridges: The New Immigration from Asia and the Pacific Islands,* ed. James T. Fawcett and Benjamin V. Carino (Staten Island, NY: Center for Migration Studies, 1987), p. 266.

10. Ibid.

11. Robert L. Boyd, "Ethnic Entrepreneurs in the New Economy" (Paper, University of North Carolina at Chapel Hill, 1988).

12. Kotkin, p. 84.

13. Ibid.

14. Wilson, "The Truly Disadvantaged."

15. Sara McLanahan, Irwin Garfinkel, and Dorothy Watson, "Family Structure, Poverty, and the Underclass," in *Urban Change and Poverty,* ed. Michael G. H. McGeary and Laurence E. Lynn, Jr. (Washington, DC: National Academy Press, 1988), pp. 102-47.

16. Eugene H. Becker, "Self-Employed Workers: An Update to 1983," *Monthly Labor Review* (1984) 107(7): 14-18; Kotkin, "The Reluctant Entrepreneurs"; Wartzman, "St. Louis Blues."

17. Kasarda, "Urban Change."

18. Roger Waldinger, et al., "Spatial Approaches to Ethnic Business," in Roger Waldinger et al., *Ethnic Entrepreneurs: Immigrant Business in Industrial Societies* (Newbury Park, CA: Sage, 1990).

19. Boyd, "Ethnic Entrepreneurs."

20. Sara McLanahan, "Family Structure and the Reproduction of Poverty," *American Journal of Sociology* 90 (1985): 873-901.

21. Irwin Garfinkel and Sara S. McLanahan, *Single Mothers and Their Children* (Washington, DC: Urban Institute Press, 1986).

22. David T. Ellwood, "Targeting the Would-Be Long-Term Recipients of AFDC: Who Should Be Served?" (Report, Harvard University, 1985).

23. James Rosenbaum, Marilyn J. Kulieke, and Leonard S. Rubinowitz, "Low-Income Black Children in White Suburban Schools: A Study of School and Student Responses," *Journal of Negro Education* 56 (1987): 35-43.

24. James Rosenbaum and Susan J. Popkin, "The Gautreaux Program: An Experiment in Racial and Economic Integration," *The Center Report: Current Policy Issues* (Center for Urban Affairs and Policy Research, Northwestern University) 2 (1).

25. John D. Kasarda, "Jobs, Migration and Emerging Urban Mismatches," in *Urban Change and Poverty,* ed. Michael G. H. McGeary and Laurence E. Lynn, Jr. (Washington, DC: National Academy Press, 1988), pp. 148-98.

26. U.S. Department of Labor, Bureau of Labor Statistics, *Establishment Data 1939-1986 Machine Readable Files* (Washington, DC).

Chapter 4

New Strategies for Inner-City Economic Development

Michael E. Porter

❋❋❋

The economic distress of America's inner cities is one of the most pressing issues facing the nation.[1] The time has come to recognize that revitalizing these areas requires a radically different approach. Today, most efforts and public resources, including the Empowerment Zone program, are still targeted toward meeting residents' immediate needs rather than generating jobs and economic opportunity that will mitigate the need for social programs in the long run. Although efforts to provide education, housing, health care, and other needed services are essential and must continue, these must be balanced with a concerted and realistic economic strategy focused on for-profit business and job development. The necessity—and the real opportunity—is to create income and wealth, by harnessing the power of market forces, rather than trying to defy them. The private sector must play a leading role and, in many ways, is already beginning to do so.

An economic strategy for inner cities is needed as a complement to (not a substitute for) the many programs designed to increase human capital and meet the basic human needs of disadvantaged populations. A successful economic strategy will result in viable businesses that can provide the employment opportunities sorely needed in, or near, distressed inner-city neighborhoods—neighborhoods in which, in most cases, African Americans and other people of color represent the majority of the population. Employment opportunities are a linchpin for the success of virtually all other programs designed to improve human capabilities, values, and attitudes in distressed communities.

This research focuses on how to create jobs and sustainable business activity that benefit

AUTHOR'S NOTE: This article has benefited from extensive research and assistance by Whitney R. Tilson, executive director of the Initiative for a Competitive Inner City. I would also like to thank Barbara J. Paige and Ronald A. Homer for their invaluable comments, as well as many other students and individuals whose research over the last several years made this article possible.

Reprinted from *Economic Development Quarterly,* Vol. 11, No. 1, February 1997.

disadvantaged inner-city residents. It grows out of a stream of research on economic development in nations, states, and cities first described in *The Competitive Advantage of Nations.*[2] The application to inner cities is based on studies of nine major inner cities;[3] hundreds of interviews nationwide with inner-city-based companies, community-based organizations (CBOs), bankers, government officials, and others; an extensive survey of the literature; and close advisory relationships with more than a dozen inner-city companies in Boston.[4] Unfortunately, reliable statistical data on business activities in inner cities are not yet available, though we have a growing body of survey evidence.

The approach to inner-city economic development my colleagues and I have developed is fundamentally different from most existing efforts. It reflects the new realities of inner-city America and the role that inner cities might occupy in the national and international economy. Although threads of this approach have appeared in previous literature and in the efforts of some organizations, the approach has not yet been implemented comprehensively in any one inner city. However, the core principles have been proven many times over throughout the world.

Part of the reason for the vigorous debate surrounding inner-city economic development is the definition of the term *economic development* itself. The American Economic Development Council (AEDC) defines economic development as "the process of creating wealth through the mobilization of human, financial, capital, physical and natural resources to generate marketable goods and services."[5] But as John Blair notes, "economic development concerns reach into numerous aspects of community life. . . . In practice, distinctions between social, political, and economic development concerns are fuzzy."[6] This has led to a tendency to widen the definition of economic development to include virtually everything and for different definitions to emerge. The result has been both confusion in communication and, we believe, unnecessary controversy. Individuals and organizations have also tended to focus on one or a few specific elements of economic development and assert their primacy.[7]

Our research concentrates on inner-city economic development in the narrow sense—the creation of jobs and sustainable business activity that benefit disadvantaged inner-city residents. This does not deny the importance of improved housing, health care, and schools to the overall revitalization of inner cities. These topics are simply not where we have concentrated and not where the greatest uncertainty in theory and practice lies.

A Strategy for Inner-City Economic Development

Our strategy begins with the premise that a sustainable economic base can be created in inner cities only as it has been elsewhere: through private, for-profit initiatives, and investments based on economic self-interest and genuine competitive advantage instead of artificial inducements, government mandates, or charity. A sound economic strategy must focus on the position of inner cities as part of regional economies, rather than treating inner cities as separate, independent economies; otherwise, economic activity there will not be sustainable. Although the changing economy, with its dual challenges of global competition and technological advances, has adversely affected inner cites, it has also created new opportunities.

There are many businesses present today in inner cities—a surprise to those who assume that little economic activity exists because of these communities' well-known problems. Our research has documented that inner-city businesses are concentrated in sectors such as food processing and distribution, printing and publishing, light manufacturing, recycling and remanufacturing, business support services for corporations, and entertainment and tourist attractions. These are all areas in which the genuine competitive advantages of inner-city locations are present.

Economic development in inner cities will come only from recognizing and enhancing the inherent advantages of an inner-city location and building on the base of existing companies, while dealing frontally with the present disadvantages of inner cities as business locations. There is genuine economic potential in inner cities that has been largely unrecognized and untapped.

The Competitive Advantages of Inner Cities

Our analysis of major cities nationwide has found that often-discussed advantages, such as low-cost labor and cheap real estate, are largely illusory. Inner cities have available workers, but wages are not lower than in rural areas or in other countries. Real estate costs may be lower than in nearby high-rent downtown areas, but cheaper real estate is available in the suburbs and elsewhere. The changing nature of the world economy means that inner cities will not be able to compete if low-cost labor and cheap real estate are their only advantages. Similarly, it is futile to try to recreate the inner-city economies of the past, with their high-wage, blue-collar manufacturing jobs, as many urban planners still hope to do. Instead, the genuine competitive advantages of inner cities fall into four areas: strategic location, integration with regional clusters, unmet local demand, and human resources.

Strategic Location

Inner cities occupy what should be some of the most valuable locations in their regions, near high-rent business centers, entertainment complexes, and transportation and communications nodes. As a result, an inner-city location can offer a competitive edge to logistically sensitive businesses that benefit from proximity to downtown, transportation infrastructure, and concentrations of companies. The just-in-time, service-intensive modern economy is only heightening the time and space advantages of these locations. Although some traditional location-sensitive businesses have been decentralized by modern technology, many others have been created (e.g., recycling, remanufacturing, value-added business services). This powerful locational advantage of inner cities, which has not been fully used or developed, explains the continued existence and growth of the many food processing, printing, business support, rapid-response warehousing and distribution, and light manufacturing companies in inner-city areas, despite the conspicuous problems.

In our surveys of cross sections of inner-city-based companies in Boston, Baltimore, Atlanta, and Oakland, strategic location was cited as the most important advantage for a significant majority of businesses. For example, in Boston, 90% of the 60 companies surveyed said proximity to customers was an important advantage, 55% cited proximity to nearby highways, and 35% cited proximity to suppliers.[8] In Atlanta, 81% of the 37 companies surveyed said strategic location was an important advantage.[9]

Integration with Regional Clusters

Longer-term development opportunities for inner cities lie in capitalizing on nearby regional clusters of firms and industries—unique concentrations of competitive companies in related fields. The notion that business development is strongly influenced by the external economies within clusters of interconnected industries has antecedents in the development literature. However, recent research has substantially developed and broadened this idea, demonstrating how it fits into a broader framework for understanding competitiveness.[10] The composition of clusters includes not only firms and suppliers but also educational institutions, specialized financial providers, and specialized research centers. The dynamic external economies of clusters and their role in new business formation are perhaps their most fundamental attributes. Although some clusters fail to grow for a wide variety of reasons, the power of a cluster-based approach to development is far greater than company-based or even sectoral efforts.[11]

An effective economic strategy for inner cities must focus on developing the clusters within inner cities, instead of isolated companies, and linking them better to those in the surrounding economy. The ability to access competitive clusters is more far-reaching in its economic implications than is the simple proximity of inner cities to the downtown or transportation in-

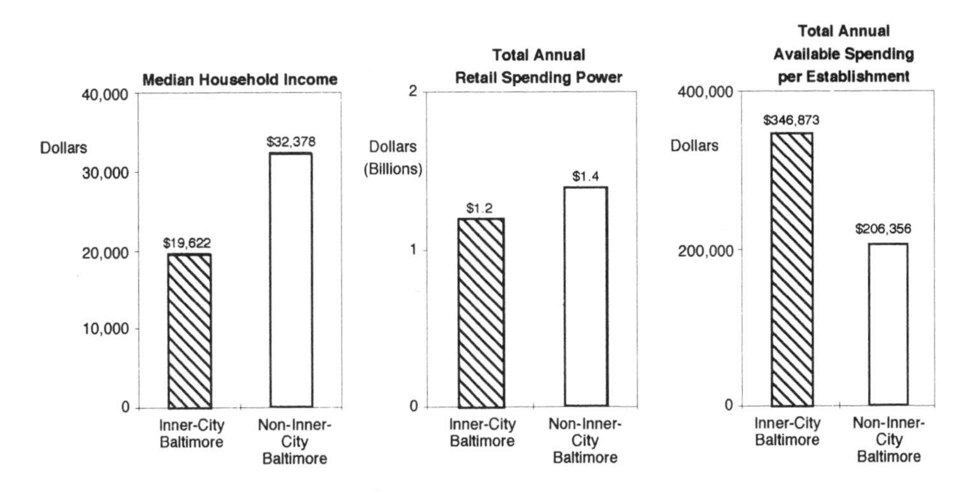

Figure 4.1. Inner-City Competitive Advantages: Baltimore's Unmet Local Demand

SOURCE: Mercer Management Consulting and Initiative for a Competitive Inner City, "The Competitive Advantages of Inner City Baltimore" (Unpublished report, Baltimore, 1995).

frastructure. Building on regional clusters involves tapping powerful external economies on information, skills, image, infrastructure, and markets. A cluster-based approach also leverages private and public investments in skills, technology, and infrastructure; for example, Boston is home to world-class clusters in health care, financial services, and tourism that surround the inner city. There are numerous opportunities to develop focused programs for training, purchasing, and business development leading to job opportunities for inner-city residents.

Unmet Local Demand

The consumer market of inner-city residents represents the most immediate opportunity for inner-city-based entrepreneurs and businesses. Despite low average incomes, high population density translates into a large local market with substantial purchasing power. Making the market even more attractive is the fact that there tend to be few competitors serving it. At a time when suburban markets are saturated, inner-city markets remain poorly served—especially in many types of retailing, financial services, and personal services. For example, Figure 4.1 shows that although the median household income in inner-city Baltimore is 39% lower than that of the rest of the city, the aggregate spending power is nearly the same, and the estimated retail spending per establishment is *two-thirds greater* in the inner city than in the rest of the city.[12] Inner-city-based businesses that serve this demand will have an advantage over more distantly located establishments because of their proximity to their customers. Inner-city-based retail and service businesses are also more likely to recognize and adapt to the fact that inner cities are distinct markets, which demand uniquely tailored product configurations, retail concepts, entertainment, and personal and business services. An opportunity is present for national retail and service chains focusing on inner cities, as well as large-scale manufacturing of tailored products to supply them.

The private sector is already waking up to the potential of inner cities. For example, supermarkets, facing market saturation in the suburbs, are launching successful stores in underserved inner cities.[13] Of the new Vons supermarkets opening in the 1990s, 25% will be urban, compared with none in the 1960s. The chief development officer of Vons said, "There are 1.7 million people within reach of our advertising, warehouses, and manufacturing who can't shop at a Vons, while the suburbs are overstored."[14] The CEO of Lucky concurs, saying, "This isn't a philanthropic exercise. There are good food customers who come out in large numbers to buy high-margin items like meat and produce that offset higher urban costs."[15] In Harlem, Fairway has opened a large, thriving supermarket that employs 170 people, 120 of whom are neighborhood residents. Pathmark has committed to open another supermarket in Harlem, in conjunction with the Local Initiatives Support Corporation's Retail Initiative program. A Pathmark company spokesman said city stores are "disproportionately profitable. Last year, they accounted for only 22% of the chain's 147 stores, but contributed 25% of its profits. In the 1990s, 50% of the chain's new stores will be urban, compared to 25% in the 1980s."[16] In Boston, seven new supermarkets have opened in the past three years, and the Stop & Shop in the South Bay Mall is one of the highest-selling stores in the 175-store chain.[17]

Other retail, service, and franchise companies are also undertaking profitable ventures in inner cities.[18] In Boston, Payless Shoes operates 7 stores that gross approximately $360 per square foot, which compares favorably with the world's best retailers.[19] In the Bronx, Caldor opened 2 stores that are already 2 of the 5 top-grossing stores in the country, and Rite-Aid announced that it will expand from 11 to more than 50 stores in the Bronx in the next few years.[20] Rite-Aid also reports that its Harlem store, opened in 1994, fills more prescriptions than does any other store in the city.[21] In South Central Los Angeles, the Magic Johnson Movie Theaters, opened in the past year, consistently rank in the top five theaters among the 21,800 surveyed nationwide, and plans are under way to open similar complexes in Atlanta, Houston, and Harlem.[22] In Boston, in addition to the hugely successful Stop & Shop mentioned above, the South Bay Mall includes the highest-volume Kmart in Massachusetts and the top-grossing Toys 'R' Us in New England.[23] A final example is Run 'N' Shoot, which is the only full-scale fitness center in Atlanta's inner city. A total of 1,000 to 1,500 people per day use the facility, which hires all of its managers and 60 part-time employees locally. A second Run 'N' Shoot will soon open nearby.

Many entrepreneurs are creating businesses that cater to the distinct needs of inner-city consumers. For example, Delray Farms, a chain of supermarkets in inner-city Chicago, specializes in meats and produce—food categories that inner-city consumers purchase in disproportionately high amounts[24]—and provides cuts of meat and types of produce that reflect the distinct ethnic mix of each neighborhood in which it operates.[25] Or consider The Lark, a chain of five men's apparel stores located in low-income neighborhoods of Gary, Indiana, and Chicago. Its product mix and marketing are uniquely tailored to local, primarily African American residents, who also comprise 90% of the workforce. Most important is first-class customer service, "because many Black shoppers say they feel slighted or even mistrusted by sales help in mainstream stores." The Lark's profits are "substantial," its theft and inventory losses are a minuscule 0.03% of sales, and employee turnover is less than 3%.[26]

Banks, initially driven by the requirements of the Community Reinvestment Act but increasingly motivated by self-interest, are making major investments in inner cities in new branches and home mortgages. Bank of America, for example, now makes half its home mortgages in California through a program, launched in 1990, called Neighborhood Advantage. The program has generated $6 billion in home loans since its inception. This program evaluates creditworthiness using nontraditional methods and requires lower down payments, yet the delinquency and foreclosure rates are two-thirds lower than Bank of America's conventional portfolio.[27]

Finally, a handful of companies have discovered lucrative niches in inner-city business lending. Medallion Funding, a private, for-profit lender and specialized small business investment corporation (SSBIC),[28] has expanded beyond its traditional lending for taxi medallions and now lends to inner-city laundromats and dry cleaners in New York City. Medallion Funding has generated average annual returns of 18% and just became a publicly traded company.[29]

Human Resources

Although inner-city populations present many workforce readiness challenges, inner-city residents can also be an attractive labor pool for businesses that rely on a loyal, modestly skilled workforce. There is the potential to build on this resource, with new approaches to education, job placement, and training. However, this requires debunking three deeply entrenched myths about the nature of inner-city residents.

The first is that inner-city residents do not want to work and opt for welfare over gainful employment. Although there is little doubt that inner cities as a whole have an undereducated, underskilled population, with a disproportionate number of people ill equipped for work for a variety of reasons, many employers report great satisfaction with their inner-city workforce. In our survey of inner-city companies in Boston, 65% of businesses cited an available workforce, 60% cited a low-cost workforce, and 50% cited a diverse workforce as critical or important advantages to their inner-city location.[30] In Atlanta, just 4% of companies were dissatisfied with their employees' skill level and 15% with their work ethic.[31] The following quotations are from our interviews:[32]

> I have no problem finding people. They come to me. I hand out 2-4 applications every day. . . . I have no problem getting them up to speed. . . . There are a lot of nice people here. . . . I've never had crime problems or seen drug problems.

> We're very devoted to our workforce. At the present time, it is the single advantage to this location.

> I have no problem finding willing and able workers. . . . I get two or three applications per day.

These perspectives are reinforced by a study in central Harlem, which concluded, "The ratio of job seekers to successful hires in the fast food restaurants studied . . . is approximately 14 to 1."[33]

However, there is clear evidence that the work ethic and qualifications of some portion of inner-city residents are sorely lacking, supporting the view of those who argue that investment in human capital is a priority. The following three quotations reflect all-too-common comments we have heard from inner-city employers:[34]

> I am dying for qualified labor, but I can't afford to hire someone who can just show up for work. . . . There are a lot of unskilled people available, but they don't meet our needs. Of the inner-city residents we try to hire for semi-skilled welding jobs, more than half flunk a drug test, and few make it through the internship period.

> The vast majority [of the inner-city workers I hired] lacked basic attitudes rather than skills. It was very difficult to find individuals who consistently arrived at work on time, followed direction, worked as a team, or showed even a modest degree of enthusiasm or ambition. It was necessary to frequently test for drug use to control this problem as well as exercise careful supervision to prevent crime in the workplace. Despite the fact that our wages and promotion opportunities were the best in each area, it was often difficult even to find willing candidates.

> Even when I hire through families or personal recommendations, I have found that 50% of the African Americans I hire don't work out. In contrast, 95% of Vietnamese I hire work out, as do 90-95% of Peruvians. Immigrants—at least certain groups of them—appreciate the job, are very reliable, and come to work every day. If you ask them to do something, they do it and don't give you any problems.

These quotations raise a number of important issues. First, they underscore the reality that the inner-city workforce has a disproportionate number of people who are problematic employees. Hence, to hire from the inner city, companies must have effective strategies for identifying good employees. Cultivating personal networks in the community and building relationships with CBOs (as discussed later) are essential.

Second, employers widely report much higher satisfaction with immigrant workers, many of whom are African American or Latino and live in inner cities. Thus perhaps a more relevant distinction than inner-city versus non-inner-city employees—or African Americans and Latinos versus other ethnic groups—is long-resident poor versus recent poor immigrants. The former tend to have greater problems entering the workforce, whether they are African Americans in the inner city or, for example, Whites in the Ozarks. Immigrants, on the other hand, tend to be a self-selecting group who find a low-income, entry-level job in the United States far better than the situation in their home country. They tend to work harder than everyone else.

Third, these quotations and our other interviews illustrate that some White inner-city business owners and managers are often quick to judge (unfairly) entire groups of people, based on only a few experiences. This is one of the most important areas in which faulty perceptions are working against economic progress in inner cities.

This discussion makes it clear that the inner-city workforce is diverse and complex and cannot easily be summarized. Although there are many problems with the workforce, however, our research reveals a meaningful proportion of unemployed or underemployed inner-city residents who are ready and able to be good employees. Rather than become paralyzed by the presence of residents unfit for work, our strategy focuses on job development for the substantial group that is ready to work. The challenge is to create more accessible jobs and better connect these people to them. Over time, more people being employed will be a virtuous cycle that will lead to others becoming more fit to be employed.

In addition to the nature of the workforce, a second myth about inner cities is that they lack entrepreneurs. In fact, there is a demonstrated capacity for entrepreneurship among inner-city residents, most of which has been channeled into microenterprises and the provision of social services. For instance, inner cities have a plethora of social service providers as well as social, fraternal, and religious organizations. Behind the creation and building of those organizations is a whole cadre of local entrepreneurs who have responded to intense local demand for social services and to funding opportunities provided by government, foundations, and private-sector sponsors. Now, the challenge is to create a climate whereby other inner-city residents, with similar talent and commitment, will build for-profit businesses that become meaningful employers and create wealth.

A third myth about inner cities is that skilled minorities—many of whom grew up in or near them—will inevitably work or create businesses in more affluent areas. Today's large and growing pool of talented minority managers represents a new generation of potential inner-city entrepreneurs. Many of these managers have developed the skills, networks, capital, and confidence to join or start entrepreneurial companies in the inner city. We know of some—including former students of mine—who are doing so. As the awareness of the economic opportunities in inner cities grows, more will follow.

The Role of the Private Sector

Many skeptics question whether the private sector is best suited, willing, or able to tap into inner cities' advantages and play a leading role in the economic revitalization of these areas. They point to the past (and, to some extent, current) behavior by the private sector—departures of companies, poor treatment of workers, and damage to the environment—and argue that the private sector is part of the problem, not the solution. Given the track record of outsiders, both in government and the private sector, such skepticism is understandable. However, if this skepticism leads to an isolationist, statist approach to economic development, inner cities will continue to decline. As Robert Kennedy once said, "To ignore the potential contribution of private enterprise is to fight the war on poverty with a single platoon, while great armies are left to stand aside."[35]

Why Aren't Market Forces Working Better?

Many people have asked, "If inner cities truly have important competitive advantages, why isn't the market working?" There are at least three answers that have been underscored by interviews both inside and outside the inner city.

First, there are many misperceptions and biases about inner cities and their opportunities—what economists call information imperfections. Fed by media coverage, many see inner cities as combat zones devoid of economic activity and populated by people with no ambition, skills, or resources. Crime is indeed a severe problem in certain inner-city areas, but not all. For example, Boston's Dorchester neighborhood is widely perceived as a high-crime area, when in fact the rates of violent crime and overall felony crime are 20% below the Boston average.[36] Such perceptions, which are often wrong or exaggerated, severely retard investment and business development in areas with obvious market opportunities.

Second, many of the inner city's competitive advantages have been diluted, if not overturned, by poor policies and leadership, whereas disadvantages remain. For example, the advantage of strategic location has often been dissipated by inadequate infrastructure investments and maintenance. Until the many disadvantages of inner cities as business locations are addressed, disadvantages will continue to outweigh advantages for many companies.

Finally, there is limited and garbled communication between entrepreneurs and companies, on the one hand, and advocates for inner cities (e.g., CBOs, foundations, and government) on the other. Advocates for inner cities often feel that companies are not doing enough for their communities, whereas businesspeople feel victimized by what they perceive to be unreasonable demands and expectations. The result is often tension, if not outright hostility, and business investment and growth are deterred. A few companies in politically sensitive or regulated industries will respond to pressure or mandates, but the effort will backfire by driving away other investors.

Inner cities, like other areas, must compete for investment and jobs. The best (and only) way to develop the economies of inner cities is to make them attractive and welcoming places in which to invest and do business, both for residents and nonresidents. We see some hopeful signs that attitudes about business in inner cities are improving.

Will the Private Sector Step Forward?

Yes, it will—not because it is forced to, but because inner cities can offer attractive markets, advantageous locations, and good employees. In response to more attention to these advantages, there is clear evidence (cited earlier) that the private sector is already investing in inner cities,[37] and the trend is building momentum. With improvements in the inner-city business environment and continued changes in perceptions, inner-city business development can accelerate.

There have been few efforts to engage the private sector through business-to-business activity, despite many organized, successful efforts to involve corporations in inner-city issues, such as housing and education. This is not due to lack of interest on the part of the private sector for programs that make economic sense. In fact, we believe so strongly in the need for better linkages between the private sector and the inner-city economy that we have founded a new organization, the Initiative for a Competitive Inner City, to develop programs to involve companies, professional service firms, and graduate business schools in assisting and creating inner-city companies in cities around the country. This effort includes close contact with local government and CBOs. Many companies—among them Bank of America, The Boston Consulting Group, Citibank, John Hancock, Lotus Development Corporation, Pacific Gas & Electric, Staples, and Textron—are actively involved. Additionally, more than 30 of the nation's leading graduate business schools have created or expanded programs linking students and faculty with inner-city companies.[38] The response gives us confidence that the business community is ready to try new approaches to urban problems, based on sound, economically driven strategies. An important role of the initiative is to lower the barriers to inner-city business development by dispelling myths about inner cities that hold back investment, publicizing successful companies, developing strategies to help enhance the competitive advantages and ameliorate the present disadvantages of inner cities, and improving communication between the private sector and inner-city advocates. The time is right for a proliferation of such efforts, in which business development is the central goal.

The Business Environment in the Inner City

As business locations, inner cities suffer many disadvantages: discrimination against residents and entrepreneurs, high taxes and utility costs, difficulty in finding affordable insurance, crime, poorly maintained logistical infrastructure, burdensome regulations and permitting requirements, environmental pollution, and a weak education and training system.

The inner city's disadvantages as a business location must be seen as an economic problem and addressed as part of an economic strategy. Too often, addressing weaknesses such as a poorly trained workforce or deficient logistical infrastructure is approached with only the social welfare of residents, not the needs of business, in mind. For example, inner-city training programs often fail to screen applicants—and even give priority to the least prepared residents in the name of fairness. Employers are then disappointed with the graduates.

Second, attempting to offset disadvantages with operating subsidies to businesses is futile. A more effective approach is to address the impediments to doing business directly. There is no substitute for reducing unneeded regulatory hurdles, simplifying permitting, and reorienting environmental cleanup requirements.

Third, our research indicates that many of the inner city's disadvantages are not inherent but are the result of poor strategies and obsolete public policies. There are many best practices nationwide that could be adopted in every inner city, as will be discussed further.

The Role of Government

Many past and present government programs have defied economic logic, distorted incentives, and failed to ease racial tensions. To quote a senior federal official, "We've put a great deal of money into inner-city economic development, and have had very few successes."[39] At the local level, state and city governments bear significant responsibility for the disinvestment by the private sector in inner cities by failing to maintain and improve schools, neglecting infrastructure, failing to provide for public safety, raising taxes excessively, and creating a morass of costly regulations.

However, this critique does not imply that all government efforts have been harmful; nor do we advocate that government should simply abandon the inner city and allow private initiatives to take over. The issue is not the importance of a continuing government role but exactly *what* that role should be. There is a continuing, vital role for government in inner-city economic development—a role focused not on direct intervention and heavy reliance on operating subsidies to attract companies, but on creating a favorable environment for business (e.g., improving the public school system, training workers, upgrading infrastructure, streamlining regulation).

Crime Prevention

Crime, with its associated fears and costs, is one of the greatest barriers to inner-city economic revitalization. Although we believe that the perception of crime is greater than the reality, our surveys reveal the costs of the reality of crime to inner-city business activity. In Boston and Atlanta, 75% and 55%, respectively, of inner-city-based companies we surveyed said that crime and security issues were serious problems.[40] In Atlanta, employers cited break-ins/vandalism, employee theft, and extra insurance and security costs as their three biggest problems.[41] In another case, the Shops at Church Square, an inner-city strip shopping center in Cleveland, spends $2 per square foot more than a comparable suburban center, because it has a full-time security guard, more lighting, and continuous cleaning. Its overall costs are thus raised by more than 20%.

There is a great deal businesses can do to mitigate the problem. Although prudent security measures are a part of the solution, we have heard again and again that, as one business owner put it, "If you become part of a neighborhood by hiring locally and participating actively in the community, the crime issue becomes virtually nonexistent."[42]

Ultimately, however, government must play a leading role, and there are signs of real progress. For example, New York City and a handful of other cities are adopting highly successful police force management and crime-fighting techniques that disproportionately benefit high-crime inner-city areas. The results have been dramatic: Although major crime has declined modestly across the country, in New York City it has fallen dramatically in the past few years (17.5% in 1995 alone) to levels not seen since the early 1970s.[43]

Discrimination

It is clear that discrimination has kept African Americans and other people of color from the educational and economic opportunities afforded many other groups in American society. The current economic weakness of inner-city communities, together with the human capital deficits that plague them, in many ways reflects the legacy of discrimination. It is also clear from our research that discrimination remains a serious problem.

However, the problems of inner cities go beyond discrimination to many factors that magnify its effects. Although continuing to work to eliminate discrimination, we must move forward with a positive strategy to address other parts of the inner-city problem. Our experience has been that a sound economic strategy based on improving competitiveness is a positive step forward that emphasizes mutual benefits and brings people together.

Affirmative Action and Government Set-Aside Programs

A central tool employed to counter discrimination has been affirmative action, which has created opportunities for minorities in a number of occupational and educational fields. In the area of business-related affirmative action programs, such as minority preferences for government contracts, further steps are needed to better tailor these programs to jobs and wealth creation for residents of distressed communities, instead of targeting minority entrepreneurs regardless of income levels and whom they employ. To quote President Clinton: "We need to do more to help disadvantaged people in distressed communities, no matter what their race or gender."[44]

Some have argued that, since minority-owned firms are more likely to hire inner-city residents than are White-owned companies, inner-city economic development efforts should focus exclusively on assisting minority-owned firms, regardless of their location.[45] Our research confirms that minority-owned companies do indeed hire a greater proportion of inner-city residents, and we are enthusiastic supporters of minority-owned businesses. However, promoting only minority-owned businesses will not be sufficient, as the record of recent decades clearly indicates. There is a need for broader strategies that will stimulate inner-city job creation by *all* types of companies. Our research reveals that there are many White-owned businesses in or near inner cities that are providing many thousands of good jobs to inner-city residents.[46] Rather than accept as a given that few White-owned businesses in inner cities will hire local residents, we need a strategy to increase their local hiring.

Business development programs also need to be refined to promote sustainable businesses, not just guarantee companies a market. Many such programs have dulled motivation and retarded cost and quality improvements. These pitfalls are not just true for minority set-asides; similar problems afflict business incentive programs around the world. It is not surprising, therefore, that a 1988 General Accounting Office report found that, within six months of graduating from the Small Business Association's (SBA's) purchasing preference program, 30% of the companies had gone out of business. An additional 58% of the remaining companies claimed that the withdrawal of the SBA's support had a devastating effect on their business. Companies benefiting from set-asides must demonstrate movement toward self-sufficiency to retain them, so that incentives are more aligned with creating sustainable, profitable businesses. By linking such incentives to distressed communities and limiting their duration, we are confident the current attacks against preference programs would diminish significantly.

The Role of Subsidies

Public funds (subsidies) will be necessary to revitalize inner cities, but they must be spent in support of an economic strategy based on competitive advantage, instead of distorting business incentives with futile attempts to lure businesses that lack an economic reason for locating in inner cities. It is appropriate for government funds to be used to prepare a site for business by assembling parcels of land, improving infrastructure, performing environmental remediation, and providing better public safety. However, the businesses that then locate in inner cities should not receive ongoing operating subsidies, or they are unlikely to become sustainable in the long run.

The Proper Use of Tax Incentives

Both at federal and state level, various tax incentives have been employed to support economic development in designated depressed areas, often called enterprise zones. Although we support measures that make inner cities more competitive (and higher taxes than exist in the surrounding region are often a significant disadvantage[47]), the record of enterprise zones is not encouraging. Again and again, businesses that locate in an area because of tax breaks or other artificial inducements, rather than genuine competitive advantages, prove not sustainable. Research is accumulating from around the world that few businesses make location decisions based on tax incentives—especially the modest ones commonly associated with enterprise zones.[48] Thus the bulk of tax breaks goes to businesses that would have been operating in the enterprise zone anyway. Enterprise zone incentives can be perverse in a number of other ways. They often fail to encourage the hiring of residents of the depressed area and to promote entrepreneurship by residents.[49] Tax breaks that do not rest on making a profit are also dangerous, because they encourage uneconomic investments.

Training

Unfortunately, the existing job training system in the United States is ineffective. Training programs are fragmented, overhead intensive, and disconnected from the needs of industry and recipients.[50] Many programs provide poor training for nonexistent jobs in industries with no projected growth; for example, a study of the Job Training Partnership Act showed that young men who had dropped out of school and enrolled in the program earned 8% *less* than those who were given no training.[51]

Inner-city job training programs have an especially difficult challenge, because they have to overcome many inner-city residents' low education levels and poor work skills. However, there are a number of extraordinary programs—such as the National Foundation for Teaching Entrepreneurship, which has helped thousands of inner-city youth start businesses. Boston's One With One and Project Protech, the Bidwell Training Center in Pittsburgh,[52] Detroit's Project: HOPE,[53] and Jobs for Youth in various cities are just a few examples of other organizations doing an outstanding job in training and placing inner-city residents in good jobs in nearby clusters. Additionally, outlined in my *Harvard Business Review* article[54] were a number of strategies both to engage the private sector to create and certify training programs, which could be built around industry clusters in the inner city and in the nearby regional economy, and to tap into existing private-sector training programs, especially corporate ones.

Access to Capital

The issue of access to capital provides an especially illuminating example of appropriate versus inappropriate government intervention. Most government efforts have focused on the creation of government loan pools and quasi-public financing entities that have produced fragmentation, market confusion, and excessive overhead costs, resulting in many uneconomic investments.

The only viable solution is to harness market forces and the resources of the private capital markets. In the area of debt financing, we argue that mainstream financial institutions must be engaged through direct incentives, such as transaction fees, to cover high overhead costs associated with small transactions and partial loan guarantees that mitigate some—though not all—of the risk. Additionally, increased disclosure of inner-city business loans would motivate efforts by private-sector lenders.

Regarding equity capital, our proposal is to *eliminate* the tax on capital gains and dividends from long-term equity investment in inner-city-based businesses (or subsidiaries) that employ a minimum percentage of inner-city residents. This proposal would, we believe, maintain a focus on genuine profit and result in new equity capital from a wide range of sources flowing into inner cites. The cost, which is likely to be modest, would be decisively outweighed by tax revenues from higher employment of inner-city residents.[55]

Deregulation

Given the history of exploitation in inner cities, some advocates resist any reform. Although I strongly oppose needless, inefficient, and destructive government regulation, I do not favor scrapping laws that govern workplace health, safety, and compensation. Instead, my primary focus is on the astounding number of rules, permits, and regulations that have accumulated at the federal, state, and especially city government levels over many years.[56] It is truly ironic that the areas in the United States that are most in need of business development are the most overregulated.

Merely registering and legally operating a small business in an inner city is daunting. For example, according to Steve Mariotti, the founder of National Foundation for Teaching Entrepreneurship, which teaches entrepreneurship to "at-risk" inner-city youth:

> The minority entrepreneur usually ends up being his own lawyer and accountant. . . . The paperwork, cost, and confusion . . . drive would-be entrepreneurs away from certainty and down a slippery slope. They develop contempt for the government, because they no longer see it as their ally. That drives people into the underground economy, where there are no contracts. Matters of dispute are settled with gun or a beating. . . . Once an entrepreneur moves into the balkanized—and chaotic—underground economy, growing the business is not a viable option.[57]

Absurd laws and regulated monopolies also plague inner-city entrepreneurs. According to a leading magazine for entrepreneurs, "about 10% of all jobs in this country require some sort of license, and many of those are low-skill, entry-level occupations such as taxicab driving, working as a street vendor, cosmetology, trash hauling, and recycling. The licensing process in these fields is often onerous and . . . preserves existing monopolies at the expense of those least able to defend themselves."[58]

Regulation also affects larger businesses and CBOs. Alan Hershkowitz, a senior scientist with the National Resources Defense Council, who is working with the Banana Kelly CDC to develop a proposed paper recycling project in the South Bronx, speaks for a large number of developers and businesspeople we have interviewed: "There is usually a lengthy, labyrinthine permitting process in cities."[59] Another example shows that for-profit businesses are not the only victims. Rev. Calvin Butts of the Abyssinian Development Corporation in Harlem laments, "If we could sit down with the city and have an ordinary negotiation, we could easily build twice the number of [housing] units, and I assure you it would cost a lot less."[60]

The regulatory process can be streamlined, as the case of Massachusetts illustrates. There, Governor William Weld's administration is reviewing all 20,000 pages of the 26-volume *Code of Massachusetts Regulations.* By January 1, 1997, it plans to rescind 19% and modify 44% of the 1,600 regulations on the books, fundamentally modifying the way the state regulates itself and its citizens.[61]

In the area of zoning, antiquated laws reflect an economy that has not existed for decades. For example, large swaths of land in New York City remain zoned for manufacturing and industrial usage, despite the fact that few such businesses remain. This land should be rezoned to allow residential and retail development.

Finally, it is ironic that current environmental policy (especially the Federal Superfund law), intended to foster cleanup of the nation's estimated 200,000 to 500,000 vacant urban industrial sites, has instead hampered such efforts. According to Henry L. Henderson, Chicago's environmental commissioner, "The shadow of crushing liability and the fear of lenders made it impossible for small businessmen to get loans to develop urban properties."[62] Liability and cleanup laws must be modified so that lenders, developers, and businesses will return to the inner city. We are pleased to see that more than half the states have passed legislation or developed policies to reduce the liability threat inherent in such properties and institute reasonable cleanup requirements. Congress is considering a host of similar proposals. Similarly, the Clinton administration initiated a $10 million grant program last year to fund cleanup and has modestly relaxed the Superfund law that governs hazardous-waste cleanups.[63]

Government as a Marketer

In the area of economic development, a critical role of government is to act as a marketer—courting, welcoming, and assisting companies. From hundreds of interviews with businesses, government officials, and others across the country, we have unfortunately found that government rarely plays this role in inner cities. For example, the CEO of a major Oakland corporation that was deciding whether to expand its plant in Oakland or open a new one in Houston recounted a story we have heard variations of many times across the country. In Houston, he was met at the airport by city officials, shown appropriate parcels of land, and given all required permits and waivers on the spot. By contrast, he was frustrated that neighborhood groups and city government in Oakland had stymied his attempts to win approval to expand his existing plant because of concerns over more traffic.[64]

The Role of Community-Based Organizations

Great effort and funding has been directed at creating and building community-based organizations (CBOs), and many have been remarkably successful in building and managing low-income housing, providing needed services, stabilizing neighborhoods, and re-creating local market demand. They deserve much credit for helping to create the conditions under which the private sector would consider investing.

Now, inner cities are ready to move to the next stage, which will require new strategies from CBOs of all types. Our model, with its focus on private, for-profit initiatives, seems threatening to some CBOs, who see themselves as advising, financing, and owning inner-city companies. It should not be. CBOs can and must play a role in inner-city economic development efforts. But choosing the proper strategy is critical, and many CBOs will have to refocus their activities. CBOs, like every other player, must identify their capabilities, resources, and limitations, and participate in economic development with the right strategy.

In the area of true business development, the record of CBOs is mixed. Although there are a number of noteworthy successes—for example, New Communities Corporation has a majority equity stake in a Pathmark supermarket in Newark's Central Ward, with sales per square foot twice the national average[65]—the strategy of advising, lending to, or operating businesses is a questionable one for most CBOs. In general, they are not equipped to provide many of the specialized inputs businesses need, and their efforts often end in failure. Moreover, CBOs can and must reach out to private-sector institutions and entities that will be essential to the ongoing growth and development of local communities. It makes little sense to attempt to re-create the expertise and resources that already exist.

Although it is difficult to generalize about such a diverse group of organizations, we believe that CBOs' economic development efforts should be guided by a business-based model. They should seek to build networks with mainstream business institutions (e.g., business schools, banks, corporations, chambers of commerce) instead of attempting to duplicate them. In many cases, CBOs already have strong relationships with the mainstream business community through, for example, efforts to develop affordable housing. These relationships can be leveraged to broker valuable business-to-business connections and resources. Thus, instead of advising businesses directly, CBOs should connect local companies with high-quality existing resources. Instead of setting up a new loan program, they should facilitate access by businesses to financing—first through banks and, failing that, through the myriad of public and quasi-public financing sources already present in most cities.

Many CBOs are also well positioned to address the vexing problem that many inner-city-based businesses hire few or no inner-city residents.[66] CBOs have, on occasion, responded by demanding that new inner-city companies hire a certain percentage of employees from the local area or by criticizing and ostracizing existing companies that do not hire enough local residents. This approach has backfired in the long run, by driving companies away and contributing to the general perception of a hostile business environment. A more productive approach is to understand the needs and perceptions of local businesses and to develop programs to address them.

The roots of this problem often go beyond poor work ethic, training, and work readiness to issues such as lack of informal networks and employer bias. CBOs, with their networks both in the community and in the private sector, can play an important role in helping overcome these barriers and connecting inner-city residents to nearby jobs. To cite a case study mentioned in my *Harvard Business Review* article, the South Brooklyn Local Development Corporation (SBLDC) played an important role in connecting local residents to jobs in the Red Hook industrial area, by developing relationships with nearby businesses and screening and referring employees to them.

Finally, CBOs with relevant experience can facilitate site improvement, development, and expansion. We have found that many businesses seeking to expand or locate in the inner city have difficulty finding suitable sites or navigating the approval and permitting process. CBOs often have significant expertise in real estate that could be applied to the development of commercial and industrial property (including site assembly, demolition, and environmental cleanup), identifying appropriate sites for expansion, and assisting companies in the permitting and approval process.

There are many examples of CBOs that are successfully collaborating with businesses. In Boston, the Dorchester Bay Economic Development Corporation was responsible for rehabilitating a building that was then occupied by a Latino-focused supermarket. The supermarket has operated successfully there and has helped revitalize the entire Uphams Corner shopping district. Another example of collaboration is cited by the CEO of First National, a leading supermarket chain:

> Cultivating close ties to community groups is the place to start. They can help identify and train local residents who will be reliable workers. Nonprofit community groups may also have development expertise and access to government financing and tax breaks. . . . It creates trust that dispels the view that outside chains come in to take advantage—a problem that Korean grocers had in Los Angeles and the Arabs in Cleveland.[67]

These examples do not involve CBOs owning and operating supermarkets but helping to create the conditions whereby private investors and entrepreneurs will invest.

Conclusion

Inner cities need new, market-oriented strategies that will build on their strengths and engage the private sector. Inner cities must and can compete. Developing a new strategy will require an understanding of what is unique about each inner city, how to build on its advantages, and a plan to eliminate or reduce the many disadvantages to conducting business. This process will require the commitment and involvement of business, government, and the nonprofit sector.

This approach has been characterized by some as out of step with the globalization of the economy and as nothing more than laissez faire. It is neither. Instead, it reflects the new shape of the postglobal, postrestructuring economy and the actual pattern of competitive success and failure of companies in inner cities. It also reflects the types of intervention by the private sector, CBOs, and government that are truly effective and needed.

The private sector must play a central role, as it has in successful economic development everywhere. The private sector is already investing in inner cities, and this trend can be accelerated by improving perceptions and addressing problems in the inner-city business environment. There is a continuing, vital role for government and for public resources in inner-city economic development to create a favorable environment for business (e.g., assembling and improving sites, training workers, upgrading infrastructure, streamlining regulation). CBOs deserve much credit for helping to create the conditions under which the private sector would consider investing. Now, however, inner cities are ready to move to the next stage, which will require new CBO strategies. CBOs should facilitate private-sector involvement, change attitudes, train residents and link them to jobs, and, when appropriate, develop sites.

As funding for traditional urban programs comes under increasing attack, those of us concerned with inner cities should not be defending failed past approaches. We must stop arguing from exceptions and recognize the rule. Instead of justifying the past, we need to turn our attention to new, market-oriented strategies that will build on strengths and engage the private sector. Although my approach has room for improvement, it is a positive way of moving forward. Despite popular perceptions, there is genuine economic opportunity in inner cities. Working together, we can unlock it.

Notes

1. There are several definitions of the term "inner city." We use the term to refer to economically distressed urban communities that we have defined, drawing on the literature and federal Empowerment Zone guidelines, as census tracts in which the median household income is no more than 80% of the median for the SMSA *and* in which the unemployment rate is more than 25% greater than the state average. The term is perhaps most widely understood to mean urban areas that commonly have large minority populations and high levels of poverty, unemployment, crime, single-parent families, high school dropout rates, and so forth. Because of these realities (and even worse perceptions), some are uncomfortable with the term. Our definition recognizes these realities and perceptions, and attempts to address them head on.

2. Michael E. Porter, *The Competitive Advantage of Nations* (New York: Free Press, 1990).

3. Atlanta, Baltimore, Boston, Chicago, Los Angeles, Newark, New York City, Oakland, and Washington, DC.

4. Given the paucity of attention to inner-city business development, there is little or no comprehensive data yet available on inner-city companies, even in individual cities. Although we have begun to compile such data, this remains an important priority for future research.

5. Richard D. Bingham and Robert Mier, eds., *Theories of Local Economic Development: Perspectives from across the Disciplines* (Newbury Park, CA: Sage, 1993), p. i.

6. John P. Blair, *Local Economic Development* (Thousand Oaks, CA: Sage, 1995), p. 22.

7. See analysis of literature of inner-city economic development in Initiative for a Competitive Inner City (ICIC), Dwight Hutchins, and Kate Moriarty, "Benchmarking Theory and Best Practices of Inner City Economic Development" (Unpublished report, Boston, 1996).

8. ICIC and The Boston Consulting Group, "Leveraging Boston's Competitive Advantage: Strategies for Inner City Business Growth" (Unpublished report, Boston, 1996).

9. ICIC, Keba Gordon, Linwood Herndon, and Rod Stovall, "A Profile of Atlanta's Inner City Competitiveness" (Unpublished report sponsored by Boral Industries, Atlanta, 1996).

10. See Michael E. Porter, *The Competitive Advantage of Nations.* A large body of literature is developing around clusters, including, for example, Michael Enright, "Organization and Coordination in Geographically Concentrated Industries," in *Coordination and Information: Historical Perspectives on the Organization of Enterprise,* ed. Daniel Raff and Naomi Lamoreaux (Chicago: Chicago University Press for the National Bureau of Economic Research, 1995), and idem, "Regional Clusters and Economic Development: A Research Agenda," working paper 94-942, Harvard Business School, Boston, MA, 1994.

11. For a partial discussion of clusters and their relationship to related concepts, see Michael E. Porter, "Comment on 'Interaction between Regional and Industrial Policies: Evidence from Four Countries,' by Markusen," in *Proceedings of The World Bank Annual Conference of Development Economics 1994,* Supplement to The World Bank Economic Review and The World Bank Research Observer, ed. Michael Bruno and Boris Pleskovic (Washington, DC: The World Bank, 1995).

12. Mercer Management Consulting and ICIC, "The Competitive Advantages of Inner City Baltimore" (Unpublished report, Baltimore, 1995).

13. For example, a study by the New York City Department of Consumer Affairs showed one supermarket for every 5,700-7,000 residents of the Upper East Side, Brooklyn Heights, and Upper West Side, but only one supermarket for every 63,818 residents of parts of Williamsburg and Bedford-Stuyvesant (cited in *The New York Times,* June 6, 1992).

14. Quoted in Susan Diesenhouse, "As Suburbs Slow, Supermarkets Return to Cities," *The New York Times,* June 27, 1993, p. 5.

15. Quoted in Dana Milbank, "Doing Well: Finast Finds Challenges and Surprising Profits in Urban Supermarkets," *The Wall Street Journal,* June 8, 1992, p. 1.

16. Quoted in Diesenhouse, "As Suburbs Slow, Supermarkets Return to Cities," p. 5.

17. Chris Reidy, "Shaw's Joins Grocery Run Back to City," *The Boston Globe,* September 28, 1996, p. 1.

18. For further information on inner-city retailing, see Jon Patricof and Willy Walker, "Inner City Retailing" (Unpublished report for ICIC, Boston, 1995). For information on franchising, see Paul Singh, "Inner City Franchising Opportunities" (Unpublished report for ICIC, Boston, 1996).

19. ICIC and The Boston Consulting Group, "Leveraging Boston's Competitive Advantage."

20. Craig Horowitz, "A South Bronx Renaissance," *New York Magazine,* November 21, 1994, p. 54.

21. Laura Bird, "Shunned No More, New York's Harlem Entices Big Chains Seeking Fresh Turf," *The Wall Street Journal,* July 25, 1996, p. B1.

22. Kenneth B. Noble, "Magic Johnson Finding Success in a New Forum," *The New York Times,* January 8, 1996.

23. Anthony Flint, "Diversity and Dollars at South Bay," *The Boston Globe,* December 3, 1995, p. 41.

24. Based on ICIC interviews with inner-city supermarket operators nationwide.

25. The management of Delray Farms, interview by ICIC, 1995.

26. Robert Berner, "Urban Rarity: Stores Offering Spiffy Service," *The Wall Street Journal,* July 25, 1996, p. B1.

27. Richard M. Rosenberg, Chairman and CEO of BankAmerica Corporation, "Banking on the New America: The Business Case for Investing in the Inner City" (Speech delivered to the 17th Annual Real Estate and Economics Symposium, U.C. Berkeley Center for Real Estate and Urban Economics, San Francisco, December 14, 1994).

28. This is a program of the Small Business Administration, which provides certain tax benefits and capital to leverage private equity investments in SSBICs, which lend to disadvantaged businesses. The program has been widely criticized, as most SSBICs are undercapitalized and lose money. See Timothy Bates, (Unpublished report to the Small Business Association, 1996), and Roy Furchgott, "Defer Capital Gains on Stock Sales? Here's the Catch," *The New York Times,* September 8, 1996, Business Section, p. 3.

29. The management of Medallion Funding, interview by ICIC, April 11, 1996.

30. ICIC and The Boston Consulting Group, "Leveraging Boston's Competitive Advantage."

31. ICIC et al., "A Profile of Atlanta's Inner City Competitiveness."

32. Quotations from ICIC interviews in Boston and Oakland, 1995.

33. Chauncy Lennon and Katherine Newman, "Finding Work in the Inner City: How Hard Is It Now? How Hard Will It Be for AFDC Recipients?" working paper, Russell Sage Foundation, New York, 1995.

34. Quotations from the following: an inner-city Baltimore manufacturer, interview by ICIC, April 1996; an inner-city manufacturing and distribution business in New Jersey and Louisiana, interviewed June 1995; and a manufacturer in Harrison, NJ, letter dated March 1995.

35. Quoted in speech by Treasury Secretary Robert E. Rubin, Los Angeles Town Hall, July 29, 1996.

36. ICIC and The Boston Consulting Group, "Leveraging Boston's Competitive Advantage." Boston Police Department data, 1995.

37. In addition to the direct investment cited earlier, "corporate philanthropy in 1993 totaled $5.3 billion, while during the same period corporations purchased $20.5 billion in goods and services from minority-owned firms; invested $2.2 billion in Low Income Housing Tax Credits; and loaned $37 billion to low and moderate income neighborhoods" ("An Exploration of Corporate Involvement in Community & Economic Development," staff paper, Office of Program-Related Investments, Ford Foundation, June 1995, p. 1).

38. In Boston, beginning in early 1994, ICIC organized a program to marshal in-depth consulting support from experienced MBA students and faculty to help inner-city-based companies realize their competitive advantages. The two- to four-person teams addressed a range of managerial issues and produced concrete benefits for client companies. Over the past three years, 69 students working on 20 teams have provided consulting services worth an estimated $350,000. In addition, the principals of ICIC and I have provided further assistance to client companies and established relationships with legal, accounting, information technol-

ogy, and financial partners in the region to provide pro bono services to clients. Overall, since the program began, our clients have created nearly 200 new jobs for inner-city residents, representing over $3,000,000 per year in wage income. ICIC is now expanding this program to other cities and business schools.

39. Anonymous, interview by ICIC, 1996.

40. ICIC and The Boston Consulting Group, "Leveraging Boston's Competitive Advantage," and ICIC et al., "A Profile of Atlanta's Inner City Competitiveness." Our findings were similar in the other cities we surveyed.

41. ICIC et al., "A Profile of Atlanta's Inner City Competitiveness."

42. An inner-city Atlanta company, interview by ICIC, 1996.

43. Eric Pooley, "One Good Apple," *Time,* January 15, 1996, p. 54; George L. Kelling, "How to Run a Police Department," *City Journal,* Autumn 1995, p. 34.

44. Quoted in Michele Galen, Susan Garland, and Catherine Yang, "Now, Affirmative Action May Help White Men," *Business Week,* July 31, 1995.

45. See survey led by Margaret Simms done by the Joint Center for Political and Economic Studies (in conjunction with *Black Enterprise* magazine and the National Minority Supplier Development Council), published in *Black Enterprise,* June 1996, p. 194. Also, see various studies by Timothy Bates, Wayne State University; and Thomas Boston and Catherine Ross, "Location Preferences of Successful African American-Owned Businesses in Atlanta," *The Review of Black Political Economy: Special Issue on Michael Porter's "Competitive Advantages of the Inner City"* (forthcoming).

46. For example, The Lark and Run 'N' Shoot, cited earlier, each of which hires more than 90% of its managers and employees locally, are White-owned. This is also true of the ICIC's clients in Boston that are White-owned.

47. For example, "The citizens of Detroit must contend with a total tax burden that is about seven times higher than the average Michigan municipality" (Stephen Moore, Director of Fiscal Policy Studies, Cato Institute, quoted in *The Detroit News,* December 19, 1993, p. 3B).

48. A General Accounting Office study found that fewer than 30% of employers cited "financial inducements" as an important location decision factor. Eleven other factors were more important.

49. "Academic studies have found that as little as 15% of the workers in some enterprise zones actually live there" (*The New York Times,* January 26, 1996). In a Louisville, Kentucky, enterprise zone, only 14% of the jobs created by companies that received tax breaks were filled by people who were unemployed or on welfare (Steven A. Holmes, "Enterprise Zone for Louisville: Past the Limit?" *The New York Times,* October 31, 1990, p. A18). In Indiana, only 6.35% of manufacturing jobs and 30% of retail jobs went to enterprise zone residents: "A survey of 155 zones in 28 states by the National Center for Enterprise Zone Research found that only 5.3% of zone businesses were minority-owned" (Dan Cordtz, "Mainstreaming the Ghetto," *Financial World,* September 1, 1992, p. 23).

50. See a multitude of studies, including ICIC, Chelli Devadutt, and Julie Fletcher, "A Survey and Analysis of the Inner City Job Training System" (Unpublished report, Boston, May 1995).

51. Jason DeParle, "Debris of Past Failures Impedes Poverty Policy," *The New York Times,* November 7, 1993.

52. See *Bidwell Training Center.* Harvard Business School Case Study No. 9-693-087. Cambridge, MA: Harvard Business School, 1993.

53. Robyn Meredith, "An Exit from the Inner City: Training for New Auto Jobs," *New York Times,* April 21, 1996, p. 10.

54. Michael E. Porter, "The Competitive Advantage of the Inner City," *Harvard Business Review* 73 (May-June 1995): 55-71.

55. Michael Porter, testimony to the House Ways and Means Committee, Subcommittee on Human Resources, Joint Hearing on H.R. 3467, "Saving Our Children: The American Community Act of 1996," July 30, 1996.

56. In Massachusetts, for example, the number of state regulations written increased from 103 in the 1960s to 375 in the 1970s to 585 in the 1980s. Today, "the complete set occupies eight feet of shelf space, and contains 15 times as many regulations as it did in 1960" (Michael Grunwald, "Quietly, Weld Aides Rewrite the State's Rulebook," *The Boston Globe,* October 3, 1996, p. 1).

57. Edward O. Welles, "It's Not the Same America," *Inc.,* May 1994, p. 85.

58. Ibid., p. 86.

59. Quoted in John Holusha, "E.P.A. Helping Cities to Revive Industrial Sites," *The New York Times,* December 4, 1995, p. A1.

60. Quoted in Philip K. Howard, *The Death of Common Sense* (New York: Random House, 1994), p. 106.

61. Grunwald, "Quietly, Weld Aides Rewrite the State's Rulebook," p. 1.

62. See John Casey, "Urban Fields of Dreams: From Contaminated Industrial Sites, Jobs and Hope," *Business Week,* May 27, 1996, p. 80.

63. Ibid.

64. Comments made by the CEO of Dreyers Ice Cream at the Holy Names College Business Symposium in October 1995.

65. New Communities Corporation, interview by ICIC, 1996; and Andrew C. Revkin, "A Market Scores a Success in Newark," *The New York Times,* April 28, 1995.

66. See study of the Red Hook area of Brooklyn by Philip Kasinitz and Jan Rosenberg, "Why Enterprise Zones Will Not Work: Lessons from a Brooklyn Neighborhood," *City Journal,* Autumn 1993, pp. 63-9. Also, see their latest unpublished work, cited by Malcolm Gladwell, "On the Waterfront, a Clash of Attitudes," *The Washington Post National Weekly Edition,* March 25-31, 1996, p. 34. Also, see study by Shorebank Corporation that shows that, of the nearly 46,000 light manufacturing jobs in the Austin neighborhood of Chicago, fewer than 10% go to local inner-city residents (senior management of Shorebank, interview by ICIC, February 12, 1996). Finally, see various enterprise zone studies cited in note 49.

67. John A. Shields, CEO, First National Supermarkets, quoted in Diesenhouse, "As Suburbs Slow, Supermarkets Return to Cities," p. 5.

PART II

Policies and Programs

Chapter 5

Adding a Stick to the Carrot

Location Incentives with Clawbacks, Rescissions, and Recalibrations

Larry C. Ledebur and Douglas P. Woodward

❋❋❋

A ... development has moved to the forefront of state and local policy in the United States, mayors and governors now measure their performance, however crudely, by plant announcements and job creation. The Reagan era's New Federalism thrust industrial recruitment to the top of the state and local policy agenda. With many federal development programs eliminated or reduced sharply, the amount of state and local money to "buy payroll" during the 1980s has spiraled upward.[1] Public officials battle fervently against one another for new plants, expansions, and relocations, armed with increasingly expensive incentive packages.

These packages tie together everything from land acquisition and job training to new roads and sewers, subsidized loans, and tax credits. Deals to entice automotive assembly plants to the Midwest and Southeast are among the best known (see Table 5.1). The interstate bidding war to subsidize the paving of the Japanese-American "auto alley" perhaps peaked in 1985 when Kentucky won a new Toyota plant with a package valued at more than $300 million.[2]

Besides these and other efforts to attract large manufacturing plants, multimillion dollar deals have been offered to retain or relocate major office facilities. In 1988, New York City bestowed Chase Manhattan with a $235 million incentive bundle to keep the bank's 5,000 jobs from moving to New Jersey. The next year, Hoffman Estates, Illinois won the bid for the relocation of 6,000 employees of the Sears Merchandising Group with a $240 million package, providing free land, worker retraining, infrastructure improvements, and tax abatements.

AUTHORS' NOTE: We thank Alice Perritt for research assistance and four anonymous referees for comments on an earlier draft.

Reprinted from *Economic Development Quarterly,* Vol. 4, No. 3, August 1990.

TABLE 5.1 Automotive Plant Incentive Packages

Company/Location	*Completion Date*	*Company's Investment (millions)*	*Estimated Annual Production*	*Estimated Employment*	*Incentives*	*Total Incentives (millions)*
General Motors (Saturn), Springhill, TN	1990	$3,500	200,000-250,000	3,000	Job training, road improvements, 40-year local tax abatement (GM makes payments in lieu of taxes)	$70 +
Toyota Motor Company, Scott County, KY	1988	$800	200,000	3,000	Land purchase assistance, site preparation, skills center, job training, highway improvements, educational programs for Japanese employees and families	$325 +
Diamond-Star (Mitsubishi/Chrysler), Bloomington/ Normal, IL	1988	$650	180,000	2,900	Road, water, sewer installation; site improvements; land purchase assistance; job training; property tax abatement; tax credits on investment state sales tax; local utility tax; pollution control bonds; water and sewer fee savings	$118.3 +
Isuzu/Fuji Motors, Lafayette, IN	1988	$500	120,000	1,700	Road, highway, sewer improvements; land acquisition assistance; job training; $1-million cultural transition fund to aid Japanese workers and families	$86 +
Mazda/Ford Motor, Flat Rock, MI	1987	$550	240,000	3,500	Road, rail, sewer, site improvements; job training; special $500,000 loan; tax abatements; Mazda will make payments in lieu of taxes	$52 +
Nissan Motor Co., Smyrna, TN	1983	$850	240,000	3,300	Job training; road, sewer, water, rail improvements; local property tax abatements; company makes payments in lieu of taxes	$66 +
Honda of America, Marysville, OH	1982	$870	330,000	4,200	Property tax abatement on buildings; previous $16.4-million grant to Honda for adjacent motorcycle factory	$16.4 +
Volkswagen AG, East Huntington, PA	1978	$236	90,000	2,500	Low-interest loans; rail and highway improvements; job training; local tax abatements; company makes payments in lieu of taxes	$86 +

SOURCE: Norman J. Glickman and Douglas P. Woodward, *The New Competitors: How Foreign Investors Are Changing the U.S. Economy* (New York: Basic Books), pp. 230-31.

A growing chorus of academics, practitioners, and politicians charges that incentives are getting out of control.[3] They contend that the economic development climate is fraught with everything from "fend-for-yourself federalism" to "fiscal fratricide."[4] A major objection is that the bargaining process often lacks any form of accountability. Even a successful effort may turn out to be a Pyrrhic victory when a prized plant is closed or scaled back, as Pennsylvania found when Volkswagen shut down in 1988.[5] Although the expected benefits of a new plant may never the reach employment, tax, and other targets that justify the costs of public subsidies, public officials rarely monitor shifting costs and benefits.

This article addresses a neglected facet of bargaining between government and business for industrial recruitment—generically called clawbacks. In addition to analyzing incentive clawbacks, we identify three related areas for policy intervention: penalties, rescissions, and recalibrations. These provisions entail financial recourse to reclaim all or some of an incentive package whenever a firm fails to meet negotiated performance requirements.

The intent of this article is to show that policymakers can avoid expensive mistakes if they tie incentives to written guarantees of job creation and other benefits. All private business transactions work within a framework of legally binding contracts. The time has come for public-private bargaining with some form of reasonable, guaranteed quid pro quo. European regional policymakers have used a carrot-and-stick approach to industrial recruitment, binding clawback provisions to incentive awards. However, to date there has been little discussion and only limited use of clawbacks in the United States.

The following is organized into four sections. First, we review the political economy of state incentive programs, focusing on their growing role in industrial recruitment. Next, we explain the economic justification for clawback provisions and other forms of policy action when governments use public money to entice new facilities. The third section presents various incentive control options in a simple cost-benefit framework. Finally, we offer guidelines for economic development policy and industrial recruitment, suggesting directions for further discussion and research.

The Proliferation of Incentive Programs

Figure 5.1 shows how state development programs have grown before and after the advent of the New Federalism in the early 1980s. Although some states already had financial assistance and tax incentive programs in place during the 1970s, they were in the minority. The slowdown in national economic activity and greater mobility of capital during the stagflation period of the 1970s intensified the interstate rivalry over economic development. Boosterism turned more aggressive, resulting in the "new war between the states" and the notorious Sunbelt-Frostbelt contest.

The New Federalism ushered in a more intense phase in this interregional competition.[6] By the late 1980s, a majority of states had programs for a broad range of incentives available to firms (see Figure 5.1). The number of states with tax incentive schemes for job creation and industrial development, including attempts to develop distressed areas, expanded significantly between 1981 and 1988.[7]

Such inventories fail to account for the size of state and local financial commitments and the way they are practically deployed. No doubt, an increase in the dollar volume of public resources invested has accompanied the proliferation of incentive programs.[8] Unfortunately, an accurate tally is not available, in part because some programs are carefully hidden or disguised by economic development agencies.

At any rate, what stands out during the Reagan era is the rise of ad hoc industrial targeting at the state and local level. Today's economic development practitioners do more than just attempt to "pick winners." Interviews disclose that they tend to "shoot anything that flies and claim anything that falls."[9]

In principle, subsidies for site location result from bargaining between government and private industry. In practice, however, private

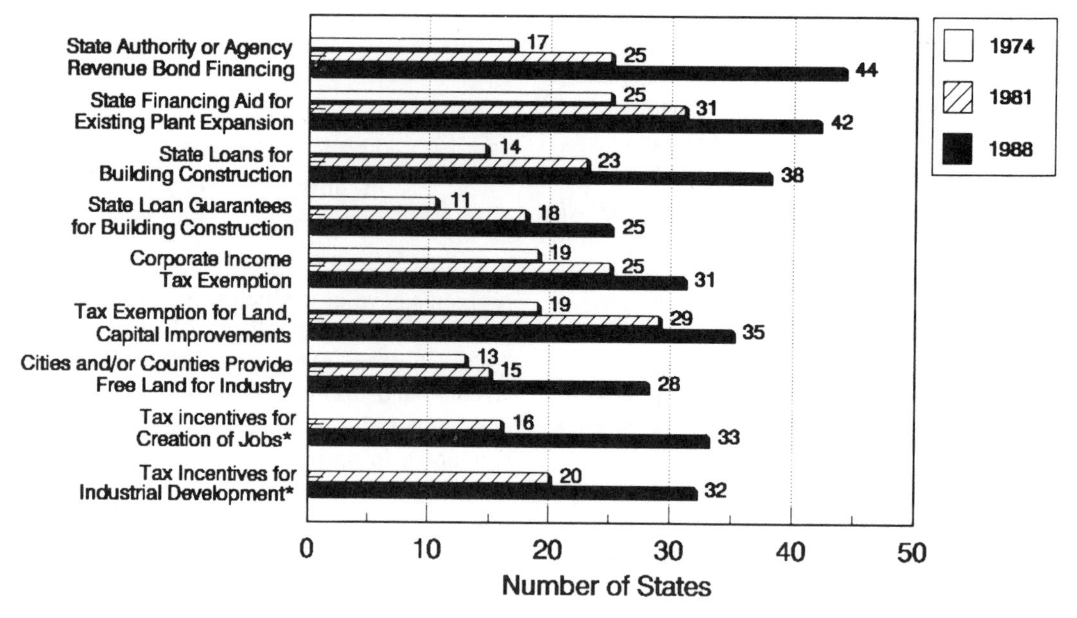

Figure 5.1. State Incentives for Industrial Development, 1974-1988

SOURCE: "Annual Survey: 50 Legislative Climates," *Industrial Development & Site Selection,* 1974, 1981, 1988.
*Figures not available before 1980.

firms have been permitted to dictate many of the terms of the final location agreement. Companies always have the advantage of choosing another location. Although there are only a few winners and many losers, states and localities enter an often counterproductive gambit to attract new capital.

The strongest argument in favor of using public money to attract new plants and expansions is straightforward: a jurisdiction would receive no new tax revenue, jobs, or other benefits if a firm chose an alternative location. The critical question then becomes whether incentives have any discernible influence on private location choices. Unfortunately, the efficacy of economic development programs is hard to evaluate systematically because they vary widely and are tailored to individual cases. Still, most studies show that incentives have little influence on location decisions. A recent report to the National Governors Association noted that "many manufacturers make the decision to locate a new facility without the benefit of state-supported incentives."[10] Empirical work dating back to the 1960s suggests that most promotional activities are insignificant determinants of site selection decisions, especially when statistical tests include access to markets, resources, labor force characteristics (quality, availability, and cost), and other important determinants as control variables.[11] Firm surveys find that most businesses consider financial incentives a minor influence on plant openings and relocations.[12]

It is not surprising that incentives carry so little weight in these studies. If a firm follows profit maximization, the location offering the greatest margin between revenues and costs will dictate the decision. Governments have little or no control over the fundamental determinants of a firm's demand and costs. Policymakers can only alter location factors like markets (on the demand side) and energy and wage costs (on the cost side) over a very long time period, if at all.

Why, then, are incentive programs proliferating? Their popularity arises in part because they provide policymakers with one of the few discretionary tools available that could sway private decisions in the short run. In other words, development officials focus their efforts on the factors they can immediately control.[13] Indeed, incentives could be effective at the margin—acting as tiebreakers when all else is the same among the communities under consideration. Empirical studies downplay the role of incentives in plant location, but this does not invalidate the argument that government policy can be influential in the final stage of a site location decision.[14]

Incentives work best after a firm has chosen a region (the Southeast or Midwest, for example) and has drawn up a final checklist of local sites. The list is often highly visible and encourages intense competition among states and localities. Some companies, Japanese multinational corporations in particular, take as long as five years to narrow their selection. Yet development agencies often have no more than a month to assemble an incentive package. There is little time for sound economic and fiscal impact analysis.

Nevertheless, states and cities believe they must act. If they seriously banned or limited the use of differential subsidies for industrial recruitment they could miss an opportunity to add hundreds or even thousands of jobs to state payrolls. Given the emphasis on job creation, action at this point appears to be politically expedient. North Carolina, for example, ostensibly refrains from using costly "giveaways."[15] Even so, both Charlotte and Raleigh, North Carolina managed to assemble heavy subsidies in their unsuccessful attempt to relocate the Sears Merchandising Group in 1989.

In fact, incentive programs naturally tend to propagate. Often, localities feel they must offer incentives to remain competitive with neighboring jurisdictions.[16] A recent study of four categories of incentives—industrial revenue bonds, loans, property tax exemptions, and income tax reductions—found empirical support for the "competitive adoption" hypothesis.[17] Changes in state incentive programs prompt neighboring states to adopt similar programs. Accordingly, one state program may quickly spread throughout a whole region.

The Justification for Incentive Controls

Recent evidence, then, suggests that incentive bidding tends to feed on itself, with more expensive items often added at the last minute in an attempt to keep ahead of the competition. These hastily assembled packages act like economic steroids—in the short run, they strengthen communities in the race for new plants and expanded payroll. Of course, that does not justify their use. Even when they are effective, incentives are open to serious objections. Not surprisingly, the opposition to incentive "giveaways" began to gather steam during the mid-1980s.

A major objection is that most locational incentive programs are highly selective and firm specific. This distinguishes incentives from economic development programs that seek to enhance the "business climate" through educational reform, research and development, support for small business, and general tax relief. State money is allocated neither uniformly nor necessarily efficiently, but on an ad hoc, case-by-case basis. It often favors large, visible projects over smaller, lower profile investments. More than anything else, this raises the question of their legitimacy. There is some evidence, largely anecdotal, that shows that tax breaks favoring individual plants can crowd out existing businesses.[18] Moreover, tax breaks and other subsidies represent lost revenue—money that cannot be spent on education, better infrastructure, and other things that most firms, and, indeed, most citizens, want.

Critics, crossing the spectrum of political opinion, now recognize that governments must carefully assess the possible benefits of an incentive package against competing uses of public money. The justification for controlling incentives stems from possible unfair state-supported advantages given to certain firms. Laissez-faire advocates argue that state-imposed controls, especially performance requirements, distort business decisions. They must agree, though, that it is industrial incentives, not these requirements, that upset private capital accumulation and allocation. With industrial subsidies, as with any other payment for product or services, failure to perform to contract specifications should result in partial or full recovery of past payments and forfeiture of future payments. Unfortunately, firms subsidized by state or local governments in the United States are rarely obliged to pay penalties for nonperformance to subsidizing governments, usually because no contract exists.

The mood is clearly changing. Some states and cities have brought lawsuits against corporations that received incentives and then closed facilities. The most significant case to date involved Duluth, Minnesota and Triangle Corporation. The city signed a contract with the company in 1982 stipulating that Triangle could not move equipment financed by tax-exempt industrial bonds. Yet seven years later, Triangle cut the Minnesota work force by one-half and moved some of its operations to Orangeburg, South Carolina. Duluth sued the company and won in February 1988, with the court ordering Triangle to return the relocated equipment. The verdict was overturned a year later, when the Minnesota Court of Appeals barred Triangle from moving any more equipment, but allowed Triangle to keep the existing equipment in South Carolina.

Another significant court challenge came in 1980, after the city of Chicago gave Playskool, Inc. a $1 million industrial revenue bond with the understanding that the company would create 400 jobs. When the plant closed after four years, city officials threatened to sue and eventually settled out of court. Playskool agreed to keep the plant open 10 more months, to set up a job placement program, and to create a $50,000 fund for displaced workers.

The lesson of both the Chicago and Minnesota cases is that communities fare better if they carefully specify a financial recovery procedure *before* a plant shuts down or scales back. Apart from these isolated legal disputes, little has been done to control the bidding war for business. Occasionally, states have proposed forming compacts to control interregional competition. In 1987, Michigan spearheaded a drive in the Midwest to limit incentives, agreeing to end tax abatements if other states reciprocated. But the initiative never took hold. Likewise, the Massachusetts legislature proposed an interstate moratorium on incentives. However, most states have shown that they are unlikely to sign such pacts and will continue to do everything they can, within their budget constraints, to entice more investment.

In lieu of an interstate compact on controlling the costs of attracting new plants and expansions, enforceable penalties imposed by individual states remain the only viable, if second-best, solution to the incentive game. Overseas such penalties are often called *clawbacks,* a comprehensive term for cancellation, reduction, and recovery of subsidies.[19] Western European countries, all with very generous incentive programs by U.S. standards, establish contractual relationships with firms they subsidize. They specify investment, employment, or output standards and penalties for failure to perform at these contractual levels. As the following examples illustrate, clawbacks are used with almost every form of industrial subsidy in these European nations.

Tax Concessions

The Netherlands subsidizes almost all types of business investment through reduced tax payments if the operation is profitable or through negative tax payments if it incurs losses. If the subsidized asset is sold within a given period, all or a portion of the subsidy must be repaid. France provides local business tax concessions to firms for qualified projects. If a firm fails to meet job targets, the concession is revoked and firms must repay the conceded tax plus interest penalties.

Employment Creation Subsidies

Italy has provided concessions on social security contributions paid by employers for new employment. Where contributions have been withheld for jobs not created, this amount can be clawed back with penalties of five times this amount. Other nations limit grant payments to employment levels as well. Great Britain and Northern Ireland both dispense grants for job creation. Yet the government pays the grants in installments, which can be reduced if job targets are not met.

Capital Grants

In addition to job-creation incentives, direct grants to stimulate investment are common in Europe. Almost all of these grants include clawback features of varying degrees of severity. In Germany and Northern Ireland, grants can be revoked and full repayment required. Belgium and France make grant payments in installments. If investment or job targets are not met, future installments can be revoked. France also allows for pro rata reductions. Italy can suspend further payments, but must sue to recover past payments.

Loans and Interest Subsidies

Several European countries administer industry subsidy programs that reduce commercial interest rates on loans for eligible projects. In Belgium, Germany, and Luxembourg these subsidies can be reduced or withdrawn if the firm fails to meet the conditions for incentive award. These nations also enforce penalties in soft loan programs. In Denmark and Great Britain, for example, loans can be called in. The state may require special payments if conditions are not fulfilled.

These examples of relatively simple clawback schemes reveal that a menu of options is available. Despite the growing interest among state and local development analysts and the increasing number of initiatives discussed earlier, so far the U.S. literature has been silent on the issue. What is needed to open the dialogue is a conceptual framework that clarifies how state and local governments can design contractual performance agreements with subsidized firms. In the next section we present the rudiments of how various options would work.

Rescissions, Clawbacks, Penalties, and Recalibrations

Just as incentive packages come in many forms, incentive controls can be tailored to individual cases. In general, however, modifications for performance deficits will fall into four basic categories:

- Rescissions: canceling a subsidy agreement;
- Clawbacks: recovery of all or part of the subsidy costs;
- Penalties: special charges for nonperformance or relocation;
- Recalibrations: subsidy adjustments to reflect changing business conditions.

To see how each of these options would work, consider the following hypothetical case. Footloose, Inc., a successful athletic shoe company, announces it will open a major production facility and promises it will move its corporate headquarters and marketing research operations to the new location. The production facility alone will hire 2,000 workers. Having made the final site selection cut for the new plant, the City of New Prosperity wants to explore the possibility of bidding for the plant with an incentive package, which may include a tax abatement.

The city then asks the Department of Economic Development to evaluate the incentive package within a cost-benefit framework. Their first step is to define the time stream of costs and benefits.

Benefits: In practice, the government may define benefits in a variety of ways, including jobs created, private investment, and tax revenues generated. For purposes of this analysis, benefits are defined by New Prosperity as the sum of the tax revenues resulting from the direct and indirect impact of new investment, output, and payroll.

Costs: All state and local development agencies must allocate limited financial resources among many competing, worthwhile projects. The city acknowledges that public resources used for subsidies, including tax expenditures, will have opportunity costs. Thus the city de-

TABLE 5.2 Hypothetical Schedule of Benefits and Costs of Subsidies (in thousands of dollars)

Year	*(1)* *Annual Benefit*	*(2)* *Annual Cost*	*(3)* *Annual Benefits minus Annual Costs*	*(4)* *Realized Benefits*	*(5)* *Realized Benefits minus Annual Costs*
1	5	52.5	–47.5	3.34	–49.16
2	10	52.5	–42.5	6.68	–45.82
3	15	52.5	–37.5	10.02	–42.48
4	20	52.5	–32.5	13.36	–39.14
5	25	52.5	–27.5	16.7	–35.8
6	30	52.5	–22.5	20.04	32.46
7	35	52.5	–17.5	23.38	–29.12
8	40	52.5	–12.5	26.72	–25.78
9	45	52.5	–7.5	30.06	–22.44
10	50	52.5	–2.5	33.4	–19.1
11	55	52.5	2.5	36.74	–15.76
12	60	52.5	7.5	40.08	–12.42
13	65	52.5	12.5	43.42	–9.08
14	70	52.5	17.5	46.76	–5.74
15	75	52.5	22.5	50.1	–2.4
16	80	52.5	27.5	53.44	0.94
17	85	52.5	32.5	56.78	4.28
18	90	52.5	37.5	60.12	7.62
19	95	52.5	42.5	63.46	10.96
20	100	52.5	47.5	66.8	14.3
Total	1050	1050	0	701.4	–348.6

cides to strive to get the greatest possible return from its budget—taking all relevant direct and indirect costs into account. The full costs of a subsidy encompass the dollar value of the subsidy, including tax revenues forgone, and any public infrastructure and service costs imposed by the anticipated activity of the firm.

Next the city identifies the upper bound of the subsidy—the cost ceiling that would not be exceeded if economic rationality guided the public decision. This maximum subsidy is determined by comparing the stream of anticipated benefits and alternative subsidy levels. The subsidy limit can be readily identified. Consider Column 1 of Table 5.2, a hypothetical stream of anticipated benefits from an annual tax abatement given to Footloose, Inc. All figures in Table 5.2 are discounted present values (DPV). Over a 20-year period, the cumulative anticipated benefits are $1,050 thousand. This is the ceiling that should not be exceeded by the discounted cost of the public subsidy over this same period. An annual tax abatement with a DPV of $52.5 thousand per year (Column 2) will result in a total subsidy cost over the 20 years equal to cumulative anticipated benefits. Note that annual subsidy costs will exceed anticipated annual benefits in years 1-10, and annual net benefits will be anticipated over the remaining 10 years of the subsidy program (Column 3).

The case where the DPV of benefits equals the DPV of subsidy costs is the breakeven, or zero rate of return, level of subsidization. Enterprising state and local government will want to achieve some positive rate of return on these public expenditures, one that exceeds the opportunity cost of the public resources. The depth of the subsidy should be established to achieve this desired rate of return on public expenditures (i.e., the ratio of DPV of benefits to DPV of costs = 1 + target rate of return).

After completing the preliminary analysis, New Prosperity presents the package to Footloose, Inc. The company accepts the offer as specified in a contract, which is to be administered by the city's Department of Economic Development. The agreement with the recipient firm is contigent upon target levels of activity (investment, employment, and output). The city also retains the ability to modify the depth of the subsidy if Footloose fails to perform at these contractual levels (i.e., when the stream of realized benefits is less than that of anticipated benefits).

The need for any adjustment would follow from monitoring and further cost-benefit analysis. Say Footloose's activity is evaluated in year 5, when it appears that the company's promises about headquarters and marketing research expansion plans will not come to pass. The new situation is presented in Table 5.2, Column 4. Here the stream of annual benefits falls below the anticipated stream used to established the threshold subsidy level. The development office of New Prosperity recognizes that the stream of realized benefits is not meeting the originally anticipated targets. Under the new stream of expected benefits, the annual benefits will not exceed annual costs until after year 15 (Column 4), and, over the 20-year period, public costs will exceed benefits by $348.6 thousand. At this point, New Prosperity development officials have several alternatives to consider.

Rescissions

As one option, the development office can simply cancel the subsidy agreement with the firm for nonperformance. In Table 5.2, cancellation to the subsidy after year 5 will result in a net loss of $212.4 thousand, the difference between cumulative costs and realized benefits over the five-year period. This may not be a net loss to the government in the long run, however, if the now unsubsidized firm continues to operate and generates future benefits.

Clawbacks

Next consider the clawback, which specifically refers to recovery of all or part of past subsidy costs. Here the government faces three choices. First, it can attempt to clawback a part of the subsidy equal to the stream of unrealized benefits over the five-year period. In Table 5.2, the amount to be clawed back is $24.9 thousand: the sum of the difference between anticipated benefits in years 1-5 (Column 1) and realized benefits (Column 4).

Second, the government can attempt to recover the subsidy costs for years 1-5 in excess of realized benefits. In Table 5.2, this amount is equal to $212.4 thousand: the sum of the difference between subsidy costs for years 1-5 (Column 2) and anticipated benefits (Column 1).

Third, if New Prosperity wished to penalize the firm for nonperformance more stringently, it could attempt to clawback the entire subsidy for the first five years, or $262.5 thousand in the example in Table 5.2. This more severe clawback option might be used, for example, if Footloose, Inc. made the decision to relocate to another jurisdiction after the initial five years.

Penalties

Disincentives for nonperformance can be increased by adding penalties to clawback provisions. For example, if the incentive is a tax abatement or tax credit, the government could clawback the taxes foregone, plus charge interest on these unpaid taxes as a penalty for nonperformance. Similar interest penalties can be applied to all forms of interest subsidies on loans (direct interest subsidies, soft loans, and loan guarantees). A variety of penalties can be devised, including foreclosure of subsidized assets (land, building, and equipment). This would be a likely option only if a plant relocated rather than scaled back its operations.

Recalibrations

An alternative to canceling the subsidy or attempting to clawback past subsidy costs is to adjust or calibrate the level of subsidy to reflect the new projected benefit stream. Three primary calibration options are available.

Option 1: De novo recalibration

The recalibration can be made in view of historical and future benefits without reference to

TABLE 5.3 Recalibration Options (in thousands of dollars)

	(1)	*(2)*	*(3)*	*(4)*	*(5)*	*(6)*	*(7)*
				Recalibrations			
Year	*Annual Benefit*	*Annual Cost*	*Realized Benefits*	*Option 1*		*Option 2 with Clawbacks*	*Option 3 Racheted*
1	5	52.5	3.34	35.07		52.5	52.5
2	10	52.5	6.68	35.07		52.5	52.5
3	15	52.5	10.02	35.07		52.5	52.5
4	20	52.5	13.36	35.07		52.5	52.5
5	25	52.5	16.7	35.07		52.5	52.5
6	30	52.5	20.04	35.07	43.4	29.28	49.6
7	35	52.5	23.38	35.07	43.4	29.28	46.7
8	40	52.5	26.72	35.07	43.4	29.28	43.8
9	45	52.5	30.06	35.07	43.4	29.28	40.9
10	50	52.5	33.4	35.07	43.4	29.28	38
11	55	52.5	36.74	35.07	43.4	29.28	35.1
12	60	52.5	40.08	35.07	43.4	29.28	32.2
13	65	52.5	43.42	35.07	43.4	29.28	29.3
14	70	52.5	46.76	35.07	43.4	29.28	26.4
15	75	52.5	50.1	35.07	43.4	29.28	23.5
16	80	52.5	53.44	35.07	43.4	29.28	20.6
17	85	52.5	56.78	35.07	43.4	29.28	17.7
18	90	52.5	60.12	35.07	43.4	28.28	14.8
19	95	52.5	63.46	35.07	43.4	29.28	11.9
20	100	52.5	66.8	35.07	43.4	29.28	9
Total	1050	1050	701.4	701.4	651	701.7	702

past subsidy costs. In this case, the recalibrated annual subsidy had been paid over the entirety of the 20-year period; the DPV of the 20-year stream of public costs would have been equal to that of the corresponding benefit stream (Column 3). However, the rate of return to the expenditure of public monies falls below the target rate of return because the subsidy level in years 1-5 exceeded the new threshold subsidy.

Alternatively, the recalibration can be made without reference to historical costs or benefits. The new subsidy level would be set to achieve the targeted rate of return on the public expenditure over the remaining 15 years. In the example of Table 5.3 (Column 5), this annual subsidy level would be approximately $43.4 thousand.

Option 2: Recalibration with clawbacks

The recalibration can also include clawback features. New Prosperity could reduce the level of the annual subsidy for years 6-20 to recover the overpayment in years 1-5. The new subsidy level would be calibrated so that the ratio of the DPV of realized benefits in years 1-5 plus that of anticipated benefits in years 6-20 over the DPV of the previously paid subsidy in years 1-5 plus the DPV of the cost stream of the subsidy for years 6-20 equaled the target rate of return. This recalibration is presented in Column 6 of Table 5.3. The annual subsidy for years 6-20 would be equal to $29.28 thousand.

Option 3: Racheted recalibration

An abrupt reduction of the annual level of the subsidy might, in itself, adversely affect a troubled company like Footloose, Inc. To provide time for the firm to adjust to the lower subsidy level, the recalibration can be accomplished by a downward rachet in the annual subsidy over the years 6-20.

The downward rachet would be calibrated to achieve the targeted rate of return by the end of the subsidy period. Thus the subsidy would be reduced in each succeeding year starting in year 6 with the extent of the annual reductions scaled to achieve the targeted rate of return, as illustrated in Column 7 of Table 5.3.

Policy Guidelines: Preparing for the Decade Ahead

By considering any one of the options described in the last section, public officials would be forced to look at how the consequences of their recruitment efforts go beyond initial job tallies advertised in the media. For example, they would be compelled to examine whether, by exempting from property taxes a plant that employs 2,000 people, they still guarantee that the community will have safe roads and drinking water, fire and police protection, parks, and a first-rate educational system.

Still, ribbon-cutting photo opportunities for major new plants often appear more politically expedient than a careful economic development plan for the future.[20] Hence the push for greater accountability is less likely to come from gubernatorial and mayoral offices than from financial officers, who are more familiar with the opportunity costs associated with development programs. Edward V. Regan, comptroller of New York State, for example, issued a stinging indictment of undisciplined development programs and the failure of public officials to make them accountable to analytical inspection. He noted, "The major players—the business community and, especially elected officials—either do not want to or do not know the costs versus benefits of playing the development game or how to change the game. Such an evaluation, by default, is now left to finance officials."[21]

Is there reason to believe that states and communities offering public incentives to the private sector will make development programs more accountable to analysis and inspection? Already, more rigorous cost-benefit analysis is being given consideration in state governments. In 1988, legislation in the State of New York proposed that every tax abatement and exemption include a statement of benefits and the costs to the taxpayers of all subsidies. The state of Illinois passed a law in 1987 requiring an economic impact study before the authorization of incentives to attract foreign investment.

The first signs that governments are moving toward clawback provisions come from Wisconsin and Vermont. Wisconsin's constitution prohibits using differential tax abatements, but to attract industry the state established assistance programs for labor training and technology development during the 1980s. The Department of Development operates under an administrative rule that requires a contract with the recipient firm. Wisconsin may void state funding when the company relocates, downgrades skill levels, or otherwise does not perform according to contract. Similarly, Vermont stipulates under contract that a company must meet employment targets and wage levels when labor training assistance is offered.

It remains to be seen whether other states and cities will follow the lead of Wisconsin and Vermont. Even so, as public officials continue to engage in subsidized industrial recruitment during the 1990s, they should at least become more skilled than they have over the last decade. Toward that end, this article has focused on one important new area for state and local policy. But the suggestions we make for improving and protecting the integrity of public funds used for industrial recruitment should be imbedded in a set of state and local economic development guidelines. Seven important steps are summarized next.

Policy Framework

State and local governments should establish clear and consistent policies on industrial subsidies and avoid operating on an ad hoc, case-by-case basis. Governments engaged in subsidizing private enterprise need to operate from a consistent framework of policies that can be effectively communicated, understood, and defended in working with potential industrial opportunities. One component of these guidelines should be a clear rationale for assisting target firms.

Subsidy Budgets

State and local governments should establish budgets for industrial subsidies, including tax concessions, that identify the amount that the unit of government is willing to invest in each fiscal year. Often, subsidy programs are administered with no clear understanding of accounting or their costs, especially those resulting from off-budget programs and tax expenditures. Administering officials should be required to conform to these budget limitations. Creation of subsidy budgets, besides controlling public costs, also encourage development authorities to establish priorities among projects based on their rate of return to the administering government.

Target Rate of Return

State and local governments should identify a target rate of return on the investment of public funds in industrial subsidies. This target rate of return should be no less than the opportunity cost of these public resources (i.e., the rate of return in the highest alternative uses). Projects that do not meet or exceed this target rate of return should not receive public subsidies.

Evaluation

State and local governments should develop the analytical capacity to evaluate the public benefits and costs of industrial subsidies. Analysis of every project considered for public subsidies should be undertaken to identify the benefits derived by the administering jurisdiction, the costs to the administering government, and the rate of return to the investment of public funds.

Cost-Effectiveness

Jurisdictions should use the most cost-effective instruments in administering industrial subsidies.[22] Tax abatements, although one of the most common subsidies, are seldom cost-effective. The more cost-effective instruments are those that leverage investment from commercial lending and investment institutions. Among these are loan guarantees, direct interest subsidies, and incentives targeted to nondepreciable assets.

Negotiating Skills

State and local governments should develop the capacity to sit with business at the negotiating table as equal partners. Business executives will respect government officials who are "businesslike," have a clear understanding of the public "bottom-line," and are skillful negotiators.

Subsidy Contracts

Governments should require legally enforceable contracts with firms as a condition of award of subsidies. These contracts should carefully specify performance criteria such as investment, employment within their jurisdiction, and public recourse for nonperformance. Subsidy contracts should include clauses that specify rescissions, clawbacks, penalties, and recalibrations that will be enforced if the subsidized firm fails to meet its promises.

These seven guidelines would help deter the often haphazard and counterproductive bargaining over incentives that has arisen inexorably over the past decade. With a comprehensive economic development program in place, states and cities could act more prudently when they make the site selection cut for a major facility. By presenting these objectives for economic development policy in this article, including specific clawback and other incentive controls, we hope to stimulate further discussion and research.

APPENDIX

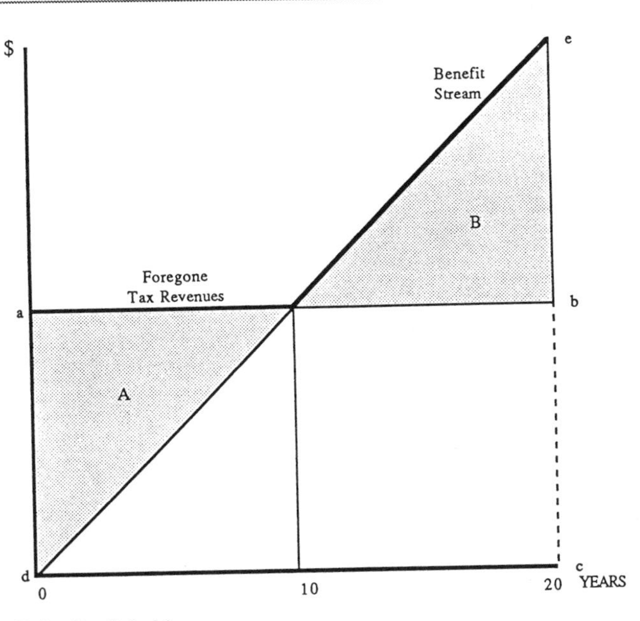

Figure 5.2. Breakeven Point for Subsidy

This appendix presents a diagrammatic exposition of the incentive control options discussed in the body of the article. Consider a time stream of anticipated benefits from a tax abatement represented by the line *de* in Figure 5.2. The time stream of costs (tax revenues forgone plus other public costs) in Figure 5.2 has constant annual costs represented by the line *ab.*[23] The relevant time period for the subsidy is assumed to be 20 years. Beyond this time horizon, all costs and benefits are discounted 100%.

Under these assumptions, the subsidy's maximum would be the level at which 20-year time stream of benefits is equal to the corresponding time stream of costs. In Figure 5.2, this will be at the annual subsidy cost of *ad* where Area *dec* = Area *abdc,* or Area A = Area B. This is the breakeven level for the unit of government administering the subsidy.

In reality, governments will (should) discount the time stream of benefits and costs of the subsidy. The threshold level of the subsidy will be that at which the discounted present value (DPV) of the 20-year stream of benefits is equal to the DPV of the corresponding time stream of costs. Note that future benefits will be discounted more heavily than future costs because of differentials in uncertainty. Benefit streams are likely to be regarded as more uncertain, whereas costs, particularly the annual tax abatement, can be identified with more certainty. In Figure 5.2, therefore, Area B would be greater than Area A at the threshold level because of the higher discount rate applied to future returns.

Consider now a situation as depicted in Figure 5.3, which is similar to the example presented earlier in Table 5.2. Here *ab* is the threshold subsidy set for the anticipated stream of benefits *cd,* whereas *ce* is the actual steam of benefits. Assume that the firm's activity is evaluated in year 5, when the administering government recognizes that the stream of realized benefits will be less than those anticipated at the beginnning of the project. This loss is shown by the Area *cde.*

Rescissions

As discussed in the body of the article, the government unit offering the incentive can now cancel the

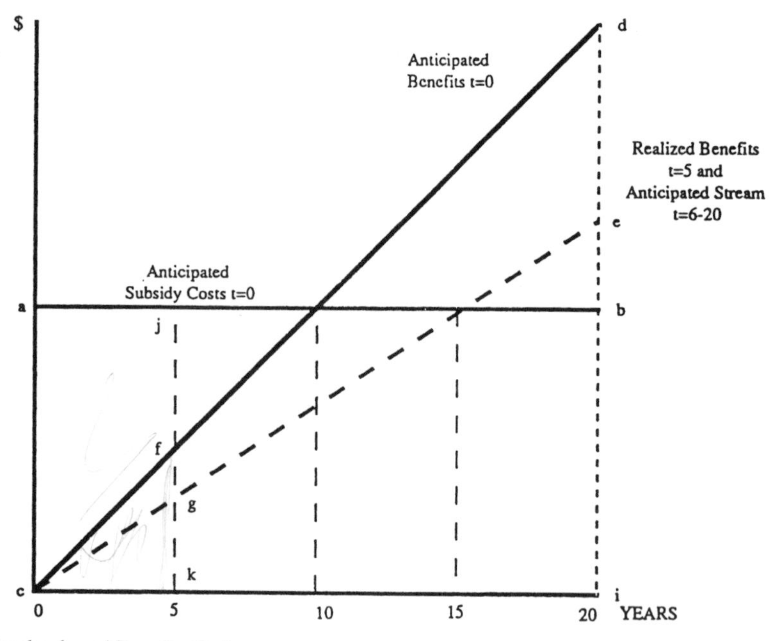

Figure 5.3. Clawback and Penalty Options

subsidy agreement with the firm for nonperformance. As shown in Figure 5.3, total costs to the government greatly may exceed realized benefits (Area *cajk* vs. *cgk*).

Clawbacks

Next consider the clawback. Here the administering government confronts three alternatives. First, it can attempt to clawback a part of the subsidy equal to the stream of unrealized benefits over the five-year period (Area *cfg* in Figure 5.3). This is the mildest form of clawback.

Second, the government can attempt to recover the subsidy costs for years 1-5 in excess of realized benefits. Here the clawback would be equal to the Area *cajg*. Third, if the unit of government can attempt to clawback the entirety of the subsidy for the first five years (Area *cajk*). In addition, as mentioned in the body of the article, disincentives for nonperformance can be increased by adding penalties to clawback provisions.

Recalibrations

Figures 5.4, 5.5, and 5.6 present the three primary calibration options.

Option 1: De novo calibration

The calibration is made without regard to past subsidy costs. In Figure 5.4, the new subsidy level will be *cl* at which Area *lcp* = Area *pen.* In this case, however, the rate of return to the expenditure of public monies falls below the targeted rate of return because total cost of the subsidy (anticipated future costs *ksni* plus historical costs *ajkc*) exceed upper threshold cost for achieving the targeted rate of return by Area *alsj.*

Alternatively, the recalibration can be made without reference to historical costs or benefits. Here the new subsidy level would be set to achieve the targeted rate of return on the public expenditure over the remaining 15 years. This level, not shown in Figure 5.4, would fall between the initial annual subsidy and *cl.*

Option 2 : Recalibration with clawbacks

This recalibration option includes clawback features. The administering government can reduce the level of the annual subsidy for years 6-20 to recover the loss in years 1-5. This would be the point at which the ratio of the DPV of the new stream of realized benefits in years 1-5 plus anticipated benefits in years 6-20 over the DPV of the subsidies of years 1-5 plus the future subsidy stream for years 6-20 equals the

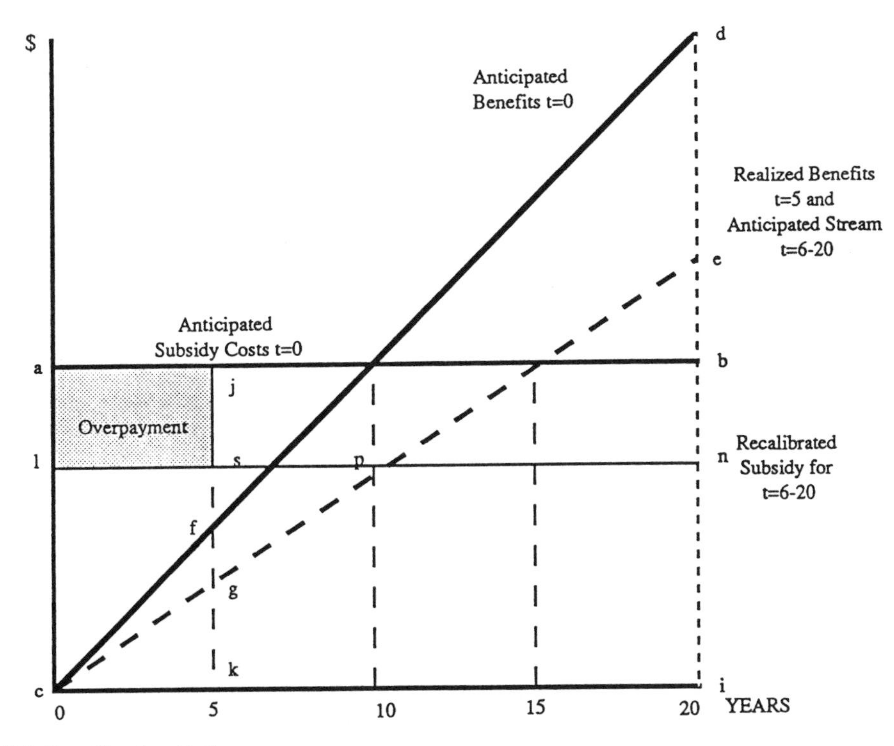

Figure 5.4. Recalibration Options

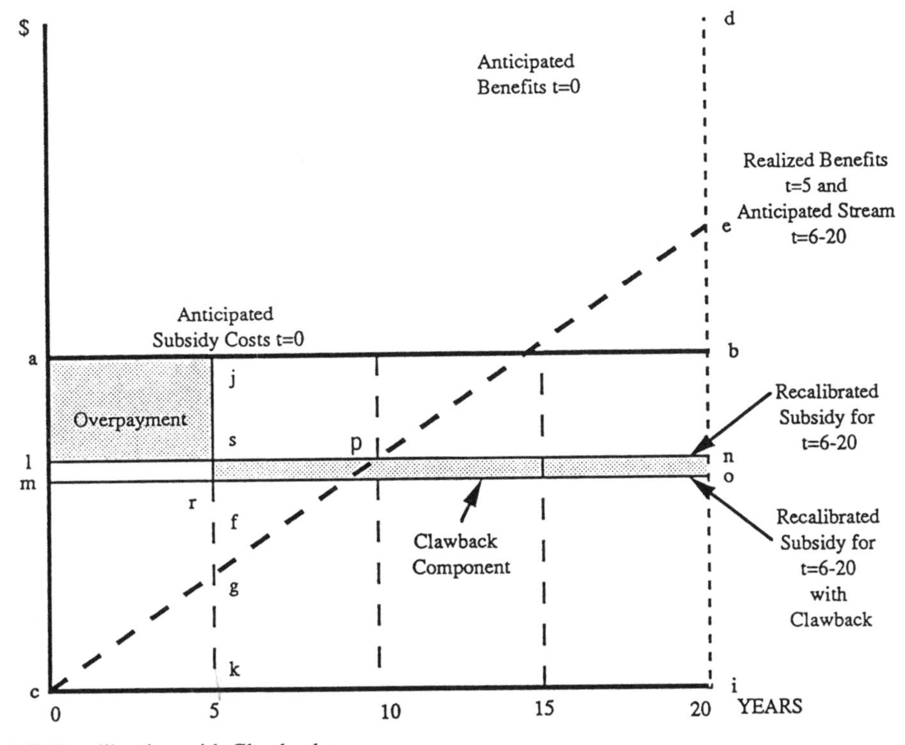

Figure 5.5. Recalibration with Clawbacks

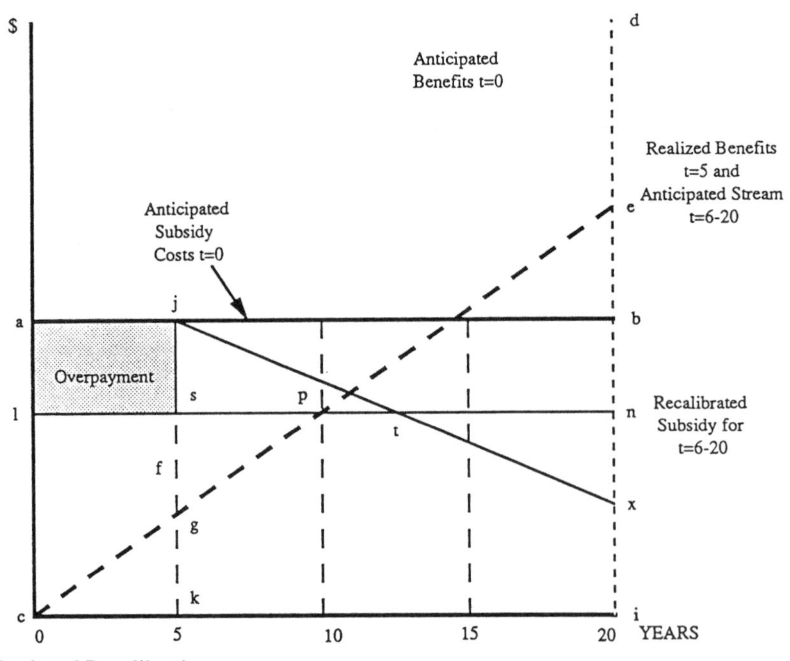

Figure 5.6. Racheted Recalibration

target rate of return on the public monies. In Figure 5.5, this annual subsidy would be *cm,* the level at which the overpayment of the subsidy in years 1-5 (Area *alsj*) is equal to the clawback component of the recalibrated subsidy (shaded Area *sron*).

Option 3: Racheted recalibration

The racheted recalibration is shown in Figure 5.6. The downward ratchet would be calibrated to achieve the targeted rate of return by the end of the subsidy period. Thus the subsidy would be reduced in each succeeding year starting in year 6 with the extent of the annual reductions scaled to achieve the targeted rate of return, as illustrated in Figure 5.6. The rate of the annual reduction (line *jx*) would be calibrated so that the overpayment of the subsidy in years 6-12.5 (Area *jst*) would equal an equivalent underpayment in years 12.5-20 (Area *tnx*).

Notes

1. A review of the literature on the use of industrial development incentives in the United States can be found in William Luker, Jr., " 'Buying Payroll': Industrial Development Incentives and the Privatization of Economic Development," (Paper presented at the Annual Meetings of the Western Social Science Association, Lyndon B. Johnson School of Public Affairs, University of Texas at Austin, April 1988, Mimeographed).

2. For details see Norman J. Glickman and Douglas P. Woodward, *The New Competitors: How Foreign Investors Are Changing the U.S. Economy* (New York: Basic Books, 1989), especially ch. 8. See also Martin Tolchin and Susan Tolchin, *Buying Into America: How Foreign Money is Changing the Face of Our Nation* (New York: Times Books, 1988), ch. 4.

3. See Georgina Fiordalishi, "How To Avoid A Bum Deal When Using Incentives to Win Businesses, Jobs," *City & State,* 19 June-2 July 1989, pp. 11-12.

4. See, for example, Tolchin and Tolchin, chap. 4.

5. A decade earlier the state offered Volkswagen a multimillion dollar incentive package following a fierce bidding war with neighboring Ohio. The package included $30 million for rail and highway connections to the plant and a $40 million loan not due until ten years after the plant closed.

6. During the 1980s, major cutbacks affected most federal programs designed to stimulate local economies. Under the Reagan administration, real spending on Urban Development Action Grants and Economic Development Administration programs fell about 53%. Even the popular Community Development Block Grant program was cut 20% (after adjusting for inflation) from 1980 through 1986. See Susan S. Fainstein and Norman Fainstein, "The Ambivalent State: Economic Development Policy in the U.S. Federal System Under the Reagan Administration," *Urban Affairs Quarterly* 25, No. 1 (1989): 41-62.

7. The best known examples are enterprise zones used to stimulate economic development in depressed urban areas. This centerpiece of Reagan's urban policy was never implemented nationally, but became popular with states and communities.

8. It is difficult to assess how the size of incentives has grown over time. For 1985, the dollar volume of state incentive programs is presented in *National Association of State Development Agencies, Directory of Incentives for Business Investment and Development in the United States: A State-by-State Guide* (Washington, DC: Urban Institute Press, 1986).

9. Herbert J. Rubin, "Shoot Anything That Flies; Claim Anything That Falls: Conversations with Economic Development Practitioners," *Economic Development Quarterly* 2, No. 3 (1988): 236-251.

10. Blaine Liner and Larry Ledebur, *Foreign Direct Investment in the United States: A Governor's Guide,* prepared for the 79th meeting of the National Governors' Association (Washington, DC: Urban Institute, July 1987), p. 5.

11. A good summary of the literature is found in John P. Blair and Robert Premus, "Major Factors in Industrial Location: A Review," *Economic Development Quarterly* 1, (1987): 72-85. For a review of early research, see Benjamin Bridges, Jr., "State and Local Inducements for Industry, Parts I and II," *National Tax Journal,* March 1965 18 No. 1, 1-14 (Part I); and June 1965 18 No. 2, 175-92. A recent empirical study of industrial development effectiveness tested a broad range of general state programs, including some of the plant location inducements considered in this article. It found that land and building subsidies, tax programs, post-secondary education assistance, advertising and other outreach programs, and research and development support are either too small to have an impact on state economic performance or are inherently ineffective. The author concludes that they tend to waste resources. See Michael I. Luger, "The States and Industrial Development: Program Mix and Policy Effectiveness," in *Perspectives on Local Public Finance and Public Policy,* Vol. 4, ed. John Quigley (Greenwich, CT: JAI, 1987), pp. 29-63.

12. Schmenner's 1980 survey of Fortune 500 plant openings and relocations demonstrated "fairly convincingly that tax and financial incentives have little influence on almost all plant location decisions." See Roger W. Schmenner, *Making Location Decisions* (Englewood Cliffs, NJ: Prentice-Hall, 1982), p. 51. More recently, the 1989 Grant Thornton survey found that manufacturers ranked state incentives 17th out of 21 factors that companies consider when deciding on locating new sites. See Grant Thornton, *The 10th Annual Manufacturing Climates Study* (Chicago: Grant Thornton, June 1989).

13. See Barry M. Rubin and C. Kurt Zorn, "Sensible State and Local Economic Development," *Public Administration Review,* March/April 1985, 333-39.

14. As for the impact of taxes on intrametropolitan and interstate location, one exhaustive review of pertinent empirical research concluded that the issue is an "open rather than a settled question." See Robert J. Newman and Dennis H. Sullivan, "Econometric Analysis of Business Tax Impacts on Industrial Location: What Do We Know, and How Do We Know It?" *Journal of Urban Economics* 23 (1988): 215-34. At the urban level, studies have shown that once available land, properly zoned, is accounted for, general tax levies may influence final manufacturing location decisions. See Michael J. Wasylenko, "Evidence of Fiscal Differentials and Intrametropolitan Firm Relocation," *Land Economics* 56, No. 3 (1980): 341-49.

15. North Carolina claims to have only four types of tax breaks, with strict limitations, and one type of financial assistance. South Carolina, by comparison, offers ten categories of tax breaks and 11 varieties of financial assistance. Greg Myers, "Bidding Wars," *Business and Economic Review,* January-March 1987, 8-12.

16. See W. E. Morgan and M. M. Hackbart, "An Analysis of State and Local Industrial Tax Exemption Programs," *Southern Economic Journal* 41 (1974): 200-5; and Bennett Harrison and Sandra Kanter, "The Political Economy of States' Job-Creation Business Incentives," *Journal of the American Institute of Planners* 44 (1978): 424-35.

17. W. Warren McHone, "Factors in the Adoption of Industrial Development Incentives by States," *Applied Economics* 19 (1987): 17-29.

18. In one case during the late 1970s, a maker of hydraulic equipment announced that it would close its Columbus, Ohio facility after the city government subsidized a West German competitor with a large tax break. See William A. Testa and David R. Allardice, "Bidding for Business," *Chicago Fed Letter* 16 (December 1988). More recently, auto parts suppliers have complained that subsidies to foreign companies are driving them out of business.

19. See Douglas Yuill and Kevin Allen, *European Regional Incentives* (Glasgow: Centre for the Study of Public Policy, University of Strathclyde, 1980) and subsequent volumes of this catalog for a discussion of clawbacks in European incentive programs. The clawback schemes discussed in this article are drawn from this catalog.

20. The economic myopia of some mayors and governors may ultimately prove to be politically short-sighted. The mayor of Flat Rock, Michigan was voted out largely because of anger about the 14-year tax holiday given to Mazda Motor Corporation.

21. Edward V. Regan, "Government Inc.: Creating Accountability for Economic Development Programs," (Washington, DC: Government Finance Research Center of the Government Finance Officers Association, April 1988; Mimeographed).

22. For information on the cost-effectiveness of alternative forms of industrial incentives see William Hamilton, Larry Ledebur, and Deborah Matz, *Industrial Incentives: Public Promotion of Private Enterprise* (Washington, DC: Aslan Press, 1985); Marc Bendick, David Rasmussen and Larry Ledebur, "Evaluating State Economic Development Incentives from the Firm's Perspective," *Journal of Business Economics* 17, No. 3, (May 1982): 23-29; and David Rasmussen, Marc Bendick, and Larry Ledebur, "A Methodology for Selecting Economic Development Incentives," *Growth and Change* 15, No. 2 (January 1984): 18-25.

23. This assumption of constant annual costs of the subsidy is probably unrealistic. If infrastructure costs are entailed in the subsidy plan, these costs are likely to be front end costs, whereas costs of maintenance and service would increase with the scale of operations of the firm. Further, if the total tax liability of the firm were abated, the cost in terms of tax revenues foregoing would increase as the value of the property increased.

Chapter 6

Ten Principles for State Tax Incentives

Keith R. Ihlanfeldt

❀❀❀

As part of our work for the Revenue Structure Study Commission of the state of Georgia, we were asked to review the state's tax incentives for economic development.[1] Specifically, we were asked to evaluate "whether the state was doing the right things and whether it was doing things right." To address these questions, we drew upon economic theory and existing empirical evidence to develop ten principles that state tax incentive programs should satisfy to maximize effectiveness and fairness. Georgia's tax incentives were then judged against each principle.

The purposes of this article are (1) to present our ten principles, in hopes that they may help other state and local governments improve their tax incentive programs; and (2) to illustrate the application of the principles by using Georgia as a case study.

Ten Tax Incentive Principles

Principle 1: Tax incentives (and other economic development policies) should be accompanied by specific programs that seek to mitigate the unwanted side effects of economic growth.

The primary objective of tax incentives and other state economic development policies is to stimulate state or local economic growth. Because both benefits and costs are associated with economic growth, it is important to couple economic development policies with auxiliary programs that minimize its unwanted side effects.

The costs of economic growth can be divided into four categories: (1) housing market effects, (2) environmental effects, (3) agglomeration diseconomies other than pollution, and (4) re-

AUTHOR'S NOTE: I wish to thank Roy Bahl for stimulating my interest in the topic and for providing literally hundreds of comments on previous drafts. I also am indebted to Matthew Murray, who patiently answered my numerous questions at the outset of this project and shared his personal library.

Reprinted from *Economic Development Quarterly,* Vol. 9, No. 4, November 1995.

ductions in the per capita provision of government services. The location of new businesses and the expansion of existing businesses within a given area stimulates population in-migration, which increases demand for housing services. Business growth also increases the earnings of existing residents, which affords them an opportunity to move upward in the housing market, and further reinforces the increase in housing demand coming from new residents.[2] The rise in the price of housing resulting from the increase in housing demand will depend on the elasticity of housing supply (i.e., the responsiveness of housing producers). Available evidence suggests that this elasticity is low over relatively long periods, which implies that economic growth will cause apartment rents and housing prices to rise.[3]

The rise in rents and housing prices accompanying economic growth can be a mixed blessing for a state. For those who benefit from employment growth, the increase in the cost of housing may be regarded as a small price to pay for having a job or earning more income. Moreover, those who own property will experience an increase in wealth, and recipients of local government services may benefit from the increased yield of the property tax. But there are those who will not benefit from the increase in labor demand: retirees, those without the skills needed by new business, or those whose labor force participation is constrained by personal or family circumstances. For them, the growth-induced rise in housing cost can be a significant burden. As a result, it is sometimes argued, economic development policies should be coupled with housing assistance for those adversely affected.[4]

A recent example of the rise in housing costs resulting from economic growth is provided by the location of General Motors Corporation's Saturn plant in Spring Hill, Tennessee. Residents of this small community have experienced significant increases in their housing rents, and first-time homebuyers have been shut out of the owner-occupied housing market.[5] Since the 6,000 Saturn jobs have gone mostly to United Automobile Workers from outside the community, the rise in the cost of housing has been particularly burdensome for existing residents. Much of their well-publicized resentment against Saturn could have been forestalled by appropriately designed housing assistance programs, financed by Saturn and/or the state of Tennessee. This illustrates the importance of forecasting likely costs of economic growth before implementing economic development policies.

The chief environmental effects of economic growth are industrial air and water pollution, automotive emissions, and soil erosion from housing construction. Policies designed to address these negative effects can be recommended on two counts: They not only enhance the quality of life but also encourage additional economic growth. Increasing empirical evidence suggests that the quality of life is an important locational determinant of businesses, especially those offering high-wage jobs.[6]

Diseconomies of agglomeration occur when concentration of population or economic activity in one place either raises the real cost of production by requiring more inputs, or reduces the real standard of living by increasing physical or social disamenities. Two good examples are crime and road congestion; researchers have found positive correlations between the magnitude of these problems and population size. Higher crime increases security and insurance costs and is certainly a significant social disamenity. Traffic congestion imposes increased travel time on both businesses and households.

A final cost of economic growth stems from the fact that the provision of government services may not keep up with the needs of a growing population. The result: Per capita provision of services declines, reducing the welfare of the community. Elementary and secondary school classes held in trailers, too few ballfields to accommodate Little Leaguers, moratoriums on sewer line tap-ins—all are manifestations of this problem, recently experienced in Georgia and other states.

Identification of the possible costs of economic growth suggests that tax incentives and other development policies cannot be initiated in a vacuum. Planning for the costs of economic growth can result in auxiliary programs to ensure economic development achieves its fundamental goal: increasing the welfare of state residents.

> Principle 2: Tax incentives should be part of a comprehensive state economic development program containing carefully crafted supply-side and demand-side policies.

Economic development policies can be broken down into two categories: supply-side and demand-side. Supply-side policies are meant to attract new firms, or encourage the expansion of existing firms, by lowering the cost of doing business within the area. Tax incentives are a supply-side policy; other examples include providing low-interest-rate loans, subsidizing the infrastructure needed by a given firm, and providing customized training to the firm's workers. The purpose of demand-side policies is to cultivate homegrown industries, based on the notion that government can intervene to help generate unique additional economic opportunities. Specific policies considered demand-side include venture capital funds, state-subsidized entrepreneurial training, subsidies for business-related applied research, and export assistance.

Ideally, empirical evidence on the relative effectiveness per dollar of government expenditure or foregone tax revenue of specific supply-side and demand-side policies should exist to guide policymakers. Unfortunately, this is not the case. Only a few studies provide evidence on the effects of specific tax incentive programs or other supply-side policies. The dearth of data on the effects of traditional incentive policies can be attributed to the fact that these programs do not easily lend themselves to statistical analysis. As Netzer has noted,

> Regression analysis works only if there are numerous observations and if the independent variables can be defined reasonably well. Both of these requirements pose special problems for the analysis of economic development policies. This is because there is considerable interstate variation in the specific provisions of similar-sounding measures, and rates of participation in similar programs differ greatly over time and place, with resulting variation in the composition of those firms that do participate.[7]

In the absence of specific evidence, studies that relate business expansion and/or location to the overall level of taxes within a given area are frequently used to gauge potential effects of tax incentive programs. The rationale here is straightforward: If incentives lower firms' taxes, and if firms respond to tax differentials in making location decisions, incentives have some effect. Over 80 studies have investigated the role of taxes in determining economic growth.[8] Those completed prior to 1975 consistently found that state and local taxes do not influence regional business growth; more recent studies, employing better data and more appropriate econometric techniques, have generally concluded that taxes do matter, though there is little concurrence on the magnitude of the effect.

Evidence on the effectiveness of demand-side policies is virtually nonexistent. However, a number of recent books (one by Scott Fosler, another by Peter Eisinger, and still another by David Osborne) enthusiastically support demand-side policies, based on the author's deductive reasoning.[9] The arguments made are that (1) small businesses create more jobs than do large businesses, (2) government can draw upon economies of scale to provide these businesses with essential services otherwise not affordable, and (3) demand-side policies prove less costly than supply-side policies, since they lend themselves to public/private partnerships and cannot result in the revenue losses endemic to tax incentives.[10]

In the absence of reliable information on the relative cost-effectiveness of specific policies, we are left with evidence that, at least weakly, supports the use of tax incentives, as well as some reasonable arguments in support of the demand-side approach. Until more conclusive research becomes available, the sensible approach is for states to "hedge their bets" by developing a comprehensive economic development plan, combining tax incentives (and other supply-side policies) with demand-side policies. Such a diversified portfolio also allows states to experiment with novel approaches.[11]

> Principle 3: Tax incentives should be general, not specific—that is, available to all firms that satisfy eligibility criteria, rather than acting as bait to lure a particular company.

Specific incentives are given to attract a large corporation or a particular business to the area; the incentive package is tailored to the company, and therefore, each one is different. General incentives apply to all businesses that satisfy certain eligibility criteria, and are analogous to an entitlement program.

Both specific and general incentives are popular with states. A recent survey, which obtained responses from 48 states, found that 25 (52%) had enacted legislation creating specific incentives over the past three years.[12] Legislation pertaining to general incentives had been passed by 45 states (94%).

Several arguments are made in favor of specific incentives. First, rates of return can be maximized, since specific incentives are targeted to those firms that have the strongest growth potential. The problem with this argument is that states typically confront the same limitations in knowledge and forecasting expertise that confront experts on the national level. There is, therefore, little reason to believe that states can successfully "pick winners" in the incentives game. Investing heavily in incentives that turn sour can prove costly. This is starkly illustrated by the state of Pennsylvania's 71-million-dollar investment in a Volkswagen plant in 1978. The state lost big after Volkswagen failed to induce any significant start-ups of local suppliers and the plant closed down after only 10 years of operation. The demise of Volkswagen in Pennsylvania was the result of unforeseen factors, such as the oil crisis, design defects, and economic recession.[13]

Another argument made in favor of specific incentives is that they can be given exclusively to companies for which the incentives will make a difference (i.e., perfect screening), thus saving the state otherwise foregone revenues. Once again, however, it is highly doubtful that states would have the requisite information to determine who would come without such incentives. Moreover, the practice of giving incentives only where they are believed to matter is obviously inequitable to businesses that come (or stay) but do not receive incentives. As a result, this strategy may backfire by contributing to the perception of a poor business climate.

There are three additional criticisms of specific incentives worth noting. First, specific incentives are generally not given to small businesses, despite the fact that these firms are an important source of employment growth. Second, specific incentives create an unlevel playing field by giving new firms a competitive advantage over existing firms. Hence the new jobs that result from specific incentives may be offset by job losses or reduced job growth by established businesses. Third, specific incentives set a precedent that can make it expensive to attract new businesses and retain existing ones. Firms may threaten not to locate or stay unless they too are given a lucrative incentive package.

> Principle 4: If the goal is economic growth as measured by employment, tax incentives should subsidize the cost of labor, not the cost of capital.

States can offset both the cost of new investment and the cost of hiring additional workers by adopting appropriate tax credits or abatements. However, if the objective is jobs creation, arguments can be made in favor of the exclusive use of labor subsidies. Roger Vaughan has argued that "reducing the tax on the income from capital to create jobs is perverse, since the policy does nothing to reduce the costs of labor."[14] If one looks only at the direct effects of a capital subsidy, Vaughan's criticism has merit; a reduction in the cost of capital, relative to the cost of labor, will cause firms to substitute capital for labor. But there are also secondary or indirect effects to be considered. To the extent that the subsidy lowers overall costs, and thereby output prices, more of the product will be demanded, causing the firm to expand production and hire more workers. The long-run effect of a capital subsidy on labor demand is therefore ambiguous and will depend on the relative strengths of primary and secondary effects. Although the recognition of secondary effects makes capital subsidies less perverse, as an employment creation policy, there remains a strong possibility that primary effects may dominate secondary effects, causing a reduction in both wages and employment.

Kesselman, Williamson, and Brandt were among the first to examine the effect of national capital subsidies on employment stimulus.[15] They studied whether capital was complementary with white- or blue-collar workers, and whether the two types were substitutes for one another. They found that capital is complementary with white-collar workers and a substitute for blue-collar employees in manufacturing. Thus a subsidy to capital will stimulate the use of capital, raise the demand for its complement (white-collar workers), and reduce the demand for substitutes or blue-collar workers. More important, for the issue at hand, they found that tax credits for employment increase the number of jobs dramatically, compared to investment tax credits (for equal revenue foregone), and that marginal employment tax credits (tax credits for new employees hired) have the largest effect on employment.[16]

In a much more recent piece, Jane Gravelle conducted simulations designed to compare the effect of labor versus capital subsidies on enterprise zone residents.[17] She simulated the effect of subsidies by relating wage and employment changes to relevant behavioral responses, as measured by elasticities.[18] Her results showed that under reasonable assumptions regarding the values of the elasticities, a capital subsidy will increase zone wages and employment much less than will a labor subsidy of the same percentage. Moreover, for products with moderate to low price elasticities of demand, a capital subsidy will actually *reduce* wages and employment. Leslie Papke's simulations of the effects of capital and labor subsidies, using a somewhat more sophisticated model than that of Gravelle, led to the same conclusions.[19]

In summary, economic theory and empirical evidence cast doubt on the effectiveness of capital subsidies in increasing employment. Whether employment expands or contracts in response to a capital subsidy depends primarily on the price elasticity of demand for the finished product. For manufacturing industries that compete in the national or international market, the price elasticity of demand is likely to be high because of opportunities for expansion. For locally marketed products, however, elasticity of demand is expected to be low. Hence, if capital subsidies (such as investment tax credits or property tax abatements) are used to stimulate employment, they should be applied selectively. However, the evidence presented above strongly favors the use of labor over capital subsidies.

Principle 5: Tax incentives should be targeted to firms in basic, rather than nonbasic, industries.

We must recognize that firms in different industries have different employment effects. Industries are sometimes classified as basic and nonbasic: basic industries are those that export their goods or services outside the state, while nonbasic industries produce for the local market. An increase in basic employment brings money into the state economy, which creates a multiplier effect by stimulating the nonbasic sector. Traditionally, manufacturing industries have been considered the basic sector. However, many service industries (such as professional, business, and health services) also export out-of-state. Targeting basic industries increases the likelihood that the benefits of incentives and other economic development programs will exceed their costs. In fact, one may argue that "when firms inherently serve local markets, subsidies to encourage their expansion or relocation can only cannibalize the businesses of firms already in the market or likely to be in it with or without subsidy."[20]

Principle 6: Tax incentives should be consistent with the state's economic development goals, and these goals should be based on careful assessments of the state's needs, strengths, and weaknesses.

The objectives of tax incentives and other state economic development policies are almost as varied as the policies themselves. Possible goals for states to prioritize before designing their development programs are considered below.

Job Creation

The most common objective is to create more jobs, which will tighten labor markets, thereby reducing unemployment and increasing wages. Job creation is especially well founded if there is slack in the local labor market, since net benefits to the community will vary directly with the number of local unemployed workers who fill the available job openings.

Change the Industry Mix

The economies of many communities are characterized by cyclical instability as a result of being nondiversified; this is particularly true where most jobs are in capital goods and durable goods manufacturing industries, since the demand for these products closely follows the business cycle. To stabilize the reaction of the local economy over the business cycle, incentives might be targeted to industries less affected by the latter (such as services) or to those that might complement existing industries, in the sense that their employment and output tend to move countercyclically.

Linkages

Reduce Intrastate Welfare Disparities

States seldom experience geographically uniform employment growth. In many, growth tends to be markedly higher in one or more urban areas, in comparison to the rest of the state. For example, over the past decade, the Atlanta metropolitan area has experienced per annum growth roughly 50% higher than the rest of Georgia. Moreover, within the state, there are significant differences in economic welfare, measured in terms of unemployment and earnings, between urban and rural counties—even within Atlanta, between northern and southern suburban counties.

What to do about these economic disparities? One approach is to let the market work; that is, if workers living in less developed areas cannot find jobs, eventually they will move where jobs are more abundant. There are two criticisms to this approach: First, people tend to have strong emotional ties to their home communities, which limits the mobility of labor and prolongs economic hardship. One classic illustration of this fact is that the shift from coal to oil occurred in the 1940s; yet the population built up in Kentucky, Ohio, West Virginia, and other eastern coal areas has largely remained in place, with persistent low wage levels and high unemployment rates. Second, Roger Bolton has argued that people attach value to a "sense of place," which he defines as "a sense of community and co-operation that is shaped by a particular geographical setting, including the natural and built environment, culture, and past history."[21]

A recognition of sense of place leads to two arguments for adopting policies to bring jobs to less developed areas. First, population out-migration from an area with insufficient employment erodes the sense of place; therefore, the exiting individual imposes negative externalities on those left behind. Second, the sense of place existent within a particular community may have value not only to the community's residents but to people living elsewhere. There may be an "option value" or "pure existence value" associated with a sense of place.[22]

Reduce Intergroup Welfare Disparities

In lieu of targeting tax incentives and other development policies to less developed areas, the objective may be to increase employment or earnings of disadvantaged groups. For example, many states now have in place special programs in support of minority-owned business development.[23] Also, it is common to find tax credits provided to businesses that hire economically disadvantaged workers.

Attract High-Technology Companies

In recent years, many states have chosen to target their economic development programs to high-technology companies. Larry Dildine lists the following advantages development officials ascribe to these companies:

- relatively high-wage employment
- relatively low environmental costs
- potentially high rates of growth
- long-term growth potential
- clustering of other high-technology firms and supplies
- export orientation
- prestige associated with business innovation.[24]

Along with these advantages, however, Dildine identifies the realities of high-risk, intensive competition among states for high-tech business, as well as the prerequisite of specialized infrastructure, especially in education. He also notes practical difficulties in defining effective incentives, targeted to high-technology companies, and evaluating their cost-effectiveness.

Fiscal Surplus

The first five objectives listed have in common an emphasis on labor market effects, but another common objective is to generate economic activity that yields tax revenues in excess of public service expenditures—that is, to attract businesses that provide a fiscal surplus that can be used to lower tax burdens of existing residents. Commerce and manufacturing are more likely to generate a fiscal surplus, whereas low- and middle-income residential development may cause public service costs to increase by more than the tax revenues generated. Hence, if the objective is fiscal improvement, policies should be adopted to encourage the former activities and discourage the latter. Fiscal surplus is most often the goal of local rather than state governments, since in a local area, there is considerable interest by elected officials in providing property tax relief to the community's residents.

Since local governments compete for new businesses, two problems may arise in economic development policies with fiscal improvement as their primary goal. First, competition may enable firms to extract tax and other financial incentives from local governments, to the extent that fiscal surplus is eliminated, at least for a number of years after the firm moves into the community. Second, interjurisdictional competition may escalate from communities offering incentives to communities choosing to keep their tax rates as low, or lower, than their competitors'. This "tax competition" has been criticized on two accounts: First, it may result in local public services, especially those disproportionately benefiting the poor, being undersupplied; second, since suburban jurisdictions surpass central cities in ability to offer lower taxes, tax competition may contribute to employment decentralization. If inner-city workers, for reasons of race or income, are unable to follow jobs to the suburbs, tax competition may heighten urban poverty. For the aforementioned reasons, such competition is considered a serious problem by many economists. Since the evidence tends to support this concern, there may be need for state intervention to curb interjurisdictional tax competition.

With the possible exception of fiscal surplus, these goals of economic development can all be considered legitimate. It should be emphasized that such goals are not necessarily mutually exclusive. Nevertheless, the success of an economic development program may clearly depend on prioritizing goals before getting started. A number of states have chosen to address goal identification by developing a strategic plan for state economic development. There is enormous variation in the quality of these plans. The better ones are based on careful assessment of the state's needs, strengths, and weaknesses, as viewed by all participating partners in the economy: businesses, local governments, and worker groups.

> Principle 7: Tax incentive programs should contain provisions to reduce potential revenue losses.

Earlier, it was argued that general incentives are superior to specific incentives; however, the former are not free of criticism, either. There is a revenue loss associated with the fact that many firms would have come to the state or expanded within the state, regardless of incentives provided, as starkly illustrated by Matthew Murray's evaluation of proposed tax incentives for the state of Tennessee.[25] From a reasonable set of assumptions regarding magnitudes of the relevant employment elasticities and multipliers, he estimated the cost per job, using various job and investment credits under the assumption of "no screening" and "perfect screening." No screening means that the credits are provided to all eligible firms, regardless of whether additional jobs are created in response to incentives; that is, even had the firm expanded employment or added capital without the credits, it still receives them. Perfect screening results when credits go only to those firms creating jobs or capital expenditures in response to incentives. The net cost per job created is defined as equaling the direct outlay of credits (i.e., the revenue foregone), minus benefits through increased tax payments resulting from the higher level of economic growth.[26]

For each of the incentives Murray evaluates, the cost per job varies widely between the no screening and perfect screening alternatives. For example, one of the job credit proposals provided a $1,000 credit ($1,500 for distressed counties) for each new job created, to manufacturers who brought in at least 35 new jobs. The credit was available for a single year, with a five-year carryforward. Under no screening, the net cost is estimated as $9,000 per job, whereas perfect screening is estimated to yield a net gain of $10,000 per job.

Development officials sometimes contend that without the tax incentive, certain jobs would not have come, and the state would not have the tax revenue in the absence of incentives; thus, incentives do not cause a true revenue loss. Murray's estimates belie this contention: Tax incentives do result in revenue losses. To reduce these losses, a number of devices might be implemented, such as requirements that a minimum number of jobs be created and maintained or dollar investment in capital be made before credits are claimed, a provision that businesses cannot use credits to reduce tax liability by more than a specified maximum, and restriction of credits to firms in those industries with large employment multipliers. Although such devices can be recommended, little hard evidence is available on their effectiveness.

Principle 8: Tax incentive programs should benefit all businesses within the targeted industry groups.

To conform to this principle, tax incentives must benefit firms new to the state as well as existent firms, regardless of their tax-year profit level. Two persuasive arguments can be made for extending tax incentives to expanding firms. First, existing business is traditionally a more important source of job growth than new or relocating firms.[27] As Michael Wasylenko puts it, "existing businesses are the 'golden geese,' and tax incentive programs that discriminate against existing businesses violate the 'golden geese' adage."[28] Second, discriminating against existent firms in granting incentives may jeopardize their competitive position, and thereby the jobs of state residents; put another way, new jobs may displace existent jobs. This argument is enhanced by Tim Bartik's finding that only about a quarter of new jobs go to original residents.[29]

Most new businesses—and many existing ones—pay few or no state income taxes. The effectiveness and fairness of tax incentives may therefore be substantially compromised unless they are structured to benefit firms both with and without income tax liabilities. Although this task can be accomplished by making credits refundable, the option may not be politically feasible, since a state budgetary appropriation would be required; it may be more practical to allow businesses to carry credits forward a number of years or to claim the credits against their sales tax or other state tax liabilities.

Principle 9: Performance evaluation methods should be adopted for periodically monitoring costs and benefits of each tax incentive.

States do little in evaluating the cost-effectiveness of their tax incentives and other economic development programs. According to Tim Bartik, there are at least three good reasons

TABLE 6.1 An Example of Minnesota's Milestones

Goal	*Milestones*
Small cities, rural, and urban areas will be economically viable	**Percentage of Twin Cities population living in census tracts with poverty rates 1.5 times the state average**
	Percentage of population living in counties with per capita income less than 70% of U.S. per capita income
	Minnesota nonmetropolitan per capita income, as percentage of U.S. nonmetropolitan per capita income
	Primary care physicians per 10,000 people in nonmetropolitan Minnesota
	Minnesota's rank in telecommunications technology
	Percentage of nonmetropolitan population in communities served by two or more options for shipping freight

SOURCE: Summary, *Minnesota Milestones,* 1993 Progress Report (St. Paul, MN: MN Planning, 1993), p. 6.

government programs are seldom evaluated: (1) Good evaluations are costly; (2) the knowledge obtained on the effectiveness of particular programs does not benefit only the state that pays for it, but can be used by other states as well; and (3) negative evaluations may be used to kill a program or agency, whereas positive evaluations will be discounted by opponents of the program.[30] Bartik considers the last reason most important in explaining this resistance to evaluation.

Although evaluation is uncommon, some efforts are under way. Eighteen states issued tax expenditure reports in 1987—the last year for which information was collected.[31] These reports are useful in assessing the direct costs of the state's tax incentives, but generally provide no information on employment benefits or increased revenues resulting from multiplier effects. Also growing in popularity is "benchmarking"; benchmarks refer to measurable indicators of progress made toward accomplishing long-run goals, or else to both goals and outcomes. Regardless of precise definition, the major difference between benchmarks and traditional assessment tools used by states is that benchmarking focuses on results and outcomes rather than activities such as dollars spent or services provided. Oregon, in 1991, was first to use the benchmarking method of performance evaluation; since then, the device has spread to many other state and local governments.

An example of benchmarking comes from Minnesota. Among its development goals is that "small cities, rural and urban areas will be economically viable." The data in Table 6.1 list the six benchmarks (Minnesota prefers to call them "milestones") used to measure progress toward this goal.

Benchmarking has a number of attractive features. First, it forces the state to envision what it wishes to become, an essential element in any long-run plan. Second, it permits progress toward long-range strategic goals to be monitored. There are also important limitations of benchmarking. First, as illustrated in Table 6.1, the relationship between milestones and goals is often unclear. Second, benchmarking does not directly link the amount of progress to specific state programs or policies; it is therefore limited in its ability to serve as an evaluation instrument.

If tax expenditure reporting and benchmarking fall short as methods of evaluation, is there no hope of objectively determining whether incentives are working? Fortunately, a couple of approaches toward evaluation hold considerable promise.

First, tax expenditure reports can be broadened to include information on firms that claim the credits, such as the number of jobs created, location of the firm by region and city size, size and industry of the firm, and whether the firm is new or expanding. West Virginia's study of its Super Credit can be considered an expanded tax expenditure report, illustrating the usefulness of knowing the characteristics of firms making use of the program.[32] The report's principal finding was that nearly 90% of the total Super Credit expenditure was attributable to taxes on coal severance activities. In effect, Super Credit enabled the coal industry to shift a portion of its inevitable modernization expenditure to the state. During the initial years of Super Credit, these same companies reduced total employment within the state by roughly 1,000 jobs. The entitlement of credit for these companies occurred, in spite of employment reductions, because of various ambiguities within the law.

Much can be learned about the effectiveness of incentives by simply administering surveys that ask whether the incentives played a role in a firm's decision making. Many economists are critical of this approach because there is incentive for "strategic behavior" on the part of respondents who believe their answers may influence the study's results—and thereby, public policy toward business. For this reason, these same economists believe, responses will be biased in favor of tax incentives. However, results from survey-based studies of locational factors lend little support to the hypothesis that firms engage in strategic behavior when answering questionnaires. If they answer the questions with the intent of influencing public policy, the expectation is that taxes would be ranked an important locational determinant; in reality, taxes show up as a marginal influence, consistent with the findings of econometric studies.

A recent study, done by the Office of the Inspector General (OIG), also contradicts economists' beliefs regarding surveys.[33] A sample of employers who had received Targeted Jobs Tax Credits against their federal income tax for hiring disadvantaged workers were asked whether they would have hired the workers anyway; 95% said they would have hired the worker without the incentive, which led OIG to recommend eliminating the program.

To minimize the possibility of bias, such surveys should be conducted by independent third parties (i.e., not by the agency that administers the program), and there should be complete client confidentiality. Bartik cites examples of high-quality survey evaluations of specific development programs conducted by various states that have adhered to these standards.[34]

Principle 10: Efforts should be made to publicize tax incentive and other economic development programs to ensure that companies are aware of them.

A survey by Michael Kieschnick found that most responding firms were unaware of the existence of state tax incentives, even though all the firms included in his survey were located in states that offered incentives.[35] Of the respondents, less than one-fifth of the new firms—a little over one-third of the expanding firms and about half of the new branch plants—were aware of state tax incentives. These findings correspond to assessments of the knowledge that firms have of federal tax incentives; for example, one study of the New Jobs Tax Credit (1977-1978) found that, one year into the program, less than half of all surveyed firms actually knew the program existed.[36]

A Description of Georgia's Tax Incentives

Georgia's tax incentives include job tax credits, investment tax credits, retraining tax credits, child care tax credits, sales tax exemptions, and property tax exemptions. All the credits apply to the state income tax. Except for the incentives related to property tax, all the individual tax incentives were created or expanded by the Georgia Business Expansion and Support Act (GBESA) of 1994. The major provisions of Georgia's tax incentives are described below, beginning with the job tax credit, which is the centerpiece of the GBESA.

Job Tax Credit

Businesses are eligible for the job tax credit if involved in manufacturing, warehousing and distribution, processing, research and development, or information processing. The amount of the credit varies by county. Each year, all the state's 159 counties are ranked using an index of economic distress, and are then divided into three tiers:

Tier 1: counties ranked 1st to 53rd, which represent the state's least developed counties
Tier 2: counties ranked 54th through 106th
Tier 3: counties ranked 107th through 159th

Companies in Tier 1 counties receive a $2,500 tax credit for each new full-time job if they create at least 10 new full-time jobs. Companies in Tier 2 counties receive a $1,500 tax credit if they create at least 25 new jobs. Tier 3 counties provide companies with a $500 tax credit if they create 50 or more new jobs.

Credits can be claimed in the second through sixth year after job creation, as long as the jobs are maintained. Credits can be carried forward for 10 years from the time the jobs were created. Taxpayers are not allowed to take more than 50% off income tax liability from the job tax credits in any one year.

Investment Tax Credit

Georgia's investment tax credit against state income taxes is based on the same three tiers as is the job tax credit program. The major provisions of the credit are listed below.

1. Eligible businesses are those operating an existing manufacturing facility or manufacturing support facility in the state for the previous three years, and who have made an investment in the taxable year in land acquisition, improvements, buildings, building improvements, machinery, or equipment.

2. The value of the credits and minimum-required dollar investments are as follows:

Companies expanding in Tier 1 counties must invest $1 million to receive a 5% credit.
Companies expanding in Tier 2 counties must invest $3 million to receive a 3% credit.
Companies expanding in Tier 3 counties must invest $5 million to receive a 1% credit.

3. The credit can be carried forward five years, from the close of the taxable year in which the qualified investment property was acquired. Taxpayers may elect to claim either the investment tax credit or the job tax credit for a given project, but not both. The total credit cannot be more than 50% of the taxpayer's total state income tax liability for that taxable year.

Retraining Tax Credit

To be eligible for the retraining tax credit, the training must be targeted to employees whose productivity has been hindered by an inability to use new technology. The major provisions of the credit are summarized below.

1. Eligible businesses have been operating in Georgia for three consecutive years and are engaged in manufacturing, warehousing and distribution, processing, tourism, or research and development.
2. The value of the credit is equal to one-quarter of the direct cost of retraining, up to $500 per full-time employee per year, for each employee who has successfully completed an approved retraining program.
3. The credit cannot exceed 50% of the taxpayer's total state tax liability for the taxable year, and it cannot be carried forward.

Child Care Tax Credit

The value of the child care tax credit is equal to 50% of the cost of operation of an employer-provided or employer-sponsored child care facility for that taxable year, minus any amount paid by employees in any taxable year.

The credit cannot exceed more than 50% of the taxpayer's total state income tax liability for that taxable year. Any unused credit may be carried forward for five years, from the close of the taxable year in which the cost of operation was incurred.

Sales Tax Exemptions

Exemptions from sales and use taxes are provided for the following:

- new machinery and equipment used directly in the manufacturing process;
- new and replacement equipment used to control air and water pollution;

- primary material handling equipment used directly for the handling and movement of property in a new or expanding warehouse, when such new facility or expansion is worth $10 million or more;
- electricity purchased for direct use in manufacturing, when the total cost of the electricity makes up 50% or more of the materials used.

Freeport Tax Exemptions

These exemptions enable cities and counties to exempt three types of inventory from property taxation:

- manufacturer's raw materials and goods-in-process
- finished goods held by the original manufacturer
- finished goods held by distributors, wholesalers, and manufacturers destined for out-of-state shipment

Each of the above classes of inventory must be separately exempted by local referendum. Exemptions may apply to 20%, 40%, 60%, 80%, or 100% of any or all of the three classes. More than three-quarters of Georgia's counties have approved freeport tax exemptions, and the majority exempt 100% of all three classes of inventory.

Property Tax Exemptions and Abatements

There is no statewide property tax abatement program in Georgia. In the case of industrial development bond financing, local communities may be able to offer companies ad valorem tax exemption, depending upon the type of Development Authority established and its legal powers. Such an exemption is possible if the Development Authority is a Constitutional Development Authority governed by a local constitutional amendment containing the power to grant the exemption and if the Development Authority is the legal owner of the property under a "sale and leaseback" arrangement. If the Development Authority is a Constitutional Development Authority without the power to grant the exemption, a Statutory Development Authority, or a Downtown Development Authority, and if legal title to the property is held by the Development Authority, the property financed by industrial development bonds is exempt from ad valorem taxation, but the leasehold estate owned by the company is subject to this tax. Under this arrangement, taxes are lower than if the company held title to the property, since the leasehold estate is assessed at a lower rate than the legal title is.

An Evaluation of Georgia's Tax Incentives

To illustrate the application of our principles, Georgia's tax incentives are judged against each principle below.

> Principle 1: Tax incentives (and other economic development policies) should be accompanied by specific programs that seek to mitigate the unwanted side effects of economic growth.

In 1987, Governor Harris created the Growth Strategies Commission. One of the goals established by the Commission was to "accommodate the inevitable growth of the future without allowing a deterioration in the quality of life." The key strategy selected to accomplish this was three-tiered planning, as legislated in the Georgia Planning Act of 1989. This act established a system of statewide planning to specifically address the unwanted side effects of economic growth. The plans of local communities are required to contain forecasts of population and employment up to 20 years in the future. Also, the effects of this growth on housing, community infrastructure, local government service, and environmental needs must be assessed, and appropriate policies must be identified to meet these needs. Communities stand to lose state loans and grants if their plans are judged inadequate, or if a plan is not submitted.

Although it may be too early to judge the effectiveness of the Georgia Planning Act, this legislation represents a major departure and can be considered a serious attempt to mitigate unwanted side effects of economic growth. However, it should be noted, communities are not required to implement their plans, which may jeopardize the effectiveness of the legislation. History has shown that land-use plans frequently end up shelved in the absence of mandatory implementation.

Principle 2: Tax incentives should be part of a comprehensive state economic development program containing carefully crafted supply-side and demand-side policies.

The Business Expansion Support Act of 1994 is one component of the governor's Development Council's Business Expansion Support Team (BEST). According to the council, BEST also offers

- enhanced coordination of assistance presently available to small and medium-sized companies;
- modernization, technology transfer, and improved productivity for Georgia business and industry;
- financing assistance for small and medium firms;
- training and human resource development assistance to existing businesses that are upgrading technology or changing processes to become globally competitive;
- a product matching system to encourage the purchase of Georgia-manufactured products;
- enhanced export assistance efforts through an export potential system, international trade shows, seminars and training at regional level, evaluation of export field staff and coordination of programs, and leveraging of federal programs.[37]

These development efforts are still evolving; six work teams throughout the state are addressing each area listed. This activity, as well as other programs the state has initiated, indicates that Georgia's tax incentives are part of a comprehensive state economic development program containing carefully crafted supply-side and demand-side policies.

Principle 3: Tax incentives should be general, not specific—that is, available to all firms that satisfy eligibility criteria, rather than acting as bait to lure a particular company.

Tax incentives offered by the state of Georgia are general in nature. Property tax abatements and exemptions provided by Local Development Authorities, however, can be considered specific incentives and are frequently employed to attract large projects. Moreover, local communities are unequal in their legal authority and financial ability to offer property tax abatements.

No information has been collected from local communities on use of property tax abatements by the state or any other entity; moreover, attempts to collect this information for our report were not successful. None of the participants (local governments, businesses, or intermediaries, such as accounting or law firms) were willing to volunteer much information. The limited information obtained suggests that (1) only nonschool taxes are typically abated, (2) poorer communities are more likely to abate than wealthier communities, (3) the communities who provide abatements consider these incentives essential to development efforts, and (4) some communities are offering abatements without legal authority to do so.

Principle 4: If the goal is economic growth, as measured by employment, tax incentives should subsidize the cost of labor, not the cost of capital.

Georgia's job tax credits, retraining tax credits, and child care credits may all be considered labor cost subsidies. The investment tax credit, property tax abatements, and sales tax exemptions on machinery and equipment are capital subsidies. As in other states, the overriding objective of Georgia's development plan is employment growth, especially in those regions that are less developed. For this purpose, labor subsidies are more cost-effective than capital subsidies.

Principle 5: Tax incentives should be targeted to firms in basic, rather than nonbasic, industries.

Job tax credits and retraining tax credits are restricted to businesses engaged in manufacturing, warehousing and distribution, processing, tourism, research and development, and selected services related to information processing. Investment tax credits and sales tax exemptions apply only to manufacturing facilities. These industries are all expected to serve markets that extend beyond state boundaries and therefore can be considered basic, as opposed to nonbasic (local). All businesses, both basic and nonbasic, are eligible for child care tax credits.

> Principle 6: Tax incentives should be consistent with the state's economic development goals, and these goals should be based on careful assessments of the state's needs, strengths, and weaknesses.

One of the primary objectives of the Governor's Development Council (GDC) was to prepare, for the first time, a state economic development strategy. This strategy, *Building a New Economic Engine for the 21st Century: Strength from Diversity,* is the result of careful assessments of the state's needs, strengths, and weaknesses.[38] As the GDC states in its overview brochure,

> The new statewide strategy is the result of six months of effort at the grassroots level. Literally thousands of business and community leaders and interested citizens participated in eleven regional meetings and intensive work sessions to offer guidance in the development of the strategy.

The incentives legislated by the GBESA are among many specific actions the GDC intends to take to realize Georgia's overall development goals.

> Principle 7: Tax incentive programs should contain provisions to reduce potential revenue losses.

All Georgia's tax incentives have provisions to reduce potential revenue losses; however, how effective these provisions will be, in actual practice, is not known. With the passage of the GBESA, tax credits are no longer strictly targeted to less developed areas, but are available to all qualifying businesses, regardless of location. Many of Georgia's urban counties have experienced strong employment growth without the incentives provided by the GBESA; these counties will continue to attract many jobs, regardless of incentives provided. The potential for revenue losses to exceed realized benefits should therefore be recognized. However, this possibility is mitigated by the fact that, in Tier 3 counties, the amount of credit per job is relatively small ($500, vs. $2,500 in Tier 1 counties), and the minimum number of jobs to be created is relatively large (50, vs. 10 in Tier 1 counties).

> Principle 8: Tax incentive programs should benefit all businesses within the targeted industry groups.

Existent companies are eligible for all tax incentives offered by the state. Although none of Georgia's tax credits is refundable, three of four programs allow carryforward. Job tax credits can be carried forward 10 years, and both investment tax credits and child care credits can be carried forward 5 years.

> Principle 9: Performance evaluation methods should be adopted for periodically monitoring costs and benefits of each tax incentive.

Each tax incentive program legislated by the GBESA requires that certification forms be completed and submitted to the Department of Revenue (DOR). The accuracy of the information on these forms will be verified only in case of an audit—historically a rare occurrence in Georgia. Businesses are therefore on their honor in claiming credits and exemptions.

The GBESA mandates that an annual report be completed by DOR on the job tax credit. The only information that must be included is the number of jobs created through the job tax credit, by county/census tract, and the amount of the tax credit used by all business enterprises. None of the other programs require any type of report to be prepared.

Georgia's strategic plan recommends that performance benchmarks be established; some examples of these are provided in the plan. Recall, however, that benchmarking has serious limitations as an evaluation tool.

In light of the above, it should be clear that Georgia's tax incentives do not conform to Principle 9; there is insufficient monitoring of both costs and benefits.

Principle 10: Efforts should be made to publicize tax incentive and other economic development programs to ensure that companies are aware of them.

Efforts are being made by the GDC to publicize its programs; for example, forums on the GBESA are being offered statewide. There is no plan, however, to assess the success of these marketing efforts.

Based on the above evaluation, Georgia's tax incentives generally conform to our principles; however, we identified four areas of concern. First, the investment tax credit and sales tax exemptions are capital subsidies, which may prove less cost-effective than the programs that subsidize the cost of labor. Second, the extension of incentives statewide may also reduce the cost-effectiveness of the overall program. The potential for revenue losses to overwhelm benefits is real for areas like Atlanta, where incentives are unlikely to play much of a role in influencing future economic development. Third, Georgia, like most states, is highly deficient in implementing effective monitoring and evaluation procedures. Finally, there is the issue of property tax abatement at the local level. Since Georgia has insufficient information about local abatements to develop sound policy, we recommended a disclosure law that would require all economic development agreements to be submitted for the state's review—but not approval. A disclosure law would reduce possible illegalities in the abatement process and would help flag those communities with abatement practices detrimental to the overall good of the community. Most important, disclosure would provide data necessary for possible statewide reform.

Conclusions

Many states have adopted tax incentives scarcely well founded on economic theory or empirical evidence. Frequently, poor incentives are contagious, as states become copycats by responding to their competitors' incentives by passing their own similar incentives.

This article has laid out ten principles that states can employ to enhance the fairness and effectiveness of their tax incentives. To illustrate their application, the principles were employed to evaluate Georgia's incentive programs.

Notes

1. The Revenue Structure Study Commission of the state of Georgia was created during the 1992 session of Georgia's General Assembly. Its purpose was to consider comprehensive reform of the state's tax system. The Policy Research Center of Georgia State University was commissioned to provide research support to the commission.

2. Economic growth increases the earnings of existing residents by lowering their nonemployment (i.e., a portion of the new jobs go to unemployed workers and workers who previously were out of the labor force) and by creating "hysteresis," which alludes to the fact that original residents benefit from growth-included movement up the job hierarchy. For more on hysteresis, see Timothy J. Bartik, *Who Benefits from State and Local Economic Development Policies* (Kalamazoo, MI: Upjohn Institute for Employment Research, 1991).

3. For evidence on the elasticity of housing supply, see Larry Ozanne and Raymond Struyk, "The Price Elasticity of Supply of Housing Services," in *Urban Housing Markets: Recent Directions in Research and Policy,* ed. C. S. Bourne and J. R. Hitchcock (Toronto: University of Toronto Press, 1978), pp. 109-38; and Frank Deleeuw and Nkanta Ekanem, "The Supply of Rental Housing," *American Economic Review* 61 (1971): 806-17.

4. Wilbur Thompson, "Economic Processes and Employment Problems in Declining Metropolitan Areas," in *Post-Industrial America: Metropolitan Decline and Inter-Regional Job Shifts,* ed. George Sternlieb and James W. Hughes (New Brunswick, NJ: Center for Urban Policy Research, Rutgers, State University of New Jersey, 1975), pp. 187-96.

5. Cynthia Mitchell, "GM a Mixed Blessing in Tennessee Town," *Atlanta Constitution,* July 31, 1990, p. A1; Darryl Fears, "GM Plant Ignites Civil War in Tennessee Town," *Atlanta Constitution,* December 18, 1991, p. A1; Mike McQueen, "Tennessee Town Fears Bust Side of Boom: GM Plant Has Stirred Some Concern," *USA Today,* November 14, 1988, p. A3.

6. See, for example, Phillip E. Graves, "A Life-Cycle Empirical Analysis of Migration and Climate by Race," *Journal of Urban Economics* 6 (1979): 135-47; and Frank W. Porrell, "Intermetropolitan Migration and Quality of Life," *Journal of Regional Science* 22 (1982): 137-58.

7. Dick Netzer, "An Evaluation of Interjurisdictional Competition through Economic Development Incentives," in *Competition among States and Local Governments*, ed. Daphne Kenyon and John Kincaid (Washington, DC: Urban Institute Press, 1991), pp. 110-32.

8. For recent reviews of the literature, see Bartik, *Who Benefits*; or R. W. Wassmer, "Local Fiscal Variables and Intrametropolitan Firm Location: Regression Evidence from the United States and Research Suggestions," *Environment and Planning C: Government and Policy* 8 (1990): 283-96; or Michael Wasylenko, "Empirical Evidence on Interregional Business Location Decisions and the Role of Fiscal Variables in Economic Development," in *Industry Location and Public Policy*, ed. Henry Herzog and Alan Schlottman (Knoxville, TN: University of Tennessee Press, 1991), pp. 13-30.

9. Scott Fosler, *The New Economic Role of American States* (New York: Oxford University Press, 1991); Peter K. Eisinger, *The Rise of the Entrepreneurial State* (Madison, WI: University of Wisconsin Press, 1988); David Osborne, *Laboratories of Democracy* (Boston, MA: Harvard Business School Press, 1988).

10. The conventional wisdom that small businesses are the primary source of employment growth has recently been attacked by Steven Davis, John Haltiwanger, and Scott Schuh, "Small Business and Job Creation: Dissecting the Myth and Reassessing the Facts" (National Bureau of Economic Research Working Paper no. 4492, 1994). They argue that prior studies have misinterpreted the data, and that in reality plants that average at least 100 employees account for 75% of gross job creation. Since their analysis is restricted to manufacturing jobs, we cannot conclude that small businesses are an unimportant source of new jobs. However, Davis et al.'s results do imply that small-business job creation has been overstated in the literature.

11. There may also be a political motivation behind the adoption of a comprehensive plan, since a larger percentage of the state's businesses may participate in the state's programs.

12. Georganna Meyer and John Hassig, "Economic Development Policy," *State Tax Notes*, November 22, pp. 1229-36.

13. Andrew Kolesar, "Can State and Local Tax Incentives and Other Contributions Stimulate Economic Development?" *Tax Lawyer* 44 (1988): 285-311.

14. Roger Vaughan, *State Taxation and Economic Development* (Washington, DC: Council of State Planning Agencies, 1979), p. 98.

15. S. R. Kesselman, S. H. Williamson, and E. R. Brandt, "Tax Credits for Employment Rather Than Investment," *American Economic Review* 67 (1977): 339-49.

16. Kesselman et al. estimate that employment levels in the United States would have been 1% higher in 1971 if employee tax credits rather than investment tax credits had been used in the 1960s and 1970s.

17. Jane Gravelle, *Enterprise Zones: The Design of Tax Incentives* (Washington, DC: Congressional Research Service, 1992).

18. The formulas Gravelle derives to measure the effect of subsidies on wages and employment contain the following elasticities: product demand elasticity, factor substitution elasticity, aggregate labor supply elasticity, and labor supply elasticity within the zone. Elasticities measure the responsiveness of one variable (e.g., product demand) to a 1% change in another variable (e.g., product price).

19. Leslie E. Papke, "What Do We Know about Enterprise Zones?" (National Bureau of Economic Research Working Paper no. 4251, 1993).

20. State Policy Research, Inc., *State Policy Reports*, August 1994, p. 6.

21. Roger Bolton, "Place Prosperity vs. People Prosperity Revisited: An Old Issue with a New Angle," *Urban Studies* 29 (1992): 185-203; and "An Economic Interpretation of a Sense of Place" (Research Paper 130, Department of Economics, Williams College, Williamstown, MA, 1989).

22. Option value refers to the value outsiders attach to a sense of place within a particular community because they want the option to move there. Pure existence value is the value attached to a sense of place simply because people believe it worthwhile.

23. An illustration is provided by South Carolina, where a corporate income tax credit against taxable income is allowed to taxpayers who hold state government contracts and subcontracts with minority businesses.

24. Larry Dildine, "Measuring State Fiscal Incentives for Economic Development of High-Technology Industries" (Paper presented at the 86th Annual Conference of the National Tax Association, November 1993).

25. Matthew N. Murray, *The Design of Economic Development Incentives for Job Creation, Capital Investment and Training* (Knoxville, TN: Center for Business and Economic Research, University of Tennessee, 1992).

26. Since public service expenditures are ignored, Murray's estimates of the cost of government in creating an additional job may be substantially underestimated.

27. This is documented by Roger Schmenner, *Making Business Location Decisions* (Englewood Cliffs, NJ: Prentice Hall, 1982).

28. Michael Wasylenko, "Empirical Evidence on Interregional Business Location Decisions," p. 15.

29. Timothy Bartik, "Who Benefits from Local Job Growth; Migrants or the Original Residents?" *Regional Studies* 27 (1993): 297-311.

30. Timothy Bartik, "Better Evaluation Is Needed for Economic Development Programs to Thrive," *Economic Development Quarterly* 8 (1994): 99-106.

31. Kimberly Edwards, "Reporting for Tax Expenditures and Tax Abatement," *Government Finance Review* 4 (1998): 13-21.

32. Mark Muchow, Alan Mierke, and Charles Lorensen, *First Report on Super Credit* (Charleston, WV: Department of Tax and Revenues, State of West Virginia, 1990).

33. Office of Inspector General, *Targeted Jobs Tax Credit Program: State of Alabama* (Washington, DC: U.S. Department of Labor, 1993).

34. Timothy Bartik, "Better Evaluation Is Needed," p. 102.

35. Michael Kieschnick, *Taxes and Growth: Business Incentives and Economic Development: Studies in Development Policy* 11 (Washington, DC: Council of State Planning Agencies, 1981).

36. Jeffrey Perloff and Michael Wachter, "The New Jobs Tax Credit—An Evaluation of the 1977-79 Wage Subsidy Program," *American Economic Review* 69 (1979): 173-79.

37. *Council Currents: Newsletter of the Governor's Development Council* (Atlanta, GA: Governor's Development Council, 1994).

38. Governor's Development Council, *Building a New Economic Engine for the 21st Century: Strength from Diversity* (Atlanta, GA: Governor's Development Council, 1994).

Chapter 7

Labor-Force-Based Development

A Community-Oriented Approach to Targeting Job Training and Industrial Development

David C. Ranney and John J. Betancur

This article reports on ongoing efforts to develop techniques that make community-based urban employment strategies feasible. Such techniques and strategies are needed because the traditional framework for labor market policy is biased in favor of firms and cannot adequately address the needs of specific pools of labor such as dislocated workers and segregated labor pools. Two common characteristics of traditional methods of labor market analysis, the nature of labor market data, and urban industrial and job training targeting policies, are particularly illustrative of these shortcomings.

The first characteristic involves the size of the labor market. Traditional frameworks are based on large labor markets comprising the space within which people can change jobs without changing residence.[1] As a result, labor market areas tend to coincide with metropolitan areas.

Second is the assumption that rational behavior on the part of firms and workers allows the market to bring the two together in a mutually satisfying result. This implies that workers are able to commute anywhere in the standard labor market area and that both workers and firms make "rational" decisions about acquiring skills, selecting the best job, or hiring the best worker. In this view, it is the market that links workers and jobs and dictates training and other priorities.

Available labor market data and most methods of labor market analysis reflect these basic assumptions. Policy intervention, therefore, also generally flows from them. There is a significant literature on the various ways in which these labor markets do not work well for all segments of the work force.[2] Explanations for this failure include insufficient information, labor market segmentation, and discrimination.

Despite this failure, economic public policy and programs such as the megaprojects developed or proposed for large cities—including airports, stadia, and waterfronts—continue operating under these assumptions. The ways in which analysis is conducted, data gathered, and policy drawn clearly assume large labor mar-

Reprinted from *Economic Development Quarterly,* Vol. 6, No. 3, August 1992.

kets operating in the forms indicated. Similarly, they maintain the bias in favor of the firms and demand all kinds of adjustments from the worker.

The authors of this article have worked with a large number of community organizations specifically concerned with the fact that many of the workers in their areas are without work. We have observed the failure of dominant labor market theory first-hand. Although skill and shortages are recorded in some parts of the metropolitan area, an abundance of unemployed and underemployed persons wishing to work are trapped in others. Neither the market nor policies and programs based on these assumptions have been very successful in bringing them together. Our experience has led us to the view that the assumptions behind the way labor market information is collected and disseminated, the techniques of labor market analysis, and the resulting policy frameworks have a bias that favors firms at the expense of many workers. Firms want the largest labor market area possible in order to be able to draw their workers from vast labor pools. Employers are not much concerned about the residence of workers as long as these can get to work every day on time. Workers, on the contrary, have to bear the costs of large labor markets. For some, there are substantial monetary and human costs in simply getting to work. Others remain unemployed due to inaccessible labor market information or job sites.

Rather than reviewing the adequacy and fairness of the assumptions underlying labor market analysis and policy, most urban employment programs today insist on trying to tinker with the present policy framework. For example, there are presently efforts to improve transportation systems and/or organize commuters for longer journeys to work. Such "reverse commuting" programs designed to bring workers to employers force many participants to travel up to two and a half hours both to and from work. As well intentioned as they are, these strategies do not address the stringent segregation restricting racial and ethnic groups to residential locations far from the jobs they could possibly get, the opportunity cost of the journey to work, or the hardships of the process.

Similarly, the division between industrial development and job training priorities fails to serve the needs of specific groups of workers. The City of Chicago, for example, like many other cities in the United States, separates economic development and job training into two departments within city government. Policies and programs in these two areas are developed and implemented with little, if any, coordination. Development efforts have tended to target firms that will meet export or import substitution objectives without direct consideration of who will be employed in these firms. Job training programs are designed primarily in relation to the guidelines of the Job Training and Partnership Act that requires short-term training and immediate placement. Thus many of these training programs target rapidly growing occupations with lower skill requirements without regard to economic development policies. Nor do training programs consider longer-term job retention and development considerations, skill transferability, and other conditions of labor pools not served by the large market. In addition, the work experience of the unemployed and underemployed and the accessibility of potential jobs to them tends to be ignored.

In short, prevailing analytical frameworks informing policy-making and program design do not include the realities of local or community-based economic development efforts and have become impediments for development of strategies addressing the employment needs of specific pools of unemployed or underemployed workers from the perspective of the workers. On the contrary, they are biased in favor of employers.

Community and other groups concerned with the employment of their residents or of specific constituencies such as dislocated or segregated labor pools need alternative frameworks for labor market data collection and analysis. Such a framework would involve other units of analysis and other forms for linking training, industrial development, and workers. One of these alternatives contemplates "shrinking" the size of the labor market, targeting specific pools of labor, and coordinating development and training priorities in locations more accessible to specific groups of workers. Shrinking the labor market involves developing intentional linkages between the needs of specific pools of workers and opportunities in local firms. Shrinking can also include development priorities for specific industrial sites that best match the experience, skills, and job training opportunities of the work force that lives nearby. Such a policy framework can reduce the costs of commuting and provide existing and potential employers with a pool of skilled workers that best meets their needs.

Community-oriented labor market policies or programs need not exclude the possibility of commuting to other areas. In fact it would be far too limiting to view a neighborhood as the sole source of resident employment.[3] But shrinking the size of the labor market needs to be part of a community-oriented approach to economic development and job training, if we are to address the needs of workers not served by the large market and the "hidden hand" process for bringing workers and firms together.

The distinction between community-based labor market strategies and those that follow from traditional labor market assumptions is this: A community-based approach targets the employment of a specific pool of workers (usually community residents) as its primary goal. More traditional strategies are focused on the workability of the market and assume that the needs of specific workers will be addressed in the process.

A problem with the community model is that techniques of labor market analysis and data are not geared to such localized strategies. Rather, both techniques and data assume the superiority of traditional conceptions of geographical labor markets. The present article reports in detail on a method, still under development, which we call "labor-force-based development." This method is an important part of ongoing efforts by the Center for Urban Economic Development of the University of Illinois at Chicago (UICUED) to respond to the need, often expressed by community groups and other strategists, for alternative frameworks of analysis that reflect the unique conditions of local communities, special labor pools, and other subunits of large cities such as Chicago.

Both the techniques themselves and the data on which they are based are in a developmental stage, but they have been used extensively by UICUED in its technical assistance work with community groups and local government. These techniques attempt to facilitate analyses of labor at a smaller scale while linking development and training priorities to the needs of local workers. A brief description of the general approach and a more detailed account of three case studies will summarize our efforts so far and illustrate some of the applications of the methodology.

Labor-Force-Based Development: The Technique

Although our experience to date is limited to the City of Chicago, the general methodology described below is applicable to any large city or urban region. The availability of the data sets that we use varies from state to state. In the State of Illinois the development of labor market information (LMI) is quite advanced. For states without some of the data capability used in our approach, the present article may suggest some directions for LMI development.

Our approach is based on the notion that the employment, experience, and skills of unemployed workers in a particular area of the city or region can become a major variable in establishing development priorities for that area. Further, with such priorities set, training programs can be designed that will make the fit between development priorities and resident work force even closer.

This approach requires two separate analyses. One is to examine the experience and skills of a specific group of workers. That group might be the residents of the "turf" of a community organization or it could be the workers in an establishment that is about to be closed. The first analysis seeks to establish the labor supply characteristics of a specific pool of labor.

A second analysis begins with the definition of a target area within which much of the labor pool identified in the first analysis could possibly be employed. The target area could be as small as a neighborhood. Once the target area is defined, an analysis of traditional labor-demand characteristics is made. Here we ask: What are the characteristics of the industrial structure in the area? How have local firms been doing relative to firms in the same industry in the city, in the region, or the nation? What kinds of people are employed by the firms in the area? Basically such an analysis seeks to determine what kinds of industries are likely to do well in the target area.

Having completed the supply and demand sides, we then determine which of the industries noted on the demand side are most likely to be able to use residents of the target area in their work force. In order to complete this approach, it is then necessary for either city or not-for-profit development groups to interview firms in the priority industries to determine more specifically what experience and background is required of their employees. Training programs can then be designed in cooperation with local job development agencies and skill-training centers so that the resident work force fits the needs of potential employers as closely as possible.

There are several difficulties in implementing this approach. Subarea statistics are not readily available for all the variables needed. Thus it is necessary to use a number of data sets that allow estimates to be made. As our description of estimating techniques will show, the resulting numbers are fairly crude. However, they are adequate for the purpose of initial targeting of industries and job training priorities.

In addition, we found that actual development opportunities and local conditions generate a variety of ways to use the general model described above. Essentially, there is no single model to achieve the objective of employing targeted groups of workers through economic development and job-training priorities. This methodology, however, has sufficient flexibility that a multiplicity of applications are possible. In the remainder of this article we will first describe the data sets and estimating methods used. Then we will present some short case studies to illustrate different ways in which the model may be applied.

Data and Methodology: Labor Supply

Determining the occupations of specific groups of employed people (i.e., the residents of a particular area) presents some difficulties. Only the decennial census collects occupational data on the *residents* of small areas. These data are available by zip code on computer tapes and include 81 different occupational categories. With 1990 data soon to be available, information will be better than it has been during the last few years. However, use of dated occupational information between census years is not as much of a problem as it would appear if one can accept the assumption that the relative distribution of occupations in a neighborhood is fairly stable over time. The validity of that assumption will vary with the community involved; one would not use the technique described subsequently for a community that has undergone rapid socioeconomic change since the census was taken. But in many instances it is legitimate to take the relative distribution of occupations (from the most recent census) and apply the percentages to estimates of the total work force that are available from a number of sources.[4]

Current data are available on the skills of unemployed people who have applied for unemployment compensation. They enables us to estimate occupational characteristics of unemployed people in specified geographical areas. These data come from state departments of employment security that administer the unemployment compensation system. In Illinois, the state is divided into service districts. An unemployed person must go to the office of the district that services his or her residence. At this office an applicant must go to a job service officer who conducts an interview and gives the individual an occupational classification code. Reports on applicant occupations and other demographic characteristics are filed with the central office monthly. The monthly reports are accumulated until there is an annual total. Using these reports, it is possible to derive occupational distributions of unemployed people within each service district.

Estimates based on these data are also not without problems. Unfortunately, boundaries of agency unemployment service districts do not follow those of any other data collection system. They cross zip codes, census tracts and even municipal boundaries. The City of Chicago has 12 of these districts. Despite the incongruity of service districts with target areas being analyzed, it is possible to adjust the target area boundaries so that there is an approximate fit. Since the analysis needs only occupational distribution, precise boundary congruity is not required. Another problem is that the data do not include people whose compensation has expired or those who have given up looking for work. Thus our estimates undercount the magnitude of the unemployment.

There are other sources of occupational information within target areas. For example, one may wish to examine labor supply in terms of a specific job development agency or the employees of a firm that is going out of business. Data from these sources are very uneven. In some cases we have examined interview files and made our own occupational classifications based on the information available. In other cases these classifications had been made but not accurately. Because the use of these data is to construct an occupational profile for development policy purposes, the lack of accuracy is not a problem. What we derive from these sources are a list of occupations and associated skills that best match the experience and training of a target population of workers. But once it comes to actually designing training programs and placing individuals, interviews that yield more precise information will be needed. That step is illustrated in one of the case studies presented below.

One further task is required because the various sources of information on labor supply use different occupational classification systems. The National Occupational Information Coordinating Committee (NOICC) has developed a crosswalk computer tape that allows comparisons to be made from one system to another. In our methodology we generally convert all occupational codes to the Dictionary of Occupational Titles (DOT) system because that yields the most information on generic skills connected with each occupation.[5] DOT occupational codes reveal the degree of sophistication needed in order to relate to data, people and things; language and math skills; physical requirements; and length of formal training.

By estimating the occupational distribution of our target labor pool and translating that distribution into DOT occupational codes, we can gain an initial insight into the work experience and skills of the targeted labor force. If that group includes residents of a community, the information would have to be supplemented with educational characteristics of those who have never been in the labor force. These estimates provide a basis for determining what kinds of industries would be most likely to hire the group of workers under consideration. Strict matching of individuals with specific jobs and the design of training programs would require actual interviews. That is a step beyond the present analysis.

Data and Methodology: Demand Side

With the profile of the available labor force completed, we are now in a position to analyze the potential of employing that labor force in a defined location. As noted earlier, the definition of that location depends on the particular application of the methodology. A community group may wish to define a target area as coterminous with its own turf boundaries. Its purpose in doing so would be to encourage economic development in their area that would be most likely to hire residents. Another possibility would be to attempt to locate, within a larger labor market area, that location closest in terms of labor demand characteristics to the skills and experience of the resident work force. A further possibility is to attempt to define redevelopment priorities for an abandoned industrial facility in terms of the objective of employing the targeted labor pool. All of these possibilities are illustrated in the case studies below. First, however, we address the problems of estimating labor demand characteristics in smaller areas.

To analyze the industrial structure of small areas, we have used two sources of information. The best source comes from state departments of employment security that uniformly gather information from firms that pay unemployment compensation taxes. The ES-202 data include addresses of firms, number of employees, wage information and the Standard Industrial Classification (SIC) up to four digits.[6] This information can be developed into a set of standard reports on an annual basis so that trends can be established. It is also possible to assemble the information by small areas defined by zip codes. In the past, these data have not been available to researchers because of confidentiality problems. Recently a number of states have worked through these problems. In the future, ES-202 data are likely to be generally available. Here too, there are some problems with accuracy. Researchers in Milwaukee have done extensive work with this data set. They enumerate a number of issues involved in the use of the ES-202 data including distinguishing between firms and establishments, summarizing monthly employment reports to get accurate figures, correcting raw data errors, changing inaccurate reporting of the location of firms.[7] Despite these problems, ES-202 is still the best source of information for localized trends in employment by industry.

Another source of the same information is not nearly as accurate but can be used if the ES-202 data is not available. Dun and Bradstreet collects the same data when doing credit checks on firms. Many states have the files known as Duns Market Indicators (DMI) and use them for analysis. The problem with the data is that there is no guarantee that a firm will report information accurately and some years there may be no report at all. Thus major trends and trend shifts need to be field checked before the data can be used to make reliable estimates.

Using these sources, it is possible to construct an industrial profile for any area that can be defined by a combination of zip codes. The profile will contain the number of firms and employees by SIC code over time. Several kinds of analysis can be done with this information. First, it is possible to determine the extent to which a particular subarea of a city or metropolitan area specializes in particular types of economic activities relative to the rest of the region, the state, and the nation. Location quotients can be calculated that make these comparisons, and shift-share techniques can be employed to determine the changes in economic specialization of a particular location.

Knowing the relative specialization of neighborhoods or other geographic locations within an urban area makes it possible to pinpoint industries that may have a geographical affinity for a particular small area. By doing shift-share on a regional basis and examining trends both in the region and the smaller area, a set of industries can be delineated that make sense in a particular location.

To translate this information into potential demand for labor with particular skills, another data set is needed. We have utilized data developed by the National Occupational Information Coordinating Committee.[8] The NOICC was established in 1976 under the Vocational Education Act to provide occupational demand information that could be used for planning purposes. Each state has a local office that is part of the NOICC network. In Illinois, that office is the Illinois Occupational Information Coordinating Committee (IOICC). Each year IOICC constructs a matrix that shows the occupational distribution within each three-digit SIC code and also shows the industrial distribution of employment in each occupation. The matrix is based in part on a survey of firms' staffing patterns that covers one-third of the SICs each year. Every industry is thus surveyed during a 3-year cycle. The survey of staffing patterns is done on a statewide basis and stepped down to each of eight regions.

The matrix also includes employment estimates for a base year (currently 1986) as well as forecasts to the year 2000. Estimates and forecasts of total employment for each industry and occupation are made by the Illinois Department of Employment Security. Although initially based on ratios relative to national data from the Bureau of Labor Statistics, they are stepped down to state, regional, and county levels. IOICC then uses the industrial staffing pattern distributions to construct matrices for the state, region, and counties.

Our methodology uses the IOICC matrix for the county to estimate the number of people employed in a smaller target area by occupation. We do this by assuming that the relative distribution of occupations within a given industrial category in the county will be similar to the smaller area. First we take the relative distribution of occupations within each industry selected by the previous location quotient and shift-share analyses (the number of workers in each occupation as a percentage of total employment). Because the IOICC data is for the county, we need to then apply these percentages to the numbers generated from ES-202 or DMI Files for the subarea of the city we are studying. This yields an estimate of the occupational demand in the selected industries in the subarea.

The estimates of occupational demand can also be converted to DOT codes as we did on the supply side. Comparing these two sets of estimates provides a basis for targeting industries for retention and attraction priority that would be most likely to hire the targeted labor pool. As noted earlier, implementation would involve more detailed investigation including interviews with both potential workers and potential employers to define the content of training programs that could enhance the fit between demand and supply. Local job development agencies could perform such tasks. The labor-force-based development analysis provides the needed first step for targeting.

We have explained in this section of the article the data sources and estimating methods used for a labor-force-based development analysis. As discussed earlier, however, labor-force-based analysis is a general idea. Different development problems require different emphases in the analysis and all the sources and methods previously described are not always appropriate. There are, in fact, numerous applications of the general approach. To illustrate some typical applications, we present some examples based on technical assistance work performed by the authors for different groups in the City of Chicago.

Applications: The Closing of Playskool

When Playskool, an old Chicago toy manufacturer, announced plans to leave the city, a controversy ensued. The city had given Playskool subsidies to create jobs. They used the subsidy and then left town as part of a corporate consolidation scheme of the parent company, Hasbro Inc. The departure was delayed, however, by considerable community opposition and a threatened lawsuit by the City of Chicago. The delay was to provide time to get the workers other jobs or to find a reuse for the facility that would hire the Playskool employees.[9]

The delay enabled the City of Chicago's Department of Economic Development (DED) to put together a team to examine the problem. The team consisted of several city departments including DED, the Finance Department, and the Mayor's Office of Employment and Training (MET). In addition, the community development corporation in the Playskool area, several community-based organizations, City Colleges, and several technical assistance centers were involved. The MET quickly went into the plant and gathered work histories of the employees at Playskool. It was determined in the process that many of the employees lived in the vicinity of the plant. Playskool, as part of its concession to the city, agreed to turn the property over to "the community." The CDC negotiated a joint venture agreement with a developer to take title to the property.

The city-organized team met to consider how to proceed. DED was prepared to offer a sizable tax break in addition to a land write-down in return for a "first source" hiring agreement that would give initial consideration for employment to the former Playskool workers. Labor-force-based development methodology was used to establish a framework for the agreement.[10] Development priorities that would fit the skills and experience of the workers were needed.

We first analyzed the work histories conducted by MET. There were 118 workers who were still unemployed and registered with MET for job placement and/or training. We immediately discovered that MET uses an occupational classification system, Standard Occupational Code (SOC), that does not match any of the three in our own system.

We decided initially that we would convert the SOC codes to the Occupational Employment Statistics (OES) system employed by IOICC so that we could use the IOICC data set to determine what kinds of industries in the Chicago area hire people with the same occupations as the Playskool workers. We made a list of the most significant OES occupations among the Playskool workers. IOICC matrix data that links occupations to industries was used to determine the following:

1. Which industries within the Chicago PMSA employ the most people in each of the occupations selected;
2. Which employ the most in relative terms (percentage of total employment in the industry); and
3. Which have the highest projected employment (to 1995)?

Lists of the top 10 industries (defined by 3-digit SIC codes) in each of the above three categories were compiled for four key occupations. The industries were each given scores based on how many times they showed up on the 12 lists.

We then put the lists and scores aside and did a more conventional analysis of the industrial structure of a grouping of zip codes that surrounded the plant. We delineated industrial specialization in those zip codes by computing location quotients that compared the target area to the remainder of the city. We used U.S. Department of Commerce publication *U.S. Industrial Outlook* to examine national trends in these industries.[11] Chicago performance was assessed using shift-share analysis. Based on these methods we prepared capsule summaries on the prospects for these industries if they were to operate at the Playskool site.

We then put the two parts of the analysis together noting which industries came up positive from both perspectives. A discussion with the Playskool work team based on this analysis resulted in a recommendation. The plastics and electronics industries seemed to be the most likely prospects. The team further refined this to target electrical connectors.

The city then moved to put together a development package that included subsidies, an agreement to attempt to find tenants in the plastics, electronics, and electrical connector industries, and other elements proposed by the team. Those included the use of part of the Playskool facility for a MET dislocated workers intake center, job training, daycare, and a legal services center. The package was rejected by the developers who had already begun to lease parts of this large industrial complex. They contended that the subsidy did not offset their need to quickly turn over their money. The prospect of the amount of time that it would take the public agencies to make the package a reality and the extra effort to target development was simply not acceptable. The CDC went along with this for fear that the Playskool property would sit vacant. Eventually the entire initiative collapsed and the Playskool workers went out on their own to seek employment where they could.

While the Playskool case illustrates the use of the labor-force-based methodology, it also presents some other lessons. Development priorities that link skills to industrial targets are not sufficient. The city moved too slowly; it was not prepared to administer such a complex arrangement on an emergency basis. Also, it may be the case that no matter how quickly the city moved it would not have been fast enough for a developer whose livelihood depends on how quickly he can turn the dollars over. Some corporate managers may well have found the package attractive, but the addition of the developer to the equation made the project difficult to implement. Institutional innovations such as public-private emergency development teams in which a public body could act as developer is needed for this type of development problem.

Applications: Development Priorities for a CDC

Another application of labor-force-based development methodologies includes establishing priorities for community development corporations. The South Austin Economic Development Corporation (SAEDC) was created in 1985 to address the development needs of the South Austin community in the far west side of Chicago. Plagued by deteriorated business strips, dilapidated and abandoned housing, concentration of un- and underemployment and poverty, manufacturing decline, and poor business activity, the community organized SAEDC to implement a community-based development program. Industrial retention and attraction and employment of residents in manufacturing jobs were among the central objectives of the corporation.

SAEDC asked the Center for Urban Economic Development (UICUED) for assistance in setting industrial priorities. UICUED and SAEDC agreed that a labor-force-based approach made the most sense as far as targeting growing or stable industries that could potentially provide the types of jobs that residents needed. UICUED then met with SAEDC and other relevant organizations in the area to discuss the approach and methodology. Based on the comments of this group and working in close cooperation with SAEDC, UICUED conducted a study of manufacturing employment and development priorities for the group of zip codes that most closely approximated SAEDC's boundaries. The study proceeded along two tracks: analysis of local manufacturing activity and analysis of the resident labor force. Both analyses followed the same methodologies used in the Playskool case to determine the needs and experiences of workers and to identify the industries most likely to provide them with the proper employment opportunities.

As in the Playskool case, we compared industries that were identified in the analysis of localized manufacturing activity with those resulting from the analysis of the labor force. Industries with an important local concentration and a regional advantage including large shares of the occupations of unemployed residents were then selected for targeting. Based on the analysis described above, UICUED organized industries into three groups: growing or stable industries with potential for development in the area, industries that would tend to hire local residents, and industries that appeared in both of the first two lists.

Considering the occupational needs of residents and the industrial outlook of industries in these lists, UICUED recommended six industries for consideration by SAEDC in setting development priorities. The study also identified occupations of unemployed residents that were declining and suggested working with residents and local institutions in the retraining of these workers for other occupations with a better projected future.[12]

SAEDC then narrowed the list to three: metal-working industries, electronic and electrical manufacturing, and plastics. In targeting these three industries, SAEDC decided to try to work with firms in their area in order to improve industrial technology, industrial management, and marketing. In addition, they engaged in efforts with the community colleges and other local organizations and individuals to develop a skills center with a capacity to respond to the specific training needs of these industrial groupings.

Listings of firms in the annotated groupings were produced from Dun and Bradstreet files. Outreach efforts were initiated. Firms were informed about the activities of SAEDC and were asked to contact SAEDC for assistance in the areas indicated. SAEDC would work with the firms to identify their needs and to secure the proper assistance. In exchange for this assistance, SAEDC asked the firms to hire area residents. As the work progressed, SAEDC began to conduct more detailed studies of the targeted industries including interviews with firms located within reasonable commuting distance of their own area. Interviews with CEOs of firms in the area identified in detail skill needs that could be related to job development for local residents.

Finding Employment Centers for Residents of a Community

The Howard Area Community Center (HACC), a nonprofit social-service agency in Chicago, suggested an application of labor-force-based development methodology that involved the identification of an accessible area outside of the neighborhood where the occupational demand matched the skills and experience of neighborhood residents. Specifically, they asked UICUED to assist their job placement efforts by developing a labor market strategy plan.

Starting with a listing of the most prevalent occupations of job applicants, UICUED targeted suburban areas adjacent to HACC's service area that were accessible by mass transportation. We first identified the main industries in these targeted areas (by size of employment). We then selected from this group those industries with a significant share of the occupations previously held by job applicants. Next, we analyzed these industries and chose those exhibiting growth or stability of employment.

Based on discussions with HACC's staff and assessments of this group of industries, we narrowed the list to five industries. We investigated this list for the names and addresses of the companies, the contact person, number of employees, and a general analysis of the types and quality of jobs that these companies could offer the job applicants serviced by HACC. Finally, we prepared a work plan on how to use this information and a marketing plan to promote the HACC's services. The plan proposed a three stage marketing effort: first, developing a package of introductory materials on the job placement services of the agency to send to the final list of firms; second, contacting the firms and announcing the mailing of the package; and third, following up with a telephone call and a visit to the firms agreeing to discuss a relationship with HACC.

Conclusions

The above cases demonstrate different applications of the labor-force-based development methodology. In the case of Playskool, the problem was to attempt to attract firms to a specific site that would be most likely to hire the people who had been employed by Playskool. In South Austin, the methods were used to help establish priorities for the activity of a community development corporation. HACC involved an effort to find an appropriate area outside of a community as a focus for local job-placement activities. In each case, the objective was to formulate development priorities that would most likely result in the direct employment of a targeted work force. The methods described earlier in this article can accomplish such objectives. The data bases we have described and the estimating methods we have developed allow economic development planners to analyze small areas and gear priorities to the needs of specific groups of workers. While the efforts described above did not preclude local groups from attempting to place unemployed workers in jobs existing in the larger labor market area, an attempt was made to "shrink" that area in order to enhance employment prospects of specific groups of workers.

This analysis does not assume or attempt a fit between local jobs and residents within a limited labor market. Rather, it focuses on the needs and opportunities of workers and employers within a smaller area. The purpose is twofold: to steer as many as possible local opportunities to those workers with limited mobility and to target development that provides the types of opportunities needed by local workers. Such an effort is justified on the grounds that local development efforts should benefit residents as much as possible, and that there is a need to help people minimize the distance between home and work, particularly when other social factors are extending it. Although larger markets provide both employers and workers with more choices, factors such as residential segregation, poor distributions of labor market information, and differential mobility work to the disadvantage of specific groups of workers. Similarly, large labor markets place unnecessary hardships and costs on workers, particularly on those workers with low skills and low wages. Meanwhile, policies reinforcing this trend pay more attention to the needs of firms than to those of the worker. This analysis proposes development and training priorities that also meet the needs of workers. Finally, prevailing analyses do not pay much attention to linkages between development and training. This tends to reinforce short-term, general trends in the market, and ignore the specific needs and potentials of particular workers.

Hence there is a bias in traditional approaches and a need for other strategies that attend to the unique needs of workers who cannot be accommodated by the natural processes of larger labor market areas. Our approach allows for targeted development based on the needs of the unemployed, the underemployed, and specific pools of workers as well as for development that links industrial and training priorities. Our cases have shown the practicality of this approach while suggesting a number of different applications.

Notes

1. William L. McKee and Richard C. Froeschle, *Where the Jobs Are: Identification and Analysis of Local Employment Opportunities* (Kalamazoo, MI: W. E. Upjohn Institute for Employment Research, 1985), p. 21; U.S. Department of Labor, Employment and Training Administration, *Directory of Important Labor Areas* (Washington, DC: U.S. Government Printing Office, 1978), pp. i-ii.

2. There is a very extensive literature on this subject. Some examples include the following: Bennett Harrison and Barry Bluestone, *The Great U-Turn: Corporate Restructuring and the Polarizing of America* (New York: Basic Books, 1988); Sar Leviatan and Isaac Shapiro, *Working but Poor: America's Contradiction* (Baltimore: Johns Hopkins University Press, 1987); Samuel Rosenberg, *The State and the Labor Market* (New York: Plenum, 1990).

3. Michael B. Teitz, "Neighborhood Economics: Local Communities and Regional Markets," *Economic Development Quarterly* 3 (1989): 111-22.

4. There are generally data services or public planning agencies that make demographic adjustments to population figures in small areas that can be used to estimate the size of a labor force. UICUED has developed a data base known as Demographer that uses estimates made by the firm, CACI Inc. to compile statistics for any group of zip codes or census tracts.

5. U.S. Department of Labor, Employment and Training Administration, *Dictionary of Occupational Titles,* 4th ed. (Washington DC: U.S. Government Printing Office, 1977), pp. xiii-xli. Illinois' NOICC, the Illinois Occupational Information Coordinating Committee (IOICC) has developed a microcomputer file of DOT codes that includes all DOT characteristics and a crosswalk against other occupational coding systems. This package, known as Micro Link, is available from IOICC, 217 East Monroe Street, Suite 203, Springfield, IL 62706.

6. Sammis B. White, John F. Zipp, William F. McMahon, Peter D. Reynolds, Jeffrey D. Osterman, and Lisa S. Binkley, "ES202: The Database for Local Employment Analysis," *Economic Development Quarterly* 3 (1990): 240-53.

7. Ibid., pp. 243-47.

8. National Occupational Information Coordinating Committee, *Vocational Preparation and Occupations,* Vol. 1, Educational and Occupational Code Crosswalk, 3d ed. (Washington, DC: U.S. Government Printing Office, 1982).

9. Robert Giloth and Robert Meir, "Democratic Populism in the U.S.A.: The Case of Playskool and Chicago," *Cities: An International Journal of Urban Policy* 3(1) (1986): 72-74.

10. John Betancur and David C. Ranney, *Labor Force Based Manufacturing and Employment Development: Priorities For the West Side of Chicago* (Chicago: Center for Urban Economic Development, December, 1986), pp. 29-46. A similar application used the methodology to assist another CDC establish target priorities for the development of a vacant site. See John Betancur and William Howard, *An Analysis of the Labor Force in Mid Southwest Area of Chicago* (Chicago: Center for Urban Econowmic Development, 1988).

11. U.S. Department of Commerce, *1990 Industrial Outlook* (Washington, DC: U.S. Government Printing Office, 1990).

12. Betancur and Ranney, *Priorities For the West Side of Chicago,* 1-29.

Chapter 8

Business Strategy and Cross-Industry Clusters

Peter B. Doeringer and David G. Terkla

❋❋❋

The history of state and local economic development policy has been marked by fads. During the 1960s, mature smokestack industries were the focus of industrial recruitment and retention efforts; attention shifted to high-technology industries during the 1970s and early 1980s, and the recruitment of foreign firms was popular during the latter half of the 1980s. The slogan of development policy during the 1990s is fast becoming "industry clusters": geographical concentrations of industries that gain performance advantages through co-location.

Industry clusters are an appealing way to organize state economic development policy. Developing clusters of multiple industries, as compared to recruiting single industries or firms, offers opportunities for leveraging industrial development incentives while diversifying a region's industrial mix. Moreover, the cluster concept opens the possibility that regional growth can be fostered by non-export- as well as export-based industries in those instances where both types of industry are part of the same cluster.[1]

State governments, ranging from Arizona and Oregon in the West to the industrial states of Michigan and Pennsylvania to high-technology Massachusetts, are now using some version of the industry cluster concept in their economic development policies. Lacking in many new policy efforts, however, are an understanding of the underlying economics of industrial clustering and a methodology for translating the cluster concept into concrete growth policies tailored to specific businesses.

This essay explores the economic foundations of business clusters that cut across different industries. Our emphasis is on business strategies, rather than the behavior of broad industry aggregates, and the way that firms (as opposed to industries) interact with one another. The findings highlight the perils of assuming universal location criteria, instead of

AUTHORS' NOTE: We are grateful to the W. E. Upjohn Institute for Employment Research for support of some of the research presented in this article.

Reprinted from *Economic Development Quarterly,* Vol. 9, No. 3, August 1995.

identifying how specific businesses locate to exploit unique regional strengths, in devising policies for nurturing clusters.

We draw on previous field research examining the role of business strategy in regional economic development in a group of industries representing a range of technologies and production methods. During 1985 and 1986, 48 interviews were conducted (covering all firms with 80 or more employees in each major three-digit Standard Industrial Classification [SIC] manufacturing industry) in the Montachusett region of Massachusetts, a local economy dominated by mature manufacturing industries such as plastics, paper, metal fabrication, and apparel. Also, during the 1991-1993 period, interviews were conducted with a sample of 50 Japanese and U.S. startups in the electronic equipment, nonelectrical machinery, and rubber and plastic products sectors in Georgia, Kentucky, New York, New Jersey, and New England.[2]

The article begins by distinguishing traditional agglomeration economies from the dynamic competitiveness concept of "competitive advantage." Uncertainty about the cause of firms' geographical clustering has led to a lack of focus in state and local economic development policies involving clustering—too little emphasis, in particular, being placed on specific causes of cluster relationships among firms across industries. We conclude with a discussion of how local economic development officials can identify the potential for clusters across industries in their local economies.

Industry Clusters, Agglomeration Economies, and Competitive Advantage

Industry clusters, a term popularized by Michael Porter as part of his study of competitive advantage, are a variation on the much older theme of local and regional agglomeration economies.[3] Economic theory posits two types of externalities that lower the costs of firms agglomerating in a common location: (1) economies based on transportation or transaction cost savings among firms in the same industry or in different industries that have input-output relationships with one another and (2) economies from shared public goods such as airports or university-based research and development (R&D) facilities.[4] The presence of positive externalities explains the clustering process, whereas specific location sites for each cluster depend on either "historical accident" or the cost advantages provided by immobile factors that attracted the firms anchoring the cluster.[5] Competitive advantage adds economic dynamics to the understanding of the clustering process. Traditional agglomeration economies are static in that the cost advantages they confer on firms remain the same through time. By contrast, competitive advantage depends on dynamic factors constantly evolving as a result of continual innovation and the transfer of new knowledge within clusters of firms.

Exploiting such dynamic externalities requires a deliberate business decision to develop efficiency-enhancing organizational relationships through direct participation in a common "production channel" or indirect relationships with governmental institutions or labor markets. Because strategic business decisions vary from firm to firm, the formation of dynamic clusters must be analyzed at the level of business units rather than on an aggregate industry-by-industry basis.

Clustering for Competitive Advantage

There are different theories on what constitutes competitive advantage. Porter, for example, postulates competitive advantages derived from (1) an innovative environment marked by vigorous competitive pressures from rival firms; (2) a customer base that pressures firms to innovate; (3) access to networks of local suppliers of specialized inputs, who themselves are constantly motivated to innovate by fierce competition with local rivals; and (4) highly specialized labor and technology that meet the needs of cluster businesses.[6] Clusters are anchored to specific locations by the immobility of specialized labor and R&D facilities, but the main source of dynamic efficiency is the interaction of competition with innovation. Aggressive competitive pressures from customers, rivals, and suppliers stimulate innovations that continuously raise total factor productivity, lower costs, and promote business growth.

A second theory of competitive advantage, advanced by Krugman, highlights the importance of large-scale firms with increasing returns to scale.[7] Clusters emerge when large firms attract supplier firms because of transportation and transaction cost savings and encourage the development of local pools of skilled labor that further contribute to agglomeration economies. Dynamic efficiency depends on continued organizational learning through exchanges of information, labor, and technology among firms within the cluster rather than on innovation induced by aggressive competition.

A third theory of competitive advantage is based on collaboration among specialized firms. This collaboration economies thesis, developed by Piore and Sabel, argues that small firms can acquire competitive advantages by using inherent flexibility to specialize in niche markets.[8] Such firms can secure cost advantages by cooperating with other specialized firms in product design, manufacturing, and marketing activities. Clusters arise because collaboration is facilitated by face-to-face interactions among cooperating producers and through local social relationships.[9]

Fuzzy Implementation

Given these different theories of the sources of competitive advantage and business clustering, it is not surprising how little agreement there is over the cluster concept implementation in specific state and local settings.[10] Some analysts equate clusters with traditional agglomeration economies and tend to focus on single-industry clusters; others define clusters to include key suppliers and customers; still others emphasize trust and collaboration among firms developed by spatial proximity.[11]

There is little consistency among states that have adopted cluster-based growth policies and little attempt to distinguish between the static and dynamic benefits available to clustering firms.[12] In practice, states tend to define clusters in ad hoc ways, using mechanical criteria: industry-based input-output relationships, broad industry growth forecasts, and untested predictions about the business potential of particular products and technologies; many clusters involve single industries to the neglect of multi-industry clustering.

The problem of translating cluster concepts into practice is nowhere more evident than it is in Massachusetts.[13] The blueprint for current Massachusetts growth policy is provided by a recent study that specifies four major industry clusters—health care, information technology, financial services, and knowledge creation services—and several minor clusters in two-digit SIC mature manufacturing industries such as plastics, machinery, and chemicals.[14]

Rather than using criteria such as the frequency and character of business relationships (or other elements of regional competitive advantage), these clusters are characterized as those that "compete nationally and internationally and have the size, sophistication, productivity, and national and international positions to drive economic upgrading."[15] There is no mention of clustering economies, no functional criteria are used to define the industrial boundaries of these clusters, and there is no indication of the degree to which the Massachusetts economy provides the rivalry and competitive pressures on the clusters that are supposed to stimulate strong business performance. Except for the information technology cluster, most industries within these clusters are from the same two- or three-digit SIC, thereby excluding most production relationships that cross industries.

Such ad hoc procedures are always questionable for defining and implementing policy. At best, these highlight the location concerns of individual industries, but they do not deal effectively with broad cluster relationships *among* industries—even though recent research finds that particular industries grow faster in cities with a more diversified industrial base than they do in those in which the base is concentrated.[16] Further, without functional procedures for determining cluster potential, it is hard to envision how either economic development analysts or business executives can determine what pieces of a cluster are already present in a region and what new clusters can be attracted to it, let alone how to recruit and nurture, the elements of any particular cluster.

Toward a Framework for Cluster-Based Development

An emerging literature provides practitioners some guidance on how to move beyond single industry clusters, based on static agglomeration economies, to enhance cluster formation among firms from different industries that share a common competitive advantages.[17] One branch of this literature has framed the analysis of clusters around specific industries and their input-output relationships.[18] Another has examined the evolution of industrial complexes ranging from those built around dominant central firms to more atomistic networks of small firms characterized by flexible specialization.[19]

Both these approaches are moving the discussion of clusters in a constructive direction, yet each has limitations. The industry-based analyses continue to view industry relationships through the distorting lens of relatively aggregate input-output relationships and obscure the possibility of competitive advantage linkages, based on factors other than product market relationships.

Studies of industrial complexes are better at capturing the microlevel details of business linkages and highlighting the importance of strategic decisions in determining the location of production. However, there is no consensus about the circumstances that lead to global networks in which production is soured wherever costs are lowest, as opposed to those that lead to localized industrial clusters, based on organizational relationships among firms or the sharing of common technologies and skills.[20]

To make sense of the diverse arrangements governing geographic division of production requires an understanding of how firm-specific strategies both define multi-industry production channels and determine the propensity of production channels to cluster within a country or region. Our research on these production and location strategies in the United States suggests three prominent reasons why firms in different industries cluster locally: (1) collaboration economies within production channels, (2) transfers of knowledge through labor market relationships, and (3) partnerships with governments and unions.

Production Channels

Production channels are useful for understanding competitive advantage clusters. Production channels are the chains of suppliers, manufacturers, and distributors that begin with basic inputs and end with marketing of the final product. They involve relationships that cut across industries (including those with an affiliated "infrastructure" of equipment vendors, information networks, transportation systems, and banking and financial institutions) as well as those within industries.

Business relationships within production channels range from exclusively market-mediated transactions to complete vertical integration. They can incorporate either dependent relationships (such as those established by dominant multinational corporations) or cooperative relationships among firms of equal power.

Although interindustry transactions incorporated within production channels can sometimes be detected in input-output tables, neither the character of relationships among firms nor the benefits of clustering can be discerned in this way. Some differences in cluster relationships are related to the type of product and the nature of production technology, but many are defined by strategic firm-specific business decisions. These decisions can vary even within industries, depending on size of firm, corporate structure, type of production process, and even nationality of ownership.

For example, even an industry like apparel and textiles, with a relatively straightforward technology and uncomplicated labor requirements, sustains both global networks with little regional or national identity and locally concentrated production relationships firmly embedded in specific social and economic settings.[21] Worldwide, the apparel and textile industry has shown a long-term trend away from clustering, as clothing manufacturers have relocated from industrial to developing countries, whereas other parts of the production channel, such as high-quality fabrics and equipment manufacturing, have remained in industrialized countries.[22]

Standard explanations identify labor costs as the key determinant of this channel dispersion. Traditional apparel manufacturing uses mass production techniques and highly simplified jobs to assemble garments from "bundles" of components, involving long production times and considerable buffer stocks between operations.[23] The labor-intensive bundle system encourages the division of labor into simple, repetitive tasks and is responsible for the shift in production to low-wage regions.

If, however, the production channel is examined as a whole, from the spinning of natural or synthetic fibers to the retailing of clothing, labor costs in apparel appear as only a part of a channelwide competitiveness equation that also includes substantial inventory costs. Scale economies in textile production, the costs of setting up production lines, the need for buffer stocks to balance die production line, and the variability of consumer tastes mean considerable inventories throughout the production channel.[24] From this broader perspective, the geographic dispersion of the production channel depends on the balance firms choose between competing on the basis of labor costs or minimizing inventories and buffer stocks through rapid response production and quick delivery.

Minimizing labor costs has been the dominant business strategy for most textile and apparel production channels in the United States and Great Britain; most such manufacturers compete on price with one another and with foreign clothing producers. In these countries, the production channel has become more and more dispersed as international differences in labor costs shift apparel manufacturing toward low-wage countries.[25] In the United States, for example, there were once strong scale economies, both within and across industries, that led to a clustering of firms in markets such as New York City.[26] However, the clusters began to disperse as the pressures of global competition from low-wage producers gradually overwhelmed cluster efficiencies.

By contrast, the apparel production channel in Italy consists of small firms that have formed cooperative alliances, which allow them to move quickly from design to production and to produce relatively short runs of product to meet replenishment demand. Apparel manufacturers work cooperatively on product design, manufacturing, and distribution with other parts of the production channel.[27] The competitive advantage comes from matching high value-added market niches to a production and marketing channel that quickly brings fresh designs to market. This channel is clustered geographically because collaboration is facilitated by overlapping social and business relationships formed by employers in close proximity.

Recently, there has been an interest in cooperation within the U.S. apparel production channel as a result of initiatives taken by clothing retailers to have apparel manufacturers speed production response times so as to reduce retail inventories and replenish stock outs. However, unlike the situation in Italy, where rapid responsiveness has led to a local clustering of the industry, electronic data interchange of sales and production information and high-speed transportation networks have reduced the need for close geographic proximity.[28]

This experience in the apparel industry is consistent with the scant literature on clustering in the United States, which suggests it is common for single industries to cluster within a region where there are transaction and transportation economies, whereas multi-industry production channels are often spatially diffused.[29] Even highly visible examples of multi-industry agglomerations—the automobile industry in Detroit, the steel industry in Pittsburgh, the high-fashion apparel manufacturing and marketing cluster in New York, and the high-technology industries of Silicon Valley and Route 128—have always sourced production outside their regions, and their production channels are becoming more and more geographically dispersed.[30] Similarly, our research has found considerable geographic dispersion but no examples of closely knit production channels.

Collaboration Economies and Geographic Clustering

Although dispersed production channels and global networking appear more common than geographic clustering in the United States, there are examples of geographic concentrations of production channels. These illustrate how dynamic competitive advantage arises from face-to-face collaboration economies.

Just-in-time production. The most well-known examples of collaboration economies are found in firms operating under just-in-time (JIT) production processes, a method pioneered by Japanese manufacturers to reduce buffer stocks and inventory costs within production channels. The rapid replenishment of inventories under JIT requires both quick delivery and close coordination of orders and production schedules among suppliers, assembly plants, and distributors. Although good transportation networks can speed delivery and electronic data can expedite communication, coordination of production and delivery schedules still benefits markedly from proximity.

JIT practices are not as widespread in American industry as they are in Japan, but suppliers in a number of industries are increasingly required by their customers to speed response and delivery in ways similar to the rapid response system in the apparel and textile production channel. JIT practices that lead to geographic clustering are particularly prevalent in production channels involving Japanese manufacturing transplants. Japanese suppliers tend to locate sufficiently close to their customers to permit JIT delivery schedules and facilitate the development of rapid response capabilities for adjusting to changing product specifications.[31] From these interactions, focused on rapid response and adjustments to changing product specifications, dynamic efficiencies emerge.

Niche markets. A second source of collaboration economies is the development of niche markets, which involve relatively low-volume, often specialized production. Such markets can be served at a distance, particularly when production is routine, but the opening of niche markets is aided by various informal contacts that occur through local business and social networks or through JIT relationships.[32]

Our study of mature manufacturing firms in central Massachusetts illustrates the process through which niche markets emerge.[33] Small- to medium-size firms in diverse product lines (such as mold making for custom-designed artificial limbs and joints, machinery and metal fabrication of prototypes for the defense industry, and customized plastic fabrication) were constantly seeking niche markets for specialized products and services; developing these markets depended on close interaction with primary customers and suppliers.

Proximity allows these niche market firms to participate in and respond rapidly to changing product designs and manufacturing practices among customers and to resolve problems quickly over product specifications and schedules. For example, custom-molding firms frequently worked with their resin suppliers to develop specialized materials for particular product runs, and they also worked closely on the designing of molds with precision fabricators of the final product. As with JIT markets, these relationships are a source of dynamic efficiency.

Labor Productivity and Technology Transfers

Labor markets are a second source of competitive advantage clusters. Conventional wisdom holds that labor markets influence business location through regional wage differentials; however, many studies have found wage differences relatively unimportant in location decisions.[34] One major reason is that wages are only one side of the labor cost equation. "True" labor costs involve adjusting wages for differences in labor productivity.

One aspect of labor productivity is investment in education and vocational skills. Schools and vocational training programs contribute to the skill base of a regional economy and can be a major asset in recruiting new firms.[35] However, education and training programs typically provide general skills that are transferable to other regions, and successful programs are easily replicated. Although such programs may contribute to clustering through traditional public good agglomeration economies, they are unlikely to confer more enduring benefits of dynamic efficiency.

A second dimension of labor productivity involves generic workforce qualities such as a high-commitment work ethic, problem-solving abilities, teamwork, and skills and knowledge of advanced technology that can be translated from one production setting to another. These qualities tend to be learned and reinforced locally, partly through socialization processes in the family and schools but primarily through various types of "learning" at the workplace.

These localized workforce qualities are well-known sources of static agglomeration economies, and our research confirms that such workforce qualities are an important source of regional differences in labor productivity.[36] However, they can foster dynamic efficiency as well.[37] Competitive advantages are available to firms drawing on local labor pools that have acquired such skills and knowledge, and the exchange of labor further enhances cluster externalities.

Workplace skills. Such social and work-based learning has a geographic identity because it is linked to an area's specific socio-economic composition and industrial mix. An example of the importance of how industrial mix affects the local skill pool is provided by our interview with a large machine tool company in the greater Cincinnati, Ohio, area. A trained and experienced labor force, this company reported, was the most important factor in its decision to locate in the area. Although employers have encouraged area vocational schools to offer courses that contribute to the supply of skilled labor, the region's basic stock of skills has been perpetuated through on-the-job training provided by its many machine tool plants.

A similar clustering of industries around a highly skilled labor force is apparent in the development of the optics and imaging industry in the Rochester, New York, area. Although Rochester accounts for only 0.4% of the nation's population, 21% of the national membership of the Society of Imaging Science and Technology lives in the area, forming a major attraction to firms in a wide range of industries.[38]

The industry specificity of such indigenous skills, however, should not be construed too narrowly. In our study of mature industries in central Massachusetts, for example, a major location factor was a set of generic skills connected with the ability to form materials with precision.[39] These generic skills were used in a range of product lines and technologies: precision molding of plastics, precision metalworking and fabrication, and precision woodworking. They seemed transferable across industries and attracted firms whose competitive advantage derived from the ability to manufacture precision products for specialized or batch production markets. Transferability of labor opens opportunities for dynamic efficiency through cross-fertilization of skills and techniques.

Workplace attitudes. In this same mature industry study, the work commitment (as well as the skills) of the local labor force was widely cited by employers as crucial to their success in customized production.[40] Several managers made direct comparisons with worker attitudes in similar facilities elsewhere in the country, emphasizing that a strong work ethic was important in maintaining a rapid response production capability and sustaining product quality.

Similar attitudes are frequently cited as critical location factors by managers of Japanese startups in the United States.[41] Japanese firms often use education criteria to screen out job candidates, but they are less concerned with education and experience than they are with recruiting workers who possess a set of traits such as adaptability, motivation, and problem-solving capacity that can contribute to dynamic efficiency.

Technology transfers. The workforce can also be a critical source of knowledge needed for technology transfers. Small firms from a wide range of industries (including telecommunications, computer peripherals, and medical equipment) that were part of the optics and imaging cluster in the Rochester area also derived competitive advantage by acquiring technology from Eastman Kodak and Xerox, the area's two largest firms.[42] These technologies spread to smaller companies, not through supplier relationships and niche markets but through the swapping of employees within a common pool of skilled and technical labor developed around the region's core technology. Again, such cross-fertilization can result in dynamic efficiency through shop floor innovations.

Governmental Partnerships

Physical infrastructure investments, such as airports and highways, are also a source of static agglomeration externalities but do not typically provide dynamic competitive advantages.[43] However, there are less tangible institutional and cultural infrastructures that can create dynamic clustering externalities, the best known examples being business-education partnerships for improving schools, integrating education with work experience, and fostering research and development activities.[44] Lesser known, but equally important, examples include labor-management climate and civic capacity.

Labor-management relations climate. The labor-management climate of a region can affect business clustering, as evidenced by business location studies showing the deterrent effects of union militancy.[45] Conversely, a positive labor relations climate can provide important positive externalities to business location.

When managers of unionized firms speak of a healthy labor-management environment, they are more concerned with the frequency of strikes and unions' willingness to accommodate day-to-day issues of work rules, safety, and flexibility in the use of labor than they are with union pressures on wage rates. Nonunion firms often cite similar advantages from their workforce. A flexible labor force enhances the ability of firms to compete in markets requiring frequent production line changes to respond rapidly to changing customer demands and be able to exploit new niches in the product market as they are discovered. Regionally, the dynamic efficiencies of a collaborative labor relations climate can reinforce other elements in the clustering process such as the formation of production channel relationships and the diffusion of technology.

Such labor relations are often easiest to achieve in small- and medium-size firms not bound by rigid organizational rules, often having personalized labor relations. By contrast, flexibility in larger firms often must be deliberately fostered to counteract the bureaucratic rigidity that comes with size; in large unionized firms, flexibility must be negotiated as well. Government can play a role in strengthening labor relations in such firms by mediating disputes and providing forums for bringing labor and management together to work on long-term economic development issues.

Civic capacity. A second type of competitive advantage partnership involves the capacity of communities (and their governments) to provide a civic climate that enhances dynamic efficiency. The most successful local economic development efforts blend business entrepreneurship with community organization and spirit.[46]

Although business strategies and other economic clustering factors are major contributors to competitive advantage, their effects can be enhanced by an effectively organized community response to development problems and opportunities.[47] Civic capacity is often highest when there are community business learning cycles that arise through public-private partnerships addressing long-term growth policies in areas such as education, labor-management relations, and regulation.

Our study of Japanese business location decisions provides particularly good evidence of the importance of civic capacity as a basis for clustering.[48] Although Japanese plants sometimes cluster because of either traditional location factors or JIT relationships, a number of plants, in states such as Georgia and Kentucky, have been attracted by the capacity of state and local government and civic organizations to work together cooperatively.

Japanese executives typically emphasize the quality and effectiveness of state government mechanisms for coordinating industrial recruiting, the degree of civic support organized by local business and government leaders, and the hospitality of government officials. However, Japanese startups may place less value on a community's capacity to promote economic development or offer development incentives per se than they do on a community's willingness to make economic and political accommodations and to ensure against unanticipated problems between the plant and the community.

Building Cluster Potential at the State and Local Levels

Industry clusters promise a new way for state and local communities to leverage economic development while avoiding traditional, zero-sum development policies. However, the trend in current state policies toward ad hoc definitions of industry clusters risks wasting development resources by neglecting important linkages among firms that cut across industries.

Many states are claiming to adopt cluster strategies but focus instead on single industries. These industries tend to be selected according to criteria such as size, sophistication, and participation in national or global markets rather than for their potential to form production channel relationships with other firms or exploit other clustering externalities.[49] Even if clusters reflect externalities, they are often those of traditional agglomeration economies rather than those of competitive advantage. At best, such efforts can capture static cost advantages by using selection criteria such as transportation and transaction costs, capital intensity, and R&D levels (see Table 8.1).

Our findings instead point to the need for state and local economic development policies based on the dynamic cluster potential of specific firms and regional economies. This means finding good partners for existing businesses, recruiting firms that have adapted alliance strategies, and taking advantage of intangible regional resources that provide externality benefits to multi-industry clusters of firms (see Table 8.1).

Finding Firms with Cluster Potential

The starting point for determining cluster potential is learning how existing firms in a region fit into larger production channels, both vertically and across product lines. The second step is determining at what points in the channel proximity matters. By mapping out channels and identifying critical "proximity points," economic development planners can begin to recruit firms that can be linked to each other, and to existing firms, in the regional economy.

A practical way of conducting such mapping is illustrated by a study conducted in the province of Alberta, Canada. Starting with detailed national input-output tables, Alberta constructed a first approximation of the production channels within which key regional industries were represented. These approximations were presented to panels of industry specialists, who further refined the channel definitions based on firsthand knowledge of their industries.[50] Gaps in the production channel were then identified and used to target recruitment efforts.

A second approach to recruiting, using production channels, is to identify firms committed to JIT production strategies. Particularly when such firms are large, as in the case of auto assembly plants, their JIT requirements are likely to attract additional suppliers in the production channel.

A third method is to assist small- and medium-size firms that already have the capacity for specialized production to develop niche markets. This can include working with large companies to identify existing supplier contracts and supplier specifications and developing a bidding capacity for small local firms. It can also involve creating networking opportunities whereby small firms can explore the production processes of large firms to identify opportunities for outsourcing production.[51]

TABLE 8.1 Agglomeration and Competitive Advantage Economies

Type of Externality	*Traditional Agglomeration Economies*	*Competitive Advantage Economies*	*Type of Cost Efficiency*	*Identifying Clusters*
Within-industry economies in production, transportation, and transaction costs	Yes	No	Static	Analyze industry characteristics (plant size, capital intensity, etc.) Match with immobile factors
Interindustry economies in transportation and transaction costs	Yes	No	Static	Input-output tables Analyze industry characteristics (product cycle, value added, etc.)
Public goods (airports, transport, university research centers)	Yes	No	Static	Analyze industry match with public goods
Highly competitive environment Specialized network of local suppliers Specialized labor and technology pools	Possibly	Yes	Dynamic • Rejuvenation of product cycle through innovation and competition • Business networks	Input-output analysis and industry focus groups for production channel gaps Analyze niche potential among existing small- and medium-sized firms Analyze match between skill base and product life cycle of firms Analyze labor-management climate and firm matches
Production and organizational learning	Possibly	Yes	Dynamic • Rejuvenation of product cycle through information and technology exchange	Analyze current environment for government-business partnerships and firm matches
Collaboration economies Transfers of knowledge (through labor market relationships) Government partnerships	Possibly	Yes	Dynamic • Specialized firms linked to one another through business and social networks • Just-in-time production • Niche markets • Workplace skills and attitudes • Technology transfer by workers • Labor-management climate • Civic capacity	Look for just-in-time linkages

With all these approaches, we must recognize that clustering forces do not necessarily conform to existing political boundaries. With good transportation networks, *close proximity* can mean being within a traveling range of two hours, as illustrated by a textile factory in Georgia that provides JIT supplies to its main customer in South Carolina or by the location requirement that the new BMW plant in South Carolina imposes on its suppliers.[52] Similarly, the labor force can commute long distances, unions can organize and bargain regionally, and social networks can operate over relatively large areas. In such circumstances, adjoining political entities that elect to pursue cluster-based growth policies should coordinate policies to take full advantage of clustering opportunities.

Identifying Generic Endowments

Industry clusters can also be formed among firms that are not part of the same production channel but share pools of labor with generic skills or generic knowledge of technologies. Such workforce- and technology-based clustering will be correlated more strongly with hard-to-observe, firm-specific factors—differentiated products, human resource management strategies, and length of production run (mass vs. customized production)—than it will with particular industries, and it will also have a product life-cycle dimension.[53] For example, firms that are engaged in prototype, customized, or batch production are likely to cluster around labor pools that are flexible and have precision skills and problem-solving capacities. By contrast, firms that supply mass markets are often more concerned with direct wage costs and the ability of workers to endure machine-paced production speeds.

Identifying generic endowments involves analyzing high-performing firms to gain a sense of the clustering strengths present in a region. Developing an industrial recruiting strategy based on matches with these endowments and with a firm's product life cycle requires a similar firm-level analysis of prospective companies.[54]

Competition and Cooperation

Porter's widely cited formulation of the cluster concept, although acknowledging a role for cooperative behavior between customer and supplier firms, emphasizes the importance of vigorous horizontal rivalry among local firms as a spur to the formation of cluster networks.[55] However, it is unclear why competition promotes the externality benefits needed for clustering—and if it does, why it has to be local competition.

In fact, intense *vertical* rivalry among firms within production channels can be counterproductive to many performance improvements, such as JIT relationships and the opening of niche markets, based on collaborative behavior. For example, the vigorous cost competition among apparel manufacturers fostered by large clothing retailers in Great Britain is inconsistent with the formation of interfirm collaboration, and the adoption of data sharing and other supportive relationships within the U.S. textile-apparel production channel was delayed, in part by vertical rivalries.

The case for achieving efficiencies through horizontal competition may be even stronger. At least one study of state-level industrial clustering found that high-performance industrial clusters operated in markets that were competitive.[56] However, this study also concluded that most of these competitors were located outside the state, usually in a different region.

On balance, promoting business rivalry within a region does not seem conducive, to cluster-based growth policies. Facilitating cooperative relationships and promoting vertical collaborations that lead to production channel clustering among suppliers and customers appear to be more constructive directions for development policy.

Mobilizing Civic Capacity

Clustering can also be encouraged through enlightened leadership by state and local institutions, such as trade unions and government agencies, whose actions affect the economic flexibility and political consistency of the business environment. For example, a number of states have sought to reduce labor-management conflict by limiting union organizing through "right to work" laws. Although such states have fewer unions, they do not always have less strike activity than more heavily unionized states, and it is conflict and militancy that influence economic development.[57] One example of a constructive regional approach to labor-management relations is found in Jamestown, New York, where strikes and difficult labor relations were threatening the industrial base of the community. Initiatives by civic leaders led to the formation of community- and plant-level labor-management committees, which succeeded in slowing job loss by reducing labor conflicts. More and more unions and companies are embarking on similar plant-level efforts aimed at reducing conflict and raising productivity, and such efforts can be further encouraged by state and local governments.[59] Similarly, Georgia's economic development strategy emphasizes enhancing the quality of governmental and civic institutions.

The biggest challenges for development policy are creating the capacity to mediate and facilitate the development process and fostering an environment that encourages collaboration among competing firms.[60] In the long run, these may be critical elements in the capacity of government to promote growth.

Notes

1. C. Tilly, "State Strategy for Developing Base Industries: A Massachusetts Case Study," *New England Journal of Public Policy* 9 (Spring/Summer 1993): 33-50.

2. For further details, see P. B. Doeringer, D. G. Terkla, and G. C. Topakian, *Invisible Factors in Local Economic Development* (New York: Oxford University Press, 1987), Appendix 4; P. B. Doeringer and D. G. Terkla, "Japanese Direct Investment and Economic Development Policy," *Economic Development Quarterly* 6 (August 1992): 255-72; and P. B. Doeringer, C. Evans, and D. G. Terkla, "Is Japanese Investment the Answer for State and Local Economic Development?" (Paper presented at the meetings of the International Regional Science Association, Chicago, November 14, 1992).

3. M. E. Porter, *The Competitive Advantage of Nations* (New York: Free Press, 1990).

4. H. O. Nourse, *Regional Economics* (New York: McGraw-Hill, 1968); A. M. Sullivan, *Urban Economics* (Homestead, IL: Irwin, 1990); J. Heilbrun *Urban Economics and Public Policy,* 3rd ed. (New York: St. Martin, 1987).

5. Nourse, *Regional Economics*, R. Vernon, *Metropolis 1985* (New York: Doubleday, 1963).

6. Porter, *The Competitive Advantage.*

7. P. Krugman, *Geography and Trade* (Cambridge, MA: MIT Press, 1991). Also, see B. Harrison, *Lean and Mean: The Changing Landscape of Corporate Power in the Age of Specialization* (New York: Basic Books, 1994).

8. M. J. Piore and C. F. Sabel, *The Second Industrial Divide: Possibilities for Prosperity* (New York: Basic Books, 1984).

9. The typical examples of these collaboration economies are found in industrial districts in Italy and Germany. Ibid. Also see Michael H. Best, *The New Competition* (Cambridge, MA: Harvard University Press, 1990).

10. N. V. Schaefer and J. A. Roy, eds., "Cluster Power: Business Networks" (Synopsis of a conference held November 18-20, 1993), Centre for International Marketing and Entrepreneurship, University of New Brunswick, 1994.

11. M. J. Enright, "The Determinants of Geographic Concentration in Industry," Working Paper 93-052, Division of Research, Harvard Business School, 1993; Porter, *The Competitive Advantage*; B. Harrison, "Industrial Districts: Old Wine in New Bottles?" *Regional Studies* 26 (1992): 460-83.

12. M. J. Waits, K. Kahalley, and R. Heffernon, "Organizing for Economic Development. New Realities Call for New Rules," *Public Administration Review* 52 (November/December 1992): 612-16.

13. Commonwealth of Massachusetts, *Choosing to Compete: A Statewide Strategy for Job Creation and Economic Growth* (Boston: Executive Office of Economic Affairs, 1993).

14. M. E. Porter (in Collaboration with Monitor Company, Inc.), *The Competitive Advantage of Massachusetts* (Cambridge, MA: Harvard Business School, 1991).

15. Ibid., p. 14.

16. E. Glaeser, H. Kallal, J. Scheinkman, and A. Shleifer, "Growth in Cities," *Journal of Political Economy* 100 (1992): 1126-52.

17. E. Sternberg, "The Sectoral Cluster in Economic Development Policy: Lessons from Rochester and Buffalo, New York," *Economic Development Quarterly* 5 (November 1991): 342-56.

18. Ibid. See also A. Markusen, "Neither Ore, nor Coal, nor Markets: A Policy-Oriented View of Steel Sites in the USA," *Regional Studies* 20 (1986): 449-61.

19. Sternberg, "The Sectoral Cluster"; M. Storper and S. Christopherson, "Flexible Specialization and Regional Industrial Agglomeration: The Case of the U.S. Motion Picture Industry," *Annals of the Association of American Geographers* 77 (1987):104-17; Piore and Sabel, *The Second Industrial Divide;* J. R. D'Cruz and A. M. Rugman, "Business Networks, Telecommunications, and International Competitiveness," mimeo, Faculty of Management, University of Toronto, October 1993; R. C. Young, J. D. Francis, and C. H. Young, "Flexibility in Small Manufacturing Firms and Regional Industrial Formations," *Regional Studies* 28 (1994): 27-38.

20. Young et al., "Flexibility in Small Manufacturing Firms"; R. B. Reich, *The Work of Nations: Preparing Ourselves for 21st-Century Capitalism* (New York: Knopf, 1991); D'Cruz and Rugman "Business Networks"; W. Lazonick, "Industrial Clusters and Global Webs: Organizational Capabilities in the American Economy," *Journal of Industrial and Corporate Change* 2 (1993): 93-105; Piore and Sabel, *The Second Industrial Divide.*

21. Ibid. Also see Best, *The New Competition,* and W. W. Powell, "Neither Market nor Hierarchy: Network Forms of Organization," in *Research in Organizational Behavior* (Greenwich, CT: JAI, 1990), pp. 295-336; Harrison, "Industrial Districts."

22. Commission of the European Communities, "Report on the Competitiveness of the European Textile and Clothing Industry," mimeo, COM(93)525, Brussels, October 27, 1993: International Labor Organization, *Textiles Committee General Report* (Geneva: ILO, 199 1).

23. F. W. Abernathy, "The U.S. Apparel Industry: Response to Global Markets" (Paper presented at the meetings of the Eastern Economic Association, March 20, 1993).

24. J. T. Dunlop and D. Weil, "Labor Productivity and Competitiveness: The Lessons of Apparel," mimeo, Harvard University/Boston University, November 1992; J. T. Dunlop and D. Weil, "Human Resource Innovations in the Apparel Industry: An Industrial Relations System Perspective," mimeo, Harvard University/Boston University, November 1992.

25. Commission of the European Communities, "Report on the Competitiveness."

26. E. M. Hoover and R. Vernon, *Anatomy of a Metropolis* (Cambridge, MA: Harvard University Press, 1959).

27. Best, *The New Competition.*

28. Dunlop and Weil, "Labor Productivity and Competitiveness"; Dunlop and Weil, "Human Resource Innovations."

29. Young et al., "Flexibility in Small Manufacturing Firms"; Enright, "The Determinants of Geographic Concentration; A. Kaufman, R. Gittell, M. Merenda, W. Naumes, and C. Wood, "Porter's Model for Geographic Competitive Advantage: The Case of New Hampshire," *Economic Development Quarterly* 8 (February 1994): 43-66.

30. P. Hall and A. Markusen, *Silicon Landscapes* (Boston: Allen & Unwin, 1985); Markusen, "Neither Ore, nor Coal, nor Markets."

31. Doeringer and Terkla, "Japanese Direct Investment"; Doeringer et al., "Is Japanese Investment"; Kangtsung Chang, "Japan's Direct Manufacturing Investment in the United States," *The Professional Geographer* 41 (August 1989): 315-28; A. Mair, "New Growth Poles? Just-in-Time Manufacturing and Local Economic Development Strategy," *Regional Studies* 27 (1993): 207-21; A. Mair, R. Florida, and M. Kenney, "The New Geography of Automobile Production: Japanese Transplants in North America," *Economic Geography,* October 1988, pp. 352-73; M. Kenney and R. Florida, *Beyond Mass Production: The Japanese System and Its Transfer to the U.S.* (New York: Oxford University Press, 1993).

32. Another example of the clustering of firms associated with both JIT and niche market characteristics is provided by Young et al. in their recent study of small manufacturing firms in two regions in New York State (see Young et al., "Flexibility in Small Manufacturing Firms"). They find that the majority of the firms in both regions were products of outsourcing from much larger regional manufacturing firms. Although many firms continued to supply their parent firms with customized products, they had also used this niche to expand their customer bases to include firms outside the region.

33. P. B. Doeringer, and D. G. Terkla, "Turning around Local Economies: Managerial Strategies and Community Assets,": *Journal of Policy Analysis and Management* 9 (1990): 487-506.

34. Timothy J. Bartik, *Who Benefits from State and Local Economic Development Policies?* (Kalamazoo, MI: W. E. Upjohn Institute for Employment Research, 1991); D. G. Terkla and P. B. Doeringer, "Explaining Variations in Employment Growth: Structural and Cyclical Change among States and Local Areas," *Journal of Urban Economics* 29 (1991): 329-48.

35. R. Gittell and P. Flynn, "Looking Forward: Massachusetts Business Climate from a Comparative and Historical Perspective," Report to the Massachusetts Industrial Finance Agency, November 1993.

36. P. B. Doeringer and M. J. Piore, *Internal Labor Markets and Manpower Analysis* (Lexington, MA: D. C. Heath, 1971); Doeringer et al., *Invisible Factors*; Doeringer et al., "Is Japanese Investment."

37. Krugman, *Geography and Trade.*

38. Sternberg, "The Sectoral Cluster."

39. Doeringer et al., *Invisible Factors.*

40. Ibid.

41. Doeringer and Terkla, "Japanese Direct Investment."

42. Sternberg, "The Sectoral Cluster."

43. Bartik, *Who Benefits.*

44. P. B. Doeringer, "Global Value Chains and Regional Economic Development" (Paper presented to International Symposium on Globalization and Competitiveness, Agadir, Morocco, May 8-11, 1995).

45. Doeringer et al., *Invisible Factors*; Doeringer and Terkla, "Japanese Direct Investment; Bartik, *Who Benefits*; Doeringer et al., "Is Japanese Investment."

46. R. Gittell *Renewing Cities* (Princeton, NJ: Princeton University Press, 1992).

47. Ibid.

48. Doeringer and Terkla, "Japanese Direct Investment"; P. B. Doeringer, "Georgia Power Company C: Japanese Business Perspectives on Georgia," mimeo, John F. Kennedy School of Government Case Program, Harvard University, 1992; R. Gittell, "Georgia Power Company B: Local Government Builds a Partnership for Economic Development," mimeo, John F. Kennedy School of Government Case Program, Harvard University, 1992.

49. Commonwealth of Massachusetts, *Choosing to Compete.*

50. P. W. Roberts and D. B. Waldron, "An Analysis of Clustering in the Alberta Economy," mimeo, Department of Organization Analysis, Faculty of Business, University of Alberta, September 1992.

51. Young et al., "Flexibility in Small Manufacturing Firms."

52. L. Riddle, "Real Estate," *New York Times,* December 8, 1993, p. D20.

53. P. M. Flynn *Technology Life Cycles and Human Resources* (Landham, MD: University Press of America, 1993); Vernon, *Metropolis 1985.*

54. Flynn, *Technology Life Cycles.*

55. Porter, *The Competitive Advantage;* Lazonick, "In Clusters and Global Webs."

56. Kaufman et al., "Porter's Model for Geographic Competitive Advantage."

57. Doeringer et al., *Invisible Factors.*

58. Gittell, *Renewing Cities.*

59. T. A. Kochan, H. C. Katz, and R. B. McKersie, *The Transformation of American Industrial Relations* (New York: Basic Books, 1986).

60. Gittell and Flynn, "Looking Forward."

PART III

Neighborhoods and Social Equity

Chapter 9

The Linkage Between Regional and Neighborhood Development

Wim Wiewel, Bridget Brown, and Marya Morris

❀❀❀

Over the past decade, much attention has been given to the issue of national and regional development and its relation to economic restructuring. By contrast, there has been relatively little serious theoretical thinking about neighborhood economic development. Regional economic theorists rarely focus on how regional changes play out at the level where people actually experience them. This gap is particularly glaring because hundreds, if not thousands, of neighborhood organizations are presently involved in neighborhood development projects.[1] Such projects are typically conceived and implemented without consideration of the economic trajectory of the region in which a neighborhood is located.

The absence of theories regarding the linkage between the region and its neighborhoods may be due to the fuzziness of the concepts involved. Does *neighborhood economic development* refer to growth and change in the economic structure of a certain place, that is, the development of the neighborhood economy? Or does it mean the access that residents of a certain place have to employment, income, and wealth generally, that is, the economic development of the neighborhood, conceived as residents? Or does it mean both?

The concept of neighborhood economy is fuzzier than that of regional economy. The latter essentially refers to the economic structure and process within a certain geographical area. Although significant trade with other regions is likely to occur, most of the labor force and economic activity occur within the area. If defined similarly to include only the economic structure and process within geographic boundaries, a neighborhood economy would be quite small. But in most cases, the neighborhood economy also implies the labor force participation, occupational distribution, and earning power of neighborhood residents. Conceptually, this may be treated as exports are treated in the regional economy, whereas the employment of nonresidents by business located in the neighborhood can be treated as imports. Although this may work conceptually, it involves such a vast proportion of the total inflow and outflow of re-

Reprinted from *Economic Development Quarterly,* Vol. 3, No. 2, May 1989.

sources as to raise doubts about the reality and integrity of the concept of a "neighborhood economy."

Because of these definitional complexities, this article is not just about the neighborhood economy, but more generally about neighborhood development within the regional context. It explores how and why changes occur in neighborhood economic activities as regional changes occur, how regional changes affect the population of neighborhoods, and how neighborhoods play different roles within the region.

The goal of this analysis is to achieve a better understanding of the economic fate of neighborhoods. We know this depends largely on what happens nationally and regionally, but just as regions have benefited differentially from recent economic changes, so are neighborhoods affected differentially by regional change. By better understanding the linkage between regions and neighborhoods, we can better assess what can be done at a neighborhood level to improve the economic conditions of residents or how specific projects or policies will affect different neighborhoods.

A full discussion of the subject would require a theory of regional development, a theory of neighborhood development, and a theory of the linkage between these two. On none of these three topics is there widespread agreement. Nevertheless, we can at least compare a number of current theories of regional and neighborhood development. These theories contain, more or less explicitly, statements about how the linkage is effectuated. The next section of this article will summarize these theories. The third section will then describe the linkage mechanisms that the theories imply. Because there is little literature that deals explicitly with the connection between regional and neighborhood development, we will offer a framework for approaching this issue that draws on existing literature on a number of related questions. The mechanisms linking the region to the national economy and the nation to the world economy may be relevant by analogy to the linkage we are exploring here.

Finally, in the conclusion, we will draw from this discussion some implications for what a theory of neighborhood development should address and what elements it should contain.

Theories of Regional and Neighborhood Development

To understand how neighborhood economies are linked to the regional economy, it is helpful to review general theories on regional or neighborhood development. Of course, if fully developed, such theories should include clear statements about this linkage. But in reality, it has rarely been the object of careful scrutiny. Nevertheless, the theories imply several alternative conceptions of the relationship.

Regional development theories can be categorized in a variety of ways, but for the present purpose we will distinguish between export-based theories, dependency theories, resource-based and cumulative causation theories, and product/profit cycle theories.

The first three of these also apply to neighborhood development, with the notion of empowerment added to the dependency theories, and a focus on self-sufficiency added to the resource-based theories. A fourth category of theories applicable to neighborhood development includes the filtering and "trickle-down" theories. We will now describe each of these in more detail, focusing on what each implies about the process by which neighborhood economies are linked to the regional economy.

Export-Based Theories

Export-based theories focus on changes in exogenous demand as the driving factor in regional growth. According to these theories, external changes are transmitted to the rest of the economy because of the input-output relations between industries and because of the relation between the basic industries (which engage in export) and the nonbasic industries that are tied to and serve the basic industries and their workers.

A primary assumption underlying most export-based models is the neoclassical economic model. Private actors are driven by competition and the desire for profit maximization; regional disparities reflect differing factor prices, with the market tending to equilibrium. Thus change occurs as factor prices change (because of new technologies or materials, transportation improvements, or productivity changes), and this change is transmitted through shifts in exogenous demand.

At the neighborhood level this model has been applied to analyze the income streams into a neighborhood.[2] In most cases the export of labor is the most significant source of neighborhood income. Export and import metaphors have also been used to describe the neighborhood's relative role in offering retail goods and services to its residents and those of other neighborhoods.[3]

The policy prescriptions emanating from such analyses revolve around substituting imports when the neighborhood is undersupplied relative to the rest of the city, or enhancing and building on a competitive position if the neighborhood appears to be exporting certain goods or services. The export model has also been used to identify industrial sectors in which a metropolitan area appears either to be undersupplied or to have a competitive advantage. Then, specific industries are selected and targeted for neighborhood economic development projects.[4]

In all of these export-based models, the mechanism linking various sectors and geographic levels is the operation of the market. In the long run, public sector intervention may attempt to affect some of the factor prices that determine relative competitiveness, but ultimately it is the market that brings about whatever linkages exist.

Dependency and Empowerment Theories

Dependency theories argue that it is the nature of capitalist development continually to expand the area from which surplus can be extracted while centralizing the surplus value and control over the process. Thus underdevelopment in poor neighborhoods or countries is not simply a result of the backwardness of productive forces or a historical lag in coming into the modern economic system. Rather, underdevelopment is evidence of a full integration into that system; it is the result of the effectively working system of exploitation.[5] Thus inequality and "backwardness" are maintained and even enhanced by the dictates of capitalist accumulation. Several authors within this perspective also expand on the role of the state in this process, and on the extent to which this process is affected by oppositional movements.[6] In moving from analysis to prescription, these theories stress the importance of empowerment, a concept developed during the urban crises of the 1960s and closely linked to the idea of black nationalism.[7] For instance, a recent manual on neighborhood economic development argues that purely economic and business development approaches will never bring about significant changes in the condition of poor people. In order to reduce dependency, the manual insists, neighborhood development must "transcend purely economic issues and enable local residents to increase their capacity to plan and coordinate the way in which their communities are run."[8]

Resource-Based, Cumulative Causation, and Self-Sufficiency Theories

In explaining regional disparities or opportunities for growth, resource-based or "staple" theories[9] focus on the existing stock of resources in a region. This may include raw materials and geography as well as the existing mix of industries and labor skills.[10]

Cumulative causation models add a dynamic element to these theories by focusing on how areas with initially higher levels of development reap further advantage because they are the most favorable place for new development. Myrdal,[11] for instance, argued that economies of scale tend to increase inequalities. Although the diffusion of innovation may spread some growth to poorer regions, the main flow of capital and skilled labor is in the other direction.

Clearly, in these models, the main mechanism is the operation of the market, especially the process of economies of scale. However, to the extent that local elites or community organizations can affect the local stock of capital and other resources, they may be able to improve their area's competitive stance.

At its extreme, this approach may lead to self-sufficiency approaches, which aim to increase local resources to minimize dependence on the outside world. For instance, the notion of poor areas as "internal colonies" suggests solutions similar to those that have been pursued by Third World countries in the process of establishing independence. This includes various economic strategies as well as attempts at separatism, such as the proposal to turn Boston's black Roxbury neighborhood into a separate city called Mandela.[12] Proposals such as this typically rest on analyses that emphasize the leakages from neighborhood economies.[13] Obviously, such an approach would require not only extensive economic changes but also considerable political action.

Product/Profit Cycle Theories

Original product cycle theory argued that products, and the producing industries go through a cycle of initial rapid growth, then stagnate and decline. As production is standardized, it becomes more important to minimize labor costs; thus, over the product cycle, locational requirements change.[14]

Markusen[15] has expanded this theory by including a greater focus on profitability. In this view, regional changes are not caused by changing factor prices or new consumer preferences, but corporate responses to different historical moments of long-term profitability cycles. Older producers need to rationalize and cheapen production, while new producers tend to locate in new areas.

The product/profit cycle consists of the following phases:

(a) In the initial period of product innovation profitability is high, with increasing employment and an emphasis on professional and technical workers.

(b) In the period of continued growth and increased competition, profits are more normal. Production is increased, there are more managerial and sales workers, and production plants are dispersed to areas of low labor costs.

(c) If some firms have been able to create an oligopoly situation, the industry may return to a stage of above-normal profits and stay in its host region. However, the high wages paid by oligopolies will retard the development of new industries and ultimately cause problems; Detroit, Buffalo, and Pittsburgh are cited as examples.

(d) If competition continues or is renewed, profits will continue to decline and force restructuring. This may include plant closings or the virtual disappearance of the industry.

Thus, in this model the process of industrial change, along with corporate decision making in response to these changes, is the main driving mechanism behind regional economic change. There is no direct equivalent to this approach in theories of neighborhood development. The closest equivalents are theories of housing filtering, which also focus on obsolescence cycles and the strategic decisions of owners to reinvest or bail out.

Filtering/Trickle-Down Theories

Anthony Downs has offered the most comprehensive formulation of filtering and trickle-down theory in regard to housing.[16] In this context, filtering and trickle down occurs when there is population and economic growth causing a continual and increased demand for better quality housing. As households trade up, the housing left behind becomes available for lower-status groups. Although this is a market process, government also plays a role in that much of the suburban growth that stimulated this process was made possible by the federal highway program. The real estate finance market with its bias to newer and single family housing similarly stimulates this process.

The process can be changed or reversed, however, when enough owners decide to invest heavily in upgrading. They may do so because of public programs and subsidies, or because of changing consumer preferences, such as the back-to-the-city movement that followed the energy crisis.

Obviously, this approach differs quite radically from the analyses of the dependency theorists, even though both deal with the same phenomena. Nevertheless, both raise similar questions regarding the degree of determinism or level of autonomy that neighborhoods have, whether within a market context or the context of capitalist accumulation. We will address this question again as we deal specifically with the linkage between the neighborhood and regional economy in the next section.

Linkage Processes

As is clear from the preceding section, these theories of regional and neighborhood development suggest several different ways in which events in the neighborhood economy are shaped by the regional economy, and vice versa. Export-based theories, resource-based theories, and filtering theories all ascribe causality to market mechanisms in which firms and households make individual choices, based on the conditions at any one time, which are then transmitted throughout the rest of the economy. The dependency and empowerment theories see the linkage as embedded in the nature of the capitalist accumulation process. However, these theories also focus on the role of the political process in either maintaining or changing the nature of these relationships, as does self-sufficiency theory. In this context, "political process" includes both government action and the attempts of different groups to influence public policy. Finally, product/profit cycle theory adds yet a fourth linkage mechanism in emphasizing the role of industrial change and corporate strategic decision making. Thus the four linkage mechanisms suggested but not elaborated by the theories are: (a) market processes, (b) the capitalist accumulation process, (c) the nature of industrial change and corporate decision making, and (d) the political process.

Clearly, a complete theory of either regional or neighborhood economic development ought to contain a theory of linkage. Each of the linkage mechanisms suggested, though insufficient by itself, enhances our understanding of how linkage occurs and thus provides an element of a comprehensive theory. In order to identify more clearly the crucial elements, the following section describes the operation of each of the linkage mechanisms in detail.

Market Mechanisms

Implicit in much regional and urban economic analysis is the notion that the market operates to link places together. For instance, Hoover and Vernon[17] see neighborhood evolution occurring in terms of expanding concentric zones, pushing out from a growing central business core. Shifts in population and job location are the dynamic forces that propel a metropolitan area through these evolutionary changes. Land prices, and the free choices of firms and households, are the operative mechanisms.

The more recent version of this model, as formulated by Downs, has been described here. In his view, the decline of inner-city and "rust-belt" neighborhoods is a result of a natural shift by residential, industrial, and commercial consumers from obsolete to more modern and efficient facilities. This process is aggravated somewhat by an oversupply of newer units, which customers prefer, thereby leaving older central city housing to become obsolete and dilapidated, often before its time. As demand declines, so do rents, and eventually owners cannot maintain or upgrade their units. Even under these conditions, though, these neighborhoods may be recycled if the market prevails and economic growth in the area as a whole continues.

In this view, the probability that any given neighborhood will experience revitalization is a function of a large number of supply and demand factors at the metropolitan and neighborhood levels.[18] Demand factors at the metropolitan level include a strong central business district with growing employment, rising real incomes, rapid in-migration of nonpoor households, and no in-migration of poor households. The supply factors include long commuting times to the CBD, restrictive suburban development, rapid increases in suburban housing prices, and loose condominium conversion regulations.

At the neighborhood level, demand side factors include proximity to local resources, good public transportation, and proximity to other revitalized neighborhoods. Supply factors include single family housing, architecturally significant structures, supportive financial institutions, and local government support for upgrading public services.

In regard to the issue of displacement, it is argued that the revitalization that is occurring is a result of macroeconomic trends in housing markets as well as large-scale demographic shifts and fundamental changes in American life-styles, which are bringing more people back to the city.[19] Sumka warns that government intervention in any market must be carefully considered for both its direct effects and its unanticipated consequences. He contends that although displacement may cause serious problems in some neighborhoods, there is little evidence to suggest that this is widespread.

Varady's[20] evaluation of the federal Urban Homesteading Demonstration program is similarly skeptical of government intervention. That program aimed to stabilize neighborhood populations and improve housing conditions by increasing the confidence level of residents. Yet Varady found that demographic and socioeconomic shifts played a more significant role in neighborhood pessimism, decisions to move, and low repair expenditures than did HUD-sponsored home improvements.

Much of the literature on rural development follows a similar market-oriented approach. Most of it focuses on changing demographics and on the changing composition of rural economies as they shift from agricultural-related economies to manufacturing, service, and high-technology industries in direct response to national-level economic restructuring. In this view, the increase in rural population between 1970 and 1980 resulted primarily from improved economic conditions in these areas, as well as from a trend among families to prefer less congested areas.[21] The dramatic expansion of manufacturing activity in rural areas in the last decade is seen as a normal stage of development in a modern nation's space economy. The forces promoting this diffusion include changes in transportation and telecommunications, which reduce the necessity of capital and labor agglomerations, and geographic differences in wage rates.[22] This "modernization" approach is of course quite different from analyses of the same phenomenon emanating from the dependency theory or product/profit cycle approaches.

Even though individual neighborhoods or rural towns cannot change the market forces that affect them, they are not necessarily seen as entirely helpless. Bradshaw and Blakely[23] argue that local policy structures can make a difference in how rural communities cope with the changes in the economy. At its extreme, this view can lead to what might be called the "positive mental attitude" approach to local development, or a belief that small areas can fully shape their own fate if only they want to. A considerable amount of the literature on small-town development, especially that written by consultants, seems to follow this line.[24]

There is a limit, though, to what can be achieved to change market conditions. Downs argues that when conditions are extremely poor in a large area, a government's limited resources will be insufficient to compensate for a lack of demand. Yet, public monies could make a difference if they are concentrated in areas where future housing demand appears probable. He also suggests a "rhetorical revival strategy," in which city officials flatly deny that a decline in housing demand exists in order to dispel any uncertainty among potential and existing homeowners.[25] Goetze similarly calls attention to the use of public relations to affect the market in specific neighborhoods.[26]

Most analysts who focus on the market as the operative mechanism linking small areas to the larger economy are, not surprisingly, more interested in the analysis of these larger economic and demographic shifts than in the details of their localized impacts. The shortcomings of this approach have been critiqued extensively, as laid out elsewhere, in an analysis of market imperfections as welt as in comprehensive attacks on the basic assumptions.[27] These critiques essentially offer alternative explanations, to which we will now turn.

Laws of Capitalist Development

Just as the invisible hand of the market presumably operates through the sovereign choices of firms and households, so are the laws of capitalist development effectuated by particular actors and institutions. However, many analysts are more concerned with formulating the general laws than with specifying how they are implemented. Accordingly, we will start with a general discussion before turning to the specific roles of industry, government, and citizens.

In the Marxian approach, the main determinant of the form urbanization takes is the mode of production, that is, capitalism and its accompanying division of labor. As Allan Scott writes:

> The sociospatial differentiation of urban neighborhoods can only be analyzed in the context of the norms and pressures of commodity-producing society . . . the social geography of the city is above all an outcome of the social division of labour.[28]

In this view, neighborhoods will differ and continue to be differentiated according to the function of their residents and places of employment within the overall system.[29] As David Harvey argues, residential differentiation reflects the class structure, which he defines as having very fine distinctions. For instance, class distinctions arise not just from the division of labor, but also from consumption classes (e.g., "yuppies"), authority relations, social mobility characteristics (e.g., professional groupings), and class consciousness and ideology (e.g., the extent to which a class becomes aware of itself as a class and sets aside internal differentiation).[30] Thus the differences between neighborhoods are not a result of the processes of social ecology as described by the Chicago School sociologists and their current representatives[31] nor of the free choices of consumers. Rather, they are a function of the nature of the capitalist production process.

In this model, change occurs because the capitalist process is a dynamic one, driven by the need for accumulation.[32] As new geographical areas are brought into the orbit of this process, or as new technologies are developed, places may become obsolete or may become reinvigorated. This process has been well documented for communities faced with the loss of their primary industries, such as the steel industry in South Chicago,[33] the milling industry in the Northwest,[34] and many others.[35] Thus poor neighborhoods (or other areas) are not poor because they are relatively underdeveloped and lagging, but because of their particular place in the system. In some cases they may well be entirely dispensable, for which Walker[36] coined the term "lumpengeography." Their only remaining function is to act as a reserve, keeping a check on the cost of doing business elsewhere, and providing a reservoir of land and workers.

The reverse can also occur—incorporation or reincorporation of places previously outside the system. Suburbanization met the demands of the working class for better living conditions, while also opening up new avenues of accumulation.[37] Gentrification or other forms of central city redevelopment similarly bring back into productive use areas that no longer function well for accumulation.[38] For instance, Giloth and Betancur, as well as Prusska-Carroll,[39] describe the change in the Chicago economy from a manufacturing to a corporate control center, and the accompanying pressures to convert industrial land to upscale residences and offices.

But the influences are not only in one direction. Once constituted as communities, these cities, towns, and neighborhoods can also become a political basis for organizing countervailing pressure. The function of residential areas may be to reproduce the social order and the labor force,[40] yet, Harvey argues, communities can also become the locus for "displaced class struggle."[41] The creation of communities that reflect the class structure and serve reproduction needs is a two-edged sword because communities also serve a very basic human need for socialization and support and thus become a basis for organizing against the accumulation process.[42] This happens not only in working-class communities. It can also happen among the bourgeoisie in response to new growth that might threaten its communities. For instance, Molotch[43] describes the opposition by real estate interests against oil exploration off the coast of Santa Barbara.

Scott and Harvey[44] probably offer the most comprehensive statements of the relation between capitalist accumulation and urban differentiation. Harvey ascribes a crucial role to financial institutions and government in regulating the specifics of the urbanization process. In his earlier work Scott seemed to suggest that the process was almost automatic, as when he writes of the "predominantly spontaneous tendency of distinctive neighborhoods to disaggregate out in urban space."[45] More recently, however, he has focused on ways in which technological changes in industry affect locational patterns.[46] Accordingly, we will now turn to what may be seen as a specification of how capitalist development effectuates the linkage between regions and neighborhoods. We will look first at the role of industries and firms, and next at that of the political realm.

Industrial Change and Corporate Decision Making

The classic statement about the importance of industrial structure in shaping urban development is Wilbur Thompson's "Tell me your industries and I will tell your fortune."[47] Behind the obvious meaning of this statement lay the implication that there was relatively little that government or anyone else could do to affect the future. This approach takes several forms.[48] One focuses on markets and transportation costs as they affect industries. With the decline in transportation costs, and their lessened importance for the rapidly growing service industries, these models apply to fewer and fewer cases. Another form of this approach centers on sector studies of particular industries, studies that are better able to take into account the complexities of industrial change. Yet firm decisions are not always made with reference to industrial sectoral considerations, they quite often are made on the basis of corporate considerations only. For instance, Ranney[49] documents several plant closings that had little to do with industry specifics but more with the corporate strategy of multinational parents. Indeed, Friedland argues that whereas cities used to thrive or decline because of the fate of the dominant industry, they now do so because of firms: "If places once connected firms, now dominant firms connect places."[50]

There are several examples of analyses focusing on industrial change and corporate decision making. Noyelle and Noyelle and Stanback[51] analyzed how major postwar changes in the economy have changed the system of cities in the United States. The main changes are the increase in the size of markets, the improvements in transportation and technology, the rise of the large corporation, and the growing importance of government and the nonprofit sector. All but the last of these are related to the nature of industries and firms.

As a consequence of these changes, cities have become sorted into four groups: diversified advanced service centers, specialized advanced service centers, production centers, and consumer-oriented centers. Each of these has several subgroups, leading to distinctions into at least 10 functional types. Given the rise of advanced services and their continuing leading role, the fate of individual cities is directly linked to their place in this hierarchy of cities and their relationship to advanced service industries.

At the regional level, Massey[52] analyzed how restructuring in the electrical engineering and electronics industries affected regions in the United Kingdom. The analysis shows that regional gains or losses in employment are related to the particular restructuring that different industries undergo at different times. More so than Noyelle, Massey places this restructuring squarely within the context of a worldwide crisis of profitability in capitalism, rather than considering it as an autonomous phenomenon.

Other examples of these kinds of analyses are various studies of boomtowns or growth industries, such as Hibbard and Davis's study of logging, Markusen's analysis of the effect of the energy crisis, or Castell's work on the spatial consequences of high technology development.[53] Here too, more detailed studies go down only to the level of cities and regions, such as Saxenian's work on Silicon Valley and Luger's[54] on North Carolina.

A more detailed analysis of the effect of industrial restructuring appears in Scott's studies of the printed circuit boards and women's dress industries in the Los Angeles region.[55] He shows how technology and the nature of the linkages between firms influence the degree of spatial concentration around specific centers. However, he does not extend this analysis to residential location of the companies' employees, nor does he really address the concept of neighborhood other than as the relative clustering of subcontractors or suppliers of the firms in his sample.

The fact that studies such as the ones discussed in this section do not adequately deal with neighborhood impacts is not just due to theoretical oversight, but also to the difficulty in obtaining relevant data. The main data sources, such as the industry censuses, go down only to municipal or county levels. Data bases that can be disaggregated further, such as Dun and Bradstreet data and ES-202 data from state offices of employment security, have only recently become available and contain many inaccuracies when used for small areas. However, some of these data problems could be overcome if some of the reporting requirements were changed.

One could also extend Noyelle and Stanback's and Massey's analyses to neighborhoods by considering both the occupations of residents and the nature of neighborhood businesses, and by using Journey To Work data. Neighborhoods could then be classified similarly to the hierarchy of cities. This would clearly link the fate of neighborhoods to that of industrial sectors, and would clarify how neighborhoods change as the regional industrial structure changes. For instance, this would certainly be true for distinguishing between neighborhoods that have significant finance sector employment and those with manufacturing employment. Even for neighborhoods with high manufacturing employment, there are bound to be differences depending on which industries provide the bulk of the employment. This approach would also address one of the shortcomings of most housing-based analyses, which take demand as largely exogenous. Indeed, some of the literature on manufacturing displacement and gentrification follows this approach, albeit implicitly.[56]

In the most detailed and comprehensive analysis of how an industry affects individual neighborhoods, David Harvey concentrates on the role of the finance industry, assisted by government institutions:

> Financial and governmental institutions . . . function to coordinate national needs (understood in terms of the reproduction of capitalist society and the accumulation of capital) with local activities and decisions. . . . These institutions regulate the dynamic of the urbanization process and also wield their influence in such a way that certain broad patterns in residential differentiation are produced. The creation of distinctive housing submarkets (largely through the mortgage market) improves the efficiency with which institutions can manage the urbanization process. But at the same time it limits the ability of individuals to make choices.[57]

Harvey elaborates this point through an extensive case study of Baltimore, showing how various financial institutions and real estate operators, along with government institutions, created and recreated housing submarkets.

What remains unclear in Harvey's analysis is where these intermediaries get their directions and how rigidly determined the outcomes are. Indeed, he dismisses protest movements against land speculators or particular government programs as ineffective in changing the nature of the process; they merely change the tools that will be used.

Harvey's approach implies that the finance, real estate, and government institutions work exactly in tandem with the accumulation needs of capital. By contrast, Edel[58] suggests that such an approach is too deterministic. He argues that not only are different social structures possible within capitalism, but some are more beneficial to the working-class than others and offer more potential for future gains. Further, he argues, there are such things as "nonreformist reforms,"[59] reforms that do more than simply remove points oil contention and thereby strengthen existing power relations. The main arena in which groups struggle over these possible outcomes in regard to neighborhoods is the realm of politics, to which we will now turn.

The Political Process

There is a long-standing debate among (neo-)Marxists about the relative autonomy of the state, and hence of the political process, vis-à-vis capitalists and the capitalist system.[60] For instance, Mollenkopf[61] argues that the analysis of communities needs to take place on four different levels, each of which is influenced by the others but also has its own autonomy. These levels are (a) the worldwide system of capitalist accumulation, (b) the national and subnational institutional social structures, (c) the domain of political alliance formation and mobilization, and (d) networks of social ties, which are partially but not fully determined by the first three levels.

Mollenkopf goes on to state that political alliances rarely follow clear-cut economic distinctions. Rather, they tend to stem from shared race, ethnicity, geography, or ideology, or are organized around specific agencies and public programs. Similarly, after an analysis of urban social movements, Castells rejects the idea that the spatial form of the city or struggles over its shape can be understood in terms of the logic of capitalism.[62] He argues that they are not determined directly by economic conditions nor by the relations between economic classes. Harvey regrets Castells's "apparent defection from the Marxist fold" and attempts to show how "relatively autonomous" urban politics can arise, which is not only compatible with but also necessary to the processes of capital accumulation.[63] Central to Harvey's formulation is the idea that capitalism is a continuous innovating mode of production, requiring constant renewal not only in its production processes but also in the accompanying structures of governance and reproduction. The more open the society, the more innovative it will be. Ultimately, however, only those innovations will survive that are compatible with the needs of accumulation.

This approach has been criticized for its basic functionalist nature: Everything that exists by definition serves to maintain the system or it would not exist (for long, in any event).[64] No true reforms, short of revolution, are possible. It may be more logical to argue that, at the very least, any given problem created, by the economic system may have multiple solutions.[65] For instance, some have seen suburbanization as only one of several possible solutions to the particular crisis of the times.[66] If this viewpoint is correct, then issues are worth fighting over because some solutions may be more or less advantageous to different groupings. Thus given the autonomy at the political level we need to have a clear analysis of its operations.

The classical statements about the operation of the political process were framed by Hunter and Dahl as a debate between elite control and pluralism.[67] Since then, others have argued that during the period Hunter and Dahl were writing about, there was in fact relatively little conflict occurring in the policy arena. Government policy was shaped, in part, by the close relationship between big business and labor. Government provided direct aid to business in the form of subsidies and through a tax system that reduced risk and enhanced profitability. It also helped diverse business interests to act collectively, and when pressure from citizen groups in the 1960s forced chances, the government mediated conflicts.

Numerous analyses have shown how government programs ostensibly directed to the public good in fact served the interests of economic growth. Federal clearance of city "slums" to make way for new development did not provide housing for people displaced in the process.[68] Housing policies later designed to provide for the displaced proved to benefit developers more than poor families.[69] Similarly, financial incentives to attract business and industry into an area often benefited the incoming firm more than the region or the neighborhood.[70]

Similarly, the roots of the gentrification issue can be traced to two other policies that reinforced the bond between government and business. The federal Highway Act encouraged urban sprawl by making access to the suburbs easier. Mass transit improvements that followed were designed to provide the suburban workers with transportation to central city jobs. The first policy aided the drain of residents and revenue from the city while providing private developers with vast opportunities for profits. The mass transit policy then robbed city residents of needed jobs.[71] Many city neighborhoods began declining until values dropped to the point that redevelopment was feasible. Government inaction in the face of this process has had the same effect as did active renewal of declining urban property in the 1960s and 1970s.[72] Low- and moderate-income families are being displaced as the housing stock in their neighborhoods is improved, and the cost of housing rises accordingly.[73]

Thus, when government cooperates with capital, policies favoring economic growth are the inevitable result. Often, growth in one region means decline in another as each region, city, or neighborhood competes with others.[74] But areas that do not strive to grow will, by default, lose to others that do.

Despite the apparent inexorability of the "growth machine," citizen groups are able to insert themselves in the policymaking process and even control it for a limited time or in specific situations. Symbolically, such action may be cast as a fight between the interests of "downtown" and those of "the neighborhoods."[75] Participation in, or control over, the political process can be achieved through political and bureaucratic enfranchisement.[76] Bureaucratic enfranchisement, in the form of advisory capacity in the government decision-making process, began in the late 1960s as a result of citizen movements and has taken a wide variety of forms, ranging from neighborhood "city halls" to community planning boards." The political power gained through the electoral process is much more significant, but without bureaucratic enfranchisement it is not well directed. In order to achieve strategic control over the planning process; citizens must have both types of enfranchisement.

For instance, Clavel[78] describes five progressive city governments in which citizen groups played a large role. These groups forced growth-oriented interests to provide significant benefits to local residents in exchange for the right to conduct business in the city. In these cities, then, developer gains were linked to citizen needs. Citizens won the power to achieve this through organizational and mobilization efforts, often targeted at a specific issue, but legitimized through the electoral process.

One large development project that was changed in this way was the proposed 1992 Chicago World's Fair. Giloth and Shlay[79] argue that the real intent behind this extravaganza on the shore of Lake Michigan, just south of downtown, was the opening up for real estate development of the poor South Side. As such, the plan was yet another sequel to the Chicago 21 plan from 1968, which laid out the "growth machine" blueprint for Chicago's redevelopment. However, a coalition of citizen groups, aided by the antagonism between Chicago and "downstate," was able to raise enough questions about the viability of the Fair to sink the project when the state legislature had to appropriate funds for it.[80] The role of local government was telling in that it shifted from full-blown support for the Fair under Mayor Byrne's administration to official neutrality under Harold Washington, presumably so as not to anger business supporters of the project.

Yet even when efforts are so mobilized, government may be quite restricted in its ability to change the course of economic development. For instance, an analysis of the administration of Harold Washington in Chicago concludes that although government made considerable changes in its own hiring and purchasing practices and opened up the policymaking process, it was far less successful in changing the nature or location of large private development projects.[81]

One of the best analyses of the link between metropolitan and neighborhood development and the role of political struggle is Gaston and Kennedy's[82] analysis of Roxbury. The low-income, black Roxbury neighborhood was not initially part of Boston's economic revival (it was part of it only in a negative sense) that started in the late 1970s: Disinvestment in Roxbury contributed to the reinvestment elsewhere. However, after playing a key role in the gentrification of the adjacent South End neighborhood, the Boston Redevelopment Authority developed plans for revitalizing Roxbury, plans involving considerable displacement and no neighborhood control. Residents organized and obtained a moratorium on continued development on the argument that the planning process had been undemocratic. Their analysis provides an example of successful resistance in a neighborhood that had been made part of the "lumpengeography" and became subject to redevelopment pressures when its land was needed to accommodate the growing downtown.

There are many other examples of organizations in individual neighborhoods or coalitions significantly affecting public policy and economic conditions in their area. Although this kind of citizen action has often been described as largely negative,[83] and indeed first emerged in opposition to urban renewal and highway construction, it has also produced many constructive proposals and many new programs. Probably the most significant examples are the Home Mortgage Disclosure Act and the Community Reinvestment Act, which were passed in response to a national organizing campaign led by National People's Action. Smaller, local efforts since then have led to the establishment of local lending pools aimed at neighborhood reinvestment, programs requiring the hiring of neighborhood residents, and a host of other state, municipal, and neighborhood programs.[84] But there are limits to the success of such efforts. For example, thanks to citizen action, Boston's "linked development" program required real estate developers to set up neighborhood projects in exchange for the right to undertake downtown projects. Yet the program was used to extend growth-oriented development beyond downtown and into Roxbury. The nature of the proposed development—high-rise office space and upscale commercial and residential space—was a direct subversion of the idea behind linked development that is explicitly justified on the grounds that new development downtown imposes costs in the form of increased demand and higher prices for housing or services.

Although linked development programs have been passed in cities that were extremely attractive to real estate developers, such as Boston, San Francisco, and Santa Monica, they have been fought and resisted elsewhere. In Chicago, no other issue galvanized the business and real estate development community as much as the proposal for a tax on new leases that would be used to finance neighborhood development.[85] Thus the experience with this policy suggests that citizens' groups or local government are limited in their ability to exact a price from the "growth machine." The Boston and Roxbury experience also shows why bureaucratic enfranchisement—the ability to oversee the actual implementation of programs—is so important, lest the intent of programs be subverted.

Several recent analyses of public sector economic development efforts have come to similar conclusions about the limits of what can be achieved. Robinson[86] distinguishes between two economic development approaches: the corporate-center approach, essentially consisting of "growth machine" policies; and "alternative economic development," emphasizing jobs for low-income residents, neighborhood development, and other redistributive issues. She argues that the mere presence of the business sector is enough to exert influence, whereas minority and low-income groups need to take power through the electoral process in order to exert influence. However, based on a survey of 141 cities, she concludes that *all* cities use the specific strategies of the corporate-center approach, and some cities (where minorities or low-income groups exert significant political power) *also* use alternative strategies.

Similarly, in an analysis of 40 cases, Wiewel and Siegel[87] found that publicly sponsored industry task forces, despite varying ideological and strategic emphases, tended to recommend very similar programs. These authors suggest that the possible realm for political discourse about economic development in American cities is so narrow that similar solutions emerge regardless of initial intentions.

Thus it is clear why the debate about the autonomy of the state continues. Although many probusiness projects have been halted because of citizen action, there is also considerable evidence that even with substantial citizen involvement, policy outcomes usually reflect the business agenda. Clearly though, the policy arena is a central focus for struggle over the linkage between regional and urban economies on the one hand, and specific neighborhoods on the other.

Conclusion

At the beginning of this article, we reviewed theories of regional and neighborhood development and pointed out how these theories implied various notions about the linkage between regions and neighborhoods. The subsequent discussion has made clear that linkage mechanisms should be an explicit part of a theory of regional and neighborhood development, rather than being treated separately. What, then, are the central elements of such a theory?

Although we may disagree with Castell's flat rejection of Marxian categories to analyze urban space, we agree with his insistence that deterministic theories and one-step explanations are insufficient to link capitalist accumulation and particular urban outcomes.[88] Even if cities and neighborhoods ultimately are shaped to serve the interests of accumulation and reproduction of the existing system, the system contains a great deal of inertia and slack. Neighborhoods usually do not change very rapidly, and the specifics of historical conditions may enable a variety of possible outcomes. Theory can develop key indicators and suggest probable results, but it will not provide full explanations or rigid predictions.

The complexity of the political process makes specificity and detail even more essential. If the urban system reflects the fine-grained class structure as discussed by Harvey,[89] then the classical categories of "capital" and "labor" include so many competing groups as to make the political process look remarkably like political pluralism. Thus the political process clearly matters and can be analyzed using categories derived from pluralism, interest-group politics, and social-movement theories. Such analysis should include an awareness of the structural basis of competing groups, whether in the production system, geographic or ethnic roots, or the political process itself.

A new element that needs to be added, especially in regard to neighborhoods, is the linkage with the process of industrial change. This should take several forms. An analysis of the occupations and places of work of residents can serve to connect the process of industrial change directly to changes in the economic fortunes of residents. Furthermore, this analysis should be expanded to develop characterizations of neighborhoods in regard to their functional role within the regional economic system. This should include the role in regard to production as well as in regard to consumption and reproduction. Such an approach would tie together one of the more promising strands in regional research—the focus on industrial restructuring—with the analysis of housing, the traditional mainstay of neighborhood research.

Finally, there remain some other factors that were not discussed in the literature review but that should be included here. The looseness of the coupling between the accumulation process and urban form is heightened by the existence of cultural and technological factors, which, although not independent, cannot be wholly subsumed under economic and political categories. For instance, the particular demographic makeup of the United States and Western Europe can be explained as a result of depression, war, and technological developments related to birth control. Yet it exerts its own independent influence in regard to total housing demand and housing preferences, the composition of the labor force, and neighborhood crime levels. It is often difficult to include these factors as an explicit component of theory, but they cannot be ignored when one seeks to explain specific urban outcomes.

Thus what is needed is a multilevel analytical model specifying how communities are shaped and what their role is in shaping the urban environment and economic system. This model would take into account the global political economy, but would connect this to the microlevel of urban politics and neighborhood change without falling into a deterministic or functionalist trap. It would allow for considerable autonomy at each level, and a multiplicity of possible outcomes. Struggle over outcomes, rather than a predetermined linkage, would be the focus of analysis.

Clearly, we need much more research before we understand this process well. The first task is to develop an understanding and typology of neighborhoods according to their role in the regional economy. This should be elaborated by case studies of the process of neighborhood development, studies that take into account local and regional political and economic factors. Many of the existing case studies (with the good exceptions noted here) either ignore these factors altogether or engage in rote acknowledgments of their importance without detailed analysis.

We also need comparative studies of how different neighborhoods in different regional economies have fared, so we can sort out the importance of regional factors.

All this work is important not just from a theoretical point of view, but for the purposes of practice as well. Economic development planners and neighborhood activists need a better sense of how their activities at different levels are linked, and to what extent their work serves neighborhood interests or merely consists of "growth machine"-oriented projects at the neighborhood level. At present, all development projects undertaken by not-for-profit organizations in a neighborhood tend to be characterized as community-oriented.[90] Yet many of these projects do not serve the needs of present residents, or do so without contributing to political enfranchisement.[91] A more thorough understanding of the nature of neighborhood economies, and their link to their surrounding area, will help us to analyze better the real impact of specific projects and strategic directions.

Notes

1. Neal R. Peirce and Carol Steinbach, *Corrective Capitalism: The Rise of America's Community Development Corporations* (New York: Ford Foundation, 1987).

2. Bennett Harrison, "Ghetto Economic Development Survey: A Review of the Literature," *Journal of Economic Literature* (March 1974): pp. 1-37; and Richard L. Schaffer, *Income Flows in Urban Poverty Areas* (Lexington, MA: Lexington Books, 1973).

3. Center for Neighborhood Technology, *Working Neighborhoods: Taking Charge of Your Local Economy* (Chicago: Center for Neighborhood Technology, 1986).

4. Wim Wiewel, Lynn McCormick, and William Howard, *Community Economic Development Strategies: A Manual for Local Action* (Chicago: Center for Urban Economic Development, University of Illinois at Chicago, 1987), especially chap. 9.

5. Candace Kim Edel, Matthew Edel, Kenneth Fox, Ann Markusen, Peter Meyer, and David Vail, "Uneven Regional Development: An Introduction to This Issue," *Review of Radical Political Economics* 10, no. 2 (Fall 1978): 1-12; and William W. Goldsmith, "Marxism and Regional Policy: An Introduction," *Review of Radical Political Economics* 10, no. 2 (Fall 1978): 13-17.

6. David Harvey, *The Urbanization of Capital. Studies in the History of Capitalist Urbanization* (Baltimore: Johns Hopkins 1985); Carla Robinson, "Economic Development Policy Approaches of City Administrations" (Paper delivered at the Annual Conference of the Association of Collegiate Schools of Planning, November 1987); and Joe R. Feagin and Michael Peter Smith, "Cities and the New International Division of Labor: An Overview," in *The Capitalist City: Global Restructuring and Community Politics,* ed. Michael Peter Smith and Joe R. Feagin (Oxford: Basil Blackwell, 1987), pp. 3-34.

7. William Goldsmith, "The Ghetto as a Resource for Black America," *Journal of the American Institute of Planners* 40 (1974): 17-30.

8. A recent manual distinguishes between business development and community empowerment approaches. See Wiewel et al., *Community Economic Development Strategies.*

9. Melville Watkins, "A Staple Theory of Economic Growth," *Canadian Journal of Economic and Political Science* (May 1963).

10. Wilbur Thompson, "Policy-Based Analysis for Local Economic Development," *Economic Development Quarterly* 2, no. 3 (August 1987): 203-13.

11. Gunnar Myrdal, *Economic Theory and Under-Developed Regions* (London: Duckworth, 1957).

12. Marie Kennedy and Chris Tilly, "Secession and the Struggle for Community Control in Boston: A City Called Mandela," *North Star,* Spring 1987, pp. 12-18. See also Harrison, "Ghetto Economic Development," and Joseph Persky, "Is the South a Colony?" *Review of Radical Political Economics* (Winter 1972).

13. Center for Neighborhood Technology, *Working Neighborhoods.*

14. Robert Kraushaar and Marshall Feldman, "Industrial Restructuring and the Limits of Industry Data," *Regional Studies* (January 1989).

15. Ann Markusen, *Profit Cycles, Oligopoly, and Regional Development* (Cambridge: MIT, 1984).

16. Anthony Downs, "Key Relationships Between Urban Development and Neighborhood Change," *Journal of the American Planning Association* 45, no. 4 (October 1979): 462-572.

17. Edgar Hoover and Raymond Vernon, *Anatomy of a Metropolis* (Cambridge: Harvard University Press, 1959).

18. Anthony Downs, *Planning with Neighborhoods* (Washington, DC: Brookings Institution, 1981).

19. Howard Sumka, "Neighborhood Revitalization and Displacement, A Review of the Evidence," *Journal of the American Planning Association* 45, no. 4 (October 1979): 480-87; and Shirley Bradway Laska and Daphne Spain, "Urban Policy and Planning in the Wake of Gentrification," *Journal of the American Planning Association* 45, no. 4 (October 1979): 523-31.

20. David Varady, "Neighborhood Confidence: A Critical Factor in Neighborhood Revitalization?" *Environment and Behavior* 18, no. 4 (July 1986): 480-501. It should be pointed out that the evaluation was conducted after the program had been in operation for only two years.

21. R. Lonsdale and H. Seyler, eds., *Nonmetropolitan Industrialization* (Washington: U. H. Winston, 1979); Kenneth Johnson, *The Impact of Population Change on Business Activity in Rural America* (Boulder, CO: Westview Press, 1985); and Jim Schwab, "Small Town, Big Dreams," *Planning* 52, no. 11 (November 1986): 4-9.

22. R. Lonsdale and H. Seyler, *Nonmetropolitan Industrialization.*

23. Ted K. Bradshaw and Edward Blakely, *Rural Communities in Advanced Industrial Society* (New York: Praeger, 1979).

24. M. Ross Boyle, *Developing Strategies for Economic Stability and Growth* (Washington, DC: National Council on Urban Economic Development, 1987). See also Raymond Lenzi and Bruce H. Murray, eds., *Downtown Revitalization & Small City Development* (Ames: Iowa State University, 1987).

25. Downs, "Key Relationships."

26. Ralph Goetze, *Building Neighborhood Confidence* (Cambridge, MA: Ballinger, 1976).

27. Alfred J. Watkins, *The Practice of Urban Economics* (Beverly Hills: Sage, 1980); Manuel Castells, *The Economic Crisis and American Society* (Princeton: Princeton University, 1980); and Allen Scott, *The Urban Land Nexus and the State* (London: Pion, 1980).

28. Scott, The *Urban Land Nexus,* p. iii.

29. Richard A. Walker, "Two Sources of Uneven Development Under Advanced Capitalism: Spatial Differentiation and Capital Mobility," *Review of Radical Political Economics* 10, no. 3 (Fall 1978): 28-37.

30. Harvey, *The Urbanization of Capital*

31. Brian Berry and John Kasarda, *Contemporary Urban Ecology* (New York: Macmillan, 1977).

32. For a full explanation see David Harvey, *The Urbanization of Capital;* and Allan Scott, *The Urban Land Nexus.*

33. Task Force on Steel and Southeast Chicago, *Building on the Basics: The Final Report of the Mayor's Task Force on Steel and Southeast Chicago* (Chicago: City of Chicago, 1986); and David Bensman and Roberta Lynch, *Rusted Dreams: Hard Times in a Steel Community* (New York: McGraw-Hill, 1987).

34. Michael Hibbard and Lori Davis, "When the Going Gets Tough: Economic Reality and the Cultural Myths of Small-Town America," *Journal of the American Planning Association* 52, no. 4 (Autumn 1986): 419-27.

35. Terry F. Buss and Steven F. Redburn, "Plant Closings: Impacts and Responses," *Economic Development Quarterly* 1, no. 2 (May 1987): 170-77.

36. Walker, "Two Sources of Uneven Development."

37. Richard Walker, "The Suburban Solution: Urban Geography and Urban Reform in the Capitalist Development of the United States" (Ph.D. diss., Johns Hopkins University, Baltimore, 1987).

38. Chester Hartman, "Comment on 'Neighborhood Revitalization and Displacement: A Review of the Evidence,'" *Journal of the American Planning Association* 45, no. 4 (October 1979): 488-91.

39. Robert Giloth and John Betancur, "Where Downtown Meets Neighborhood: Industrial Displacement in Chicago, 1978-1987," *Journal of the American Planning Association* 54 (1988): 279-90; and Marika Pruska-Carroll, *The Printers Row: A Case of Industrial Displacement* (Chicago: Center for Urban Economic Development, University of Illinois at Chicago, 1987).

40. Scott, *The Urban Land Nexus*; and Harvey, *The Urbanization of Capital.*

41. David Harvey, "The Urban Process Under Capitalism: A Framework for Analysis," in *Urbanization and Urban Planning in Capitalist Society,* ed. Michael Dear and Allan Scott (London and New York: Methuen, 1981), pp. 91-122.

42. John Mollenkopf, "Community and Accumulation," in Dear and Scott, pp. 319-338; and Harry C. Boyte, *The Backyard Revolution* (Philadelphia: Temple University Press, 1981).

43. Harvey Molotch, "Capital and Neighborhood in the United States: Some Conceptual Links," *Urban Affairs Quarterly* 14 (1979): 289-312.

44. Harvey, *The Urbanization of Capital*; and Scott, *The Urban Land Nexus.*

45. Scott, ibid., pp. 125.

46. Allen Scott, "Industrial Organization and the Logic of Intra-Metropolitan Location, II: A Case Study of the Printed Circuits Industry in the Greater Los Angeles Region," *Economic Geography* 59, no. 4 (October 1983): 343-67; and Allen Scott, "Industrial Organization and the Logic of Intra-Metropolitan Location, III: A Case Study of the Women's Dress Industry in the Greater Los Angeles Region," *Economic Geography* 60, no. 1 (January 1984): 3-27.

47. Wilbur Thompson, "Internal and External Factors in the Development of Urban Economies," in *Issues in Urban*

Economics, ed. Harvey Perloff and Lowden Wingo, Jr. (Washington: Johns Hopkins, 1968), p. 46.

48. Kraushaar and Feldman, "Industrial Restructuring."

49. David Ranney, "Manufacturing Job Loss and Early Warning Indicators," *Journal of Planning Literature* (Winter 1988): 22-35.

50. Roger Friedland, "The Politics of Profit and the Geography of Growth," *Urban Affairs Quarterly* (September 1983): 41-54.

51. Thierry Noyelle, "The Rise of Advanced Service: Some Implications for Economic Development in U.S. Cities," *Journal of the American Planning Association* 49, no. 3 (Spring 1983): 280-90; and Thierry Noyelle and Thomas Stanback, *The Economic Transformation of American Cities* (Totowa, NJ: Rowman and Allanheld, 1984).

52. Doreen Massey, "The UK Electrical Engineering and Electronics Industries: The Implications of the Crisis for the Restructuring of Capital and Locational Change," in *Urbanization and Urban Planning in Capitalist Society,* ed. Dear and Scott.

53. Manuel Castells, ed., *High Technology, Space, and Society* (Beverly Hills: Sage, 1985); Hibbard and Davis, "When the Going Gets Tough"; Ann Markusen, "Class, Rent and Sectoral Conflict: Uneven Development in U.S. Boomtowns," *Review of Radical Political Economics* 10, no. 3 (1978): 117.

54. AnnaLee Saxenian, "Silicon Valley and Route 128: Regional Prototypes or Historic Exceptions?" in *High Technologies,* ed. Castells; Michael Luger, "Does North Carolina's High-Tech Development Program Work?" *Journal of the American Planning Association* 50, no. 3 (Summer 1984): 280-89.

55. Scott, "Industrial Organization and the Logic of Intra-Metropolitan Location" (Vols. II and III).

56. Robert Giloth and John Betancur, "Where Downtown Meets Neighborhood." See also, Edward Soja, Rebecca Morales, and G. Wolff, "Urban Restructuring: An Analysis of Social and Spatial Changes in Los Angeles," *Economic Geography* 59, no. 2 (1983): 195-230; and Hill and Bier in this issue.

57. Harvey, *The Urbanization of Capital,* p. 121.

58. Matthew Edel, "Capitalism, Accumulation and the Explanation of Urban Phenomena," in *Urbanization and Urban Planning,* ed. Dear and Scott.

59. Andre Gorz, *Strategy for Labor* (Boston: Beacon Press, 1964).

60. Nicos Poulantzas, *Political Power and Social Classes* (London: New Left Books, 1973); Ralph Miliband, "The State and Revolution," *Monthly Review* 11 (1970): 11; and Erik O. Wright, *Class, Crisis and the State* (London: Verso, 1979).

61. Mollenkopf, "Community and Accumulation," in *Urbanization and Urban Planning,* ed. Dear and Scott. For a similar approach, see Michael P. Smith and Richard Tardanico, "Urban Theory Reconsidered: Production, Reproduction and Collective Action," in *The Capitalist City,* ed. Smith and Feagin.

62. Manuel Castells, *The City and the Grassroots* (Berkeley: University of California Press, 1983).

63. Harvey, *The Urbanization of Capital,* p. 125.

64. Feagin and Smith, "Cities and the New International Division of Labor," in *The Capitalist City* ed. Feagin and Smith.

65. M. Edel, "Capitalism, Accumulation and the Explanation of Urban Phenomena."

66. Richard Walker, *The Suburban Solution*; Matthew Edel, "Rent Theory and Labor Strategy: Marx, George and the Urban Crisis," *Review of Radical Political Economics* 9, no. 4 (Winter): 1-15.

67. Robert A. Dahl, *Who Governs* (New Haven: Yale, 1961); and Floyd Hunter, *Community Power Structures: A Study of Decision Makers* (Chapel Hill: University of North Carolina Press, 1959).

68. Samuel Bowles, David M. Gordon, and Thomas E. Weisskopf *Beyond the Waste Land* (Garden City, NY: Anchor Press/Doubleday, 1983); Susan S. Fainstein, Norman I. Fainstein, Richard Child Hill, Dennis R. Judd, and Michael Peter Smith, *Restructuring the City: The Political Economy of Urban Development* (New York: Longman, 1983).

69. Nancy Kleniewski, "From Industrial to Corporate City: The Role of Urban Renewal," in *Marxism and the Metropolis: New Perspectives in Urban Political Economy,* ed. William Tabb and Larry Sawers (New York: Oxford University Press, 1978); Martin Anderson, *The Federal Bulldozer* (Cambridge: MIT, 1969); Deborah A. Auger, "The Politics of Revitalization in Gentrifying Neighborhoods: The Case of Boston's South End," *Journal of the American Planning Association* 45, no. 4 (October 1979): 515-22.

70. David Ranney, "Manufacturing Job Loss and Early Warning Indicators."

71. Mark I. Gelfand, *A Nation of Cities* (New York: Oxford University Press, 1975); and Alan Lupo, Frank Colcord, and Edmund P. Fowler, *Rites of Way: The Politics of Transportation in Boston and the U.S. Cities* (Boston: Little, Brown, 1971).

72. John R. Logan, "The Disappearance of Communities from National Urban Policy," *Urban Affairs Quarterly* 19, no. 1 (September 1983): 75-90.

73. Thomas J. Lenz, "Neighborhood Development: Issues and Models," *Social Policy* (Spring 1988): 24-30; and Auger, "The Politics of Revitalization."

74. Harvey Molotch and John Logan, *Urban Fortunes: The Political Economy of Place* (Berkeley: University of California Press, 1987); and Robert Goodman, *The Last Entrepreneurs: America's Regional Wars for Jobs and Dollars* (Boston: South End Press, 1979).

75. Robert Giloth and Robert Mier, "Developing Leaders for Community Service," in *Handbook for Community Service,* ed. Susan Lourenco (San Francisco: Jossey-Bass, forthcoming).

76. Fainstein et al., *Restructuring the City.*

77. Patricia Wright, *Neighborhood Planning Boards* (Chicago: Center for Urban Economic Development, University of Illinois at Chicago, 1986).

78. Pierre Clavel *The Progressive City* (New Brunswick, NJ: Rutgers University Press, 1986).

79. Anne Shlay and Robert Giloth, "Whose World Moves the 1992 Fair," *The Neighborhood Works* (May 1984): 18-20.

80. Robert McClory, *The Fall of the Fair* (Chicago: Chicago 1992 Committee, 1986).

81. Wim Wiewel and Nicholas Rieser, "The Limits of Progressive Municipal Economic Development: Job Creation in Chicago, 1979-1983," *Community Development Journal,* forthcoming.

82. Mauricio Gaston and Marie Kennedy, "Capital Investment or Community Development? The Struggle for Land Control by Boston's Black and Latino Community," *Antipode* 19, no. 2 (1987): 178-209.

83. John McCarron, " 'Reform' takes a costly toll." *Chicago Tribune,* August 28, 1988, p. 1.

84. For just a few examples, see Harry Boyte, *The Backyard Revolution* (Philadelphia: Temple, 1981); Lenz, "Neighborhood Development"; and Wiewel et al. *Community Economic Development Strategies.*

85. Wiewel and Rieser, "The Limits of Progressive Municipal Economic Development." On linked development more generally, see W. D. Keating, "Linking Downtown Development to Broader Community Goals: An Analysis of Linkage Policy in Three Cities," *Journal of the American Planning Association* 52, no. 2 (Spring 1986): 133-44.

86. Robinson, "Economic Development Policy Approaches."

87. Wim Wiewel and Wendy Siegel, "Industry Taskforces as Pragmatic Planning: The Effect of Ideology, Sponsorship and Economic Context on Strategy Selection" (Paper delivered at the annual meeting of the Association of Collegiate Schools of Planning, 1987).

88. Castells, *The City and the Grassroots.*

89. Harvey, *Urbanization of Capital.*

90. Peirce and Steinbach, *Corrective Capitalism*; and Richard Taub, *Community Capitalism* (Cambridge, MA: Harvard Business School, 1988).

91. Patricia Wright, " 'Community Capitalism'—Or Just 'Capitalism' ?" *Chicago Enterprise,* July 1988, pp. 12-13.

Chapter 10

Economic Restructuring

Earnings, Occupations, and Housing Values in Cleveland

Edward W. Hill and Thomas Bier

❋❋❋

Employment in the Cleveland Primary Metropolitan Statistical Area (PMSA) peaked in the third quarter of 1979 (1979:3) at nearly 903,000 and then began a slide that did not stop until the first quarter of 1983 (1983:1).[1] (See Figure 10.1.) The region lost 13% of total employment and 14% of its annualized earnings over this time period, amounting to $2.4 billion per year 1982-84 dollars.[2]

A gradual recovery began in 1983. In 1987:2 total employment stood at 871,000. But the composition of employment is very different than it was in 1979. The direction of the change is similar to other rust belt metropolitan areas. One-quarter (68,422) of all manufacturing jobs present in 1979 disappeared by 1983; another 5,000 were lost between 1983:1 and 1987:2. Manufacturing employment accounted for 30.3% of total employment in 1979:1, but in 1987:2 it was 23%. Three sectors increased their shares of local employment over this time period: services from 18.6% to 25.2%; retail trade from 16.2% to 17.6%; and finance, insurance, and real estate from 5.2% to 6.0%.

There are several facts that should be stressed in any quick accounting of the economic performance of Cleveland's economy. The nation experienced two business cycles from 1979 to 1987—Cleveland never saw the recovery between the 1979 and 1981 recessions. The drop in manufacturing employment is part of a long-term secular change in the economy that began before 1979: manufacturing's share of employment dropped 3.4% from 1975:1 to 1979:1. Lower demand for domestic steel, automobiles and auto parts, and capital

---

AUTHORS' NOTE: This research was funded with support from the Urban University Program of the Ohio Board of Regents; a University Technical Assistance Grant from the U.S. Department of Commerce, Economic Development Administration; the Cleveland Foundation; the George Gund Foundation; and BP America. We thank Richard Bingham, Lawrence Keller, Norman Krumholz, Harry Margolis, George Zeller, and two reviewers of the *Quarterly* for their comments and criticisms. Wim Wiewel added to the article with his detailed comments and encouragement. Mark Salling, Daniel Meany, Nell Ann Shelley, and Mark Hoffman helped with data collection and analysis.

Reprinted from *Economic Development Quarterly,* Vol. 3, No. 2, May 1989.

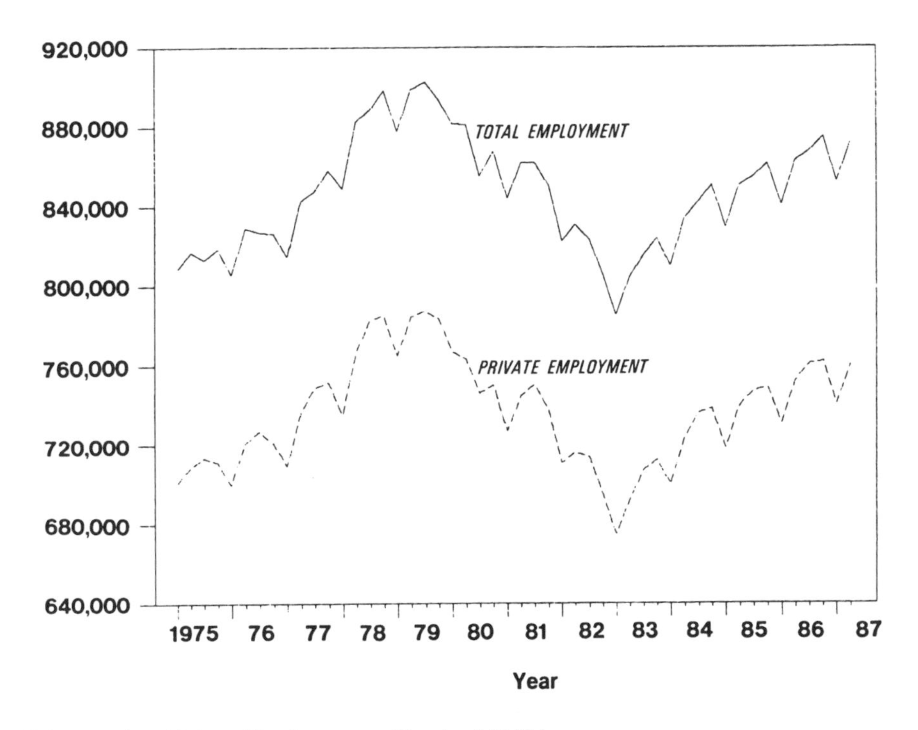

Figure 10.1. Total and Private Employment—Cleveland PMSA

goods became evident in the mid-1970s. The change in demand was a result of three forces. First, there was a long-term drop in demand for these goods. Second, the high value of the dollar resulted in erosion of demand for products manufactured in Cleveland. Third, domestic auto and steel firms made bad strategic decisions about pricing, quality, and design, which resulted in the manufacture of inferior goods, further eroding their market positions.

Structural shifts and cyclical fluctuations in the economy were matched by population losses in both the city of Cleveland and the PMSA. The PMSA lost 8% of its population, or 165,000 people, between 1970 and 1980. It is expected that an additional 33,000 to 42,000 people moved out of the region from 1980 to 1985. The city of Cleveland lost 23.6% of its population from 1970 to 1980. Recent projections indicate that out-migration is continuing. The 1985 population of the city is estimated to be 539,700.[3]

The city's loss of population in and of itself is less troubling than the fact that the out-migration appears to be selective. The remaining population is increasingly poor. Despite the fact that the unemployment rate has been dropping since 1983:1 the number of poor in the city and surrounding county has increased. Nearly 40% of the city's population was below the poverty line in 1987, an increase from 27% in 1980.[4]

A short list of anomalies in Cleveland is instructive: the unemployment rate is down but the poverty rate is up; neighborhood retailers are going out of business but a high-end "festival" retail mall opened up downtown in 1987; *The Cleveland Plain Dealer* reports that there are 5,000 homeless families in Cuyahoga County (which includes Cleveland and 80% of the PMSA's population), yet housing abandonment is accelerating in the inner neighborhoods and rents are extremely low by national standards.

The conclusion we reach in this article is that there is a consistent pattern to these observations. The economy of old-order Cleveland rested on blue-collar occupations. Neighborhoods developed to meet the demands of this class of residents, which in turn formed a complex and long-lasting pattern of social relationships, which were thrown out of kilter when old-order Cleveland suddenly passed away.

The economy of the new-order favors different occupations and classes of labor. In addition, the distribution of earnings in this new economy is very different from that of the old. We will show that not only are earnings more unequal than in the mid-1970s but that they are lower in real terms. The inequality is due to two factors: a reduction in the number of earners in the middle of the distribution and the addition of low-paying jobs.

These ideas suggest that the spatial distribution of winners and losers from 1979 to 1987 was extremely unequal. The impact of structural changes in employment is much more than an individual phenomenon—it has a detrimental impact on specific neighborhoods. Cleveland's neighborhoods are fairly homogeneous in terms of occupation, income, and race. It stands to reason that if the impact of economic restructuring is felt disproportionately by specific groups of workers it will be transmitted to where they live. Some neighborhoods will benefit, others will lose. We use data on poverty and housing prices as indicators of spatial change.

Theory

Most studies of neighborhood change, succession, status, and structure are either longitudinal case studies of a single neighborhood or cross-sectional models of many neighborhoods developed with secondary data from the decennial census. Neither approach is suitable for this article. We are interested in the spatial impact of regional economic events that began in the third quarter of 1979 and continued until the first quarter of 1983. Data on incomes and occupations from the 1980 Census miss much of the impact because it reports 1979 annual incomes and occupations. In fact, the Census is a monument to old-order Cleveland, and we use it as such.

Developing fully specified time series models about the linkage between regional economic and neighborhood change requires more data than are available. We attempt to circumvent the data problem and address the questions at hand by using theory as a guide in selecting two sets of variables for study. One is an indicator of the direct impact of regional economic change: the quarterly distribution of earnings. The other serves as an indirect indicator of neighborhood vitality: poverty rates and the value of housing. We concentrate on telling two stories that are related by time and theory. The link between earnings and the dependent variables, poverty rates and housing values, is made with 1980 census tract data on the residential location of operatives, fabricators and laborers, and professional and technical workers.

We do not directly analyze the impact of change on neighborhoods; we use census tracts as the spatial unit of analysis.[5] We do not have data to identify neighborhoods, which would require information on the social and economic status of residents, the ages and life cycle stage of the local population, and also the racial composition of communities and their home ownership pattern.

The Model

A fully specified model of the link between regional economic change and neighborhoods has four components: external economic demand, the regional labor market, the housing market, and the resulting structure of neighborhoods (Figure 10.2). Demand for labor in any regional market is predicated on the external demand for its products. A change in the level and composition of demand or in technology will change the occupational demand for workers in the region.

Demand for labor will interact with existing supply, and movement toward a new equilibrium will begin, resulting in changes in the distribution of earnings, rate of unemployment, and in the number of poor. Two types of migration decisions will result from major changes in the regional economy. People will weigh opportunities inside and outside of the region and make decisions about interregional moves. They will also examine their work locations, future earnings prospects, housing values, the status and racial composition of neighborhoods in the region, and the cost of moving and decide

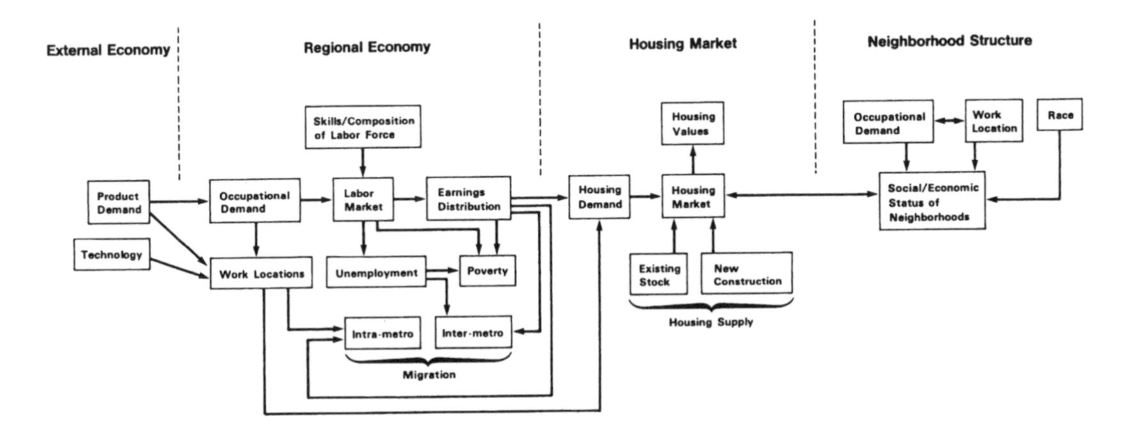

Figure 10.2. Links Between the Regional Economy and the Neighborhood

whether they should migrate within the region. The distribution of earnings is a major link between the regional labor market and the housing market.

The social and economic status of neighborhoods is a critical component in determining who will be attracted to different areas and serves as a vital link between the housing market and the ecological interrelationship of neighborhoods. The occupation of neighborhood residents is one of the major determinants of social status. Occupations also serve as a link between the changing fortunes of the regional economy and the health of neighborhoods. Because people tend to choose residences based on the status of neighborhoods and housing prices, there are broad similarities in the occupations and incomes of residents. If demand for these occupations deteriorates in an evolving economy, the neighborhood can be downgraded in status as housing values deteriorate. The opposite is true if occupational demand strengthens.

Two ways in which the degradation in value and status can be offset are by rapid immigration from outside of the region and large-scale family formation taking place without an offsetting increase in housing supply. There are other structural and institutional elements that are important to fully understand how neighborhoods evolve. For example, racism and redlining prevent the exercise of choice and inhibit adjustment in the housing stock. In Cleveland, race limits neighborhood choice. Blacks do not have access to the entire spectrum of neighborhoods.[6]

It is possible to create a reduced form of the elaborate model presented in Figure 10.2. However, the reduced form is not a quantitative model; it uses quantitative information to construct a qualitative picture of the social and economic transformation of a region. The foundation is an examination of the occupational structure of census tracts before the economic transformation took place. Data on the earnings distribution and housing values will be examined after presenting the hypothesized connections between regional economic and neighborhood change in Cleveland.

Exogenous demand for locally manufactured goods declined rapidly in 1979:3, and it appears to be permanent. At the same time, surviving manufacturers introduced labor-saving technologies to reduce the labor content of their production processes. Both of these movements adversely affected the employment prospects of laborers and operatives. At the same time, there has been an increase in service sector employment, but at a rate that does not fully offset losses in manufacturing employment. Increased demand in the service sector resulted in an expansion of employment opportunities for white-collar professionals as well as for the unskilled.

We expect that the earnings distribution became increasingly skewed as a result of regional economic restructuring. Employment was added at the high and low ends of the distribution through expansion of the service sector. At the same time, the number of jobs contracted in the middle of the distribution, resulting from a decline in the manufacturing sector. This is the declining-middle hypothesis.[7] We also expect to see deterioration in the two indicators of neighborhood health in the census tracts where blue-collar workers lived in 1980: an increase in poverty rates and deterioration of housing values. Conversely, those areas of the region that are in demand by workers in higher status occupations should see a relative increase in value and a decrease in poverty rates. We test the hypotheses about the earnings distribution in the next section.

Examining the Trend in Earnings

Constructing the Earnings Distribution

A direct test of the hypothesis that the earnings distribution has changed over time requires inspecting quarterly or annual individual and family data on incomes and earnings. Unfortunately, these are not available for metropolitan areas. Instead, we took quarterly average earnings data at the two-digit level of the Standard Industrial Classification (SIC) from the Bureau of Labor Statistic's ES-202 records and created a synthetic earnings distribution.

The earnings distribution was synthesized in two respects. First, the 69 two-digit industries present in the Cleveland PMSA were rank ordered by their real earnings, for each quarter from 1975:1 through 1987:2.[8] Then each industry's portion of total employment was added to the percentage of persons employed in all industries with lower average earnings, creating a cumulative density function of employment based on earnings. This was done for each quarter. Each industry's average earnings was then treated as grouped data in calculating the mean and standard deviation of the earnings distribution. The reported average quarterly earning for the PMSA is a true mean, but the standard deviation is synthesized.[9] These parameters were then used to calculate the skewness statistic and Gini coefficient of each quarter's distribution of earnings.

Shifts in the Earnings Distribution

Quarterly average and median real earnings in the Cleveland PMSA from 1975:1 to 1987:2 are plotted in Figure 10.3. The average earnings of the industry employing the median worker is called the median per capita earning of the distribution. There are three distinct parts to the figure. The portion of the graph from 1975:1 to 1979:1 shows the recovery from the 1975 recession to the 1979 peak. The second portion is from 1979:1 to 1983:1. This is the movement from peak-to-trough. The third section depicts the current recovery phase, from 1983:1 to 1987:2.

An important transition occurred near the 1983 trough. Median earnings became greater than average earnings, implying that more jobs were added at the far left of the earnings distribution—the low end—than at the far right during the last recovery. This is in contrast to earlier distributions. The addition of low-wage jobs drags down average earnings. Low-wage employment became relatively more important after the 1983 trough was reached.

This point is reinforced when the skewness statistic of the earnings distribution is examined (Figure 10.4). The earnings distribution would be symmetrical if the skewness statistic equals 0.0; a positive number indicates that the distribution is skewed to the right—toward high-wage employment. All of the skewness statistics since 1983:1, with the exception of fourth quarter calculations, are negative. The only other time this occurred was during the 1975 recession. This means that since the 1983 trough the modal job added to Cleveland's economy has paid a relatively low wage.

A third picture of change in the earnings distribution is provided by plotting the Gini coefficient[10] (Figure 10.5). The Gini coefficient measures the spread of the distribution; larger Gini scores indicate wider distributions. The result is unmistakable. Gini scores have been on

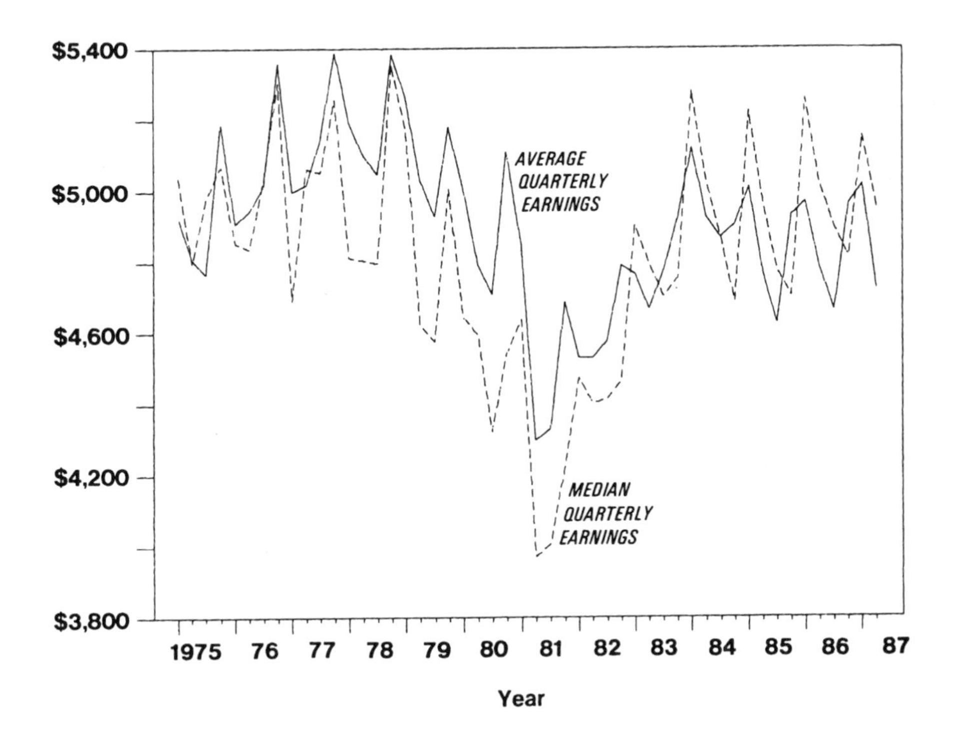

Figure 10.3. Average and Median Quarterly Earnings (in 1982–1984 Dollars)

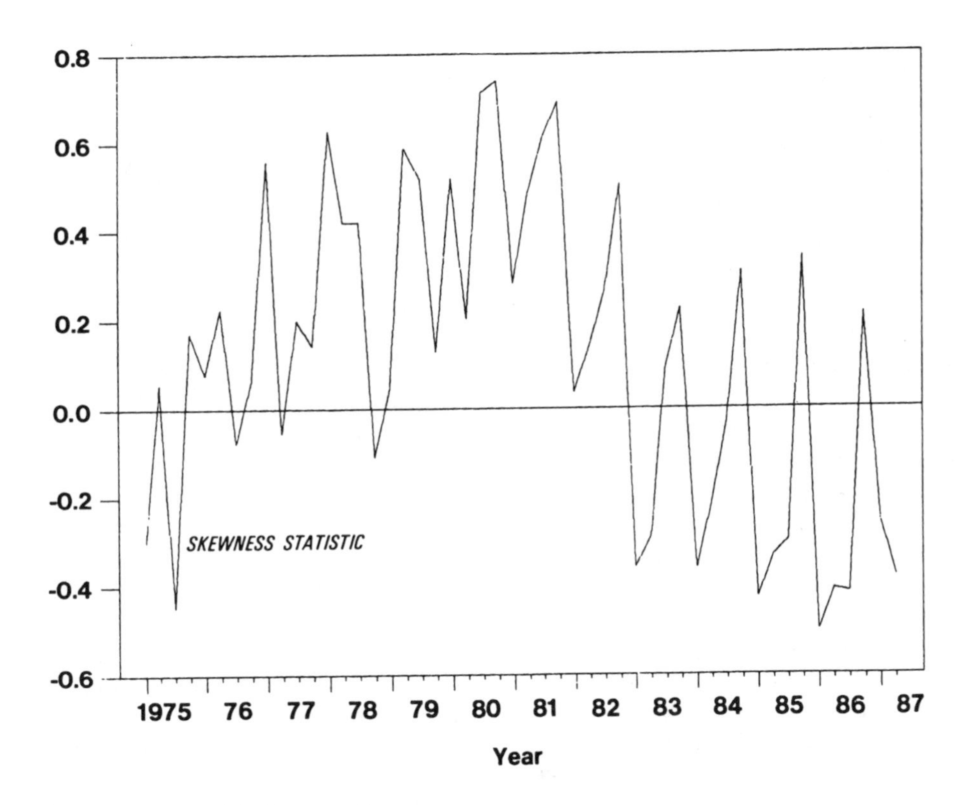

Figure 10.4. Skewness of the Earnings Function

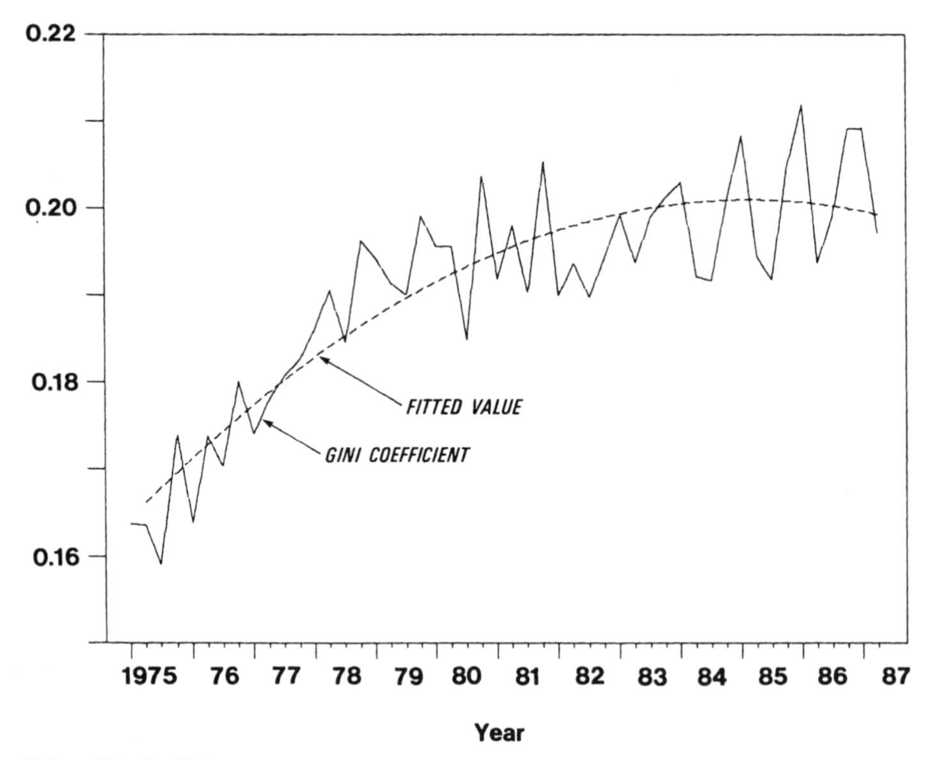

Figure 10.5. Gini Coefficient

GINI	is the quarterly Gini coefficient of the synthetic earnings distribution;
TIME	is a time-trend indicator variable, it has a value of 1.0 in 1975:1 and is increased by 1.0 in each succeeding quarter;
TIME**2	is the square of TIME;
RHO	is the correction factor for first order autocorrelation;

	GINI = B1	+B2(TIME)	+B3(TIME**2)	RHO	
COEFFICIENT	0.162561	0.001857	–0.000022	0.01466	(1)
t-STATISTIC	(55.59)*	(7.18)*	(4.64)*	(0.90)	

R**2 = 76.3 Durbin-Watson = 1.99

*Significantly different from 0.0 at the 0.01 significance level.

an unremitting upward climb since the recovery from the 1975 recession. Two lines are drawn on Figure 10.5. The solid line traces the path of Gini scores from 1975:1 to 1987:2. The dashed line is a plot of the fitted values to a time trend regression equation, where

[SEE BOX ABOVE]

The equation indicates that the Gini scores have increased in a nonlinear fashion, resembling a parabola. This means that the rate of increase has subsided in recent quarters. The function reached a turning point in the first quarter of 1985; the Gini scores have shown signs of moderating since then. It appears that the rate of increase in the Gini coefficient was a leading indicator of structural change in the regional economy. There was a marked shift in the distribution of earnings throughout the 1976

GINI	is defined in equation (1);
PCTMFG	is the percentage of private employment in the manufacturing sector;
PCTSERV	is the percentage of private employment in the service sector;
RHO	is defined in equation (1);

GINI =	B1	+B2(PCTMFG)	RHO	
COEFFICIENT	0.2609	−0.00245	0.495	(2)
t-STATISTIC	(13.75)*	(3.66)*	(3.94)*	
R**2 = 63.9 Durbin-Watson = 2.26				

GINI =	B1	+B2(PCTSERV)	RHO	
COEFFICIENT	0.1345	0.00278	0.449	(3)
t-STATISTIC	(8.27)*	(3.55)*	(3.35)*	
R**2 = 62.0 Durbin-Watson = 2.30				

*Significantly different from 0.0 at the 0.01 significance level.

to 1979 recovery that preceded the widely recognized structural change in this particular part of the rustbelt economy. The rate at which inequality increased has diminished since 1981, and has shown signs of marginal reversal from 1985 to 1987. This means that the regional economy is generating a fairly stable distribution of earnings.

We performed two tests to determine if change in the Gini coefficient is attributable to shifts in the structure of employment since 1975. We could not specify a complete test of this hypothesis because of the high correlation over time between the percentage of people employed in the manufacturing sector and in the service sector. We expected to see an inverse relationship between the portion of the working population employed in manufacturing and the Gini coefficient, and a positive correlation between the portion employed in the service sector and the Gini coefficient. Two equations were estimated, where

[SEE BOX ABOVE]

In each case the expectation was supported by the data.

The information provided in Figures 10.3 through 10.5 indicates that the earnings distribution has become less tight over time. This by itself tells little about the direction of inequality in earnings. However, the skewness statistics inform us that the modal wage has also shifted from the right side of the distribution to the left, from higher quarterly earnings to lower quarterly earnings. We also know that the average and median real quarterly wage remains, substantially below that of the late 1970s. These facts imply that the modal wage in the Cleveland economy has deteriorated significantly. At the same time, the path of the Gini coefficients indicates that the earnings distribution has become wider. We interpret this to mean that although jobs have been added to the high end of the distribution, more have been added at the low end. We deduce that the middle has deteriorated over time both in absolute and relative terms.

There are three caveats. First, the results are optimistic. The data are biased in an upward direction because they do not show the impact that nonemployment has had on the income distribution. Second, the synthetic earnings distribution assumes that there is no variance in earnings within a two-digit industry. This will also understate inequality. Third, these are earnings data, not income data. Income distributions are much more unequal than are earnings distributions in the nation as a whole.[11] This is a result of the concentration of unearned income among the top quintile and decile of the population. We see no reason to expect that the relationship between these two distributions will be any different in Cleveland.

In the remainder of the article we examine the spatial distribution of occupations, poverty,

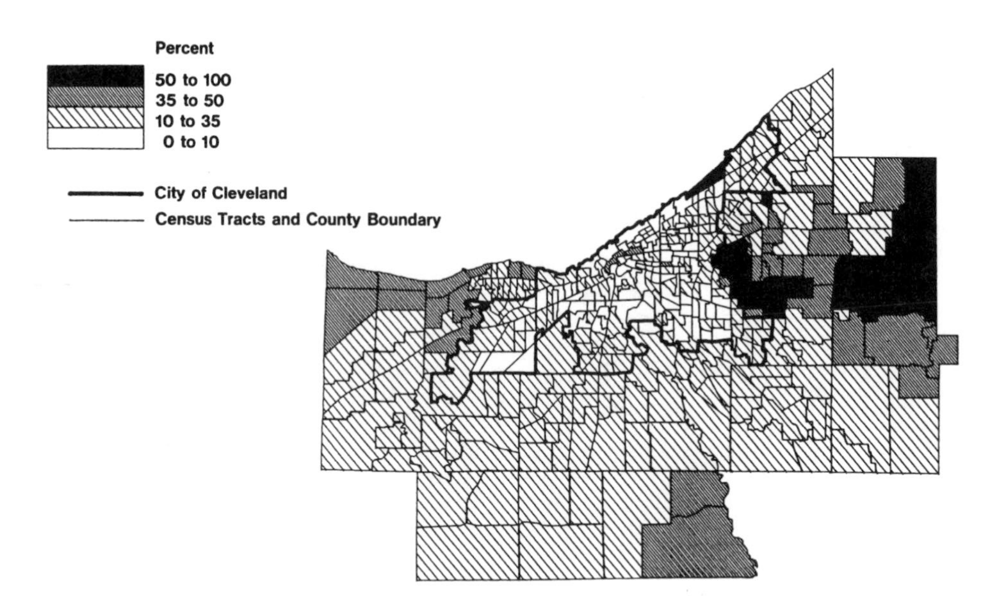

Map 10.1. Percentage Employed in Managerial and Professional Occupations, 1980—Cuyahoga County, Ohio

SOURCE: 1980 Census of Population and Housing. Prepared by Northern Ohio Data and Information Service, College of Urban Affairs, Cleveland State University.

and housing values—all indicators of the strength or weakness of neighborhoods. We assume that these changes reflect the shift in the distribution of earnings and the decline in the average purchasing power of real earnings. We expect to find a decline in those parts of the metropolitan area where the blue-collar middle class, and those who aspired to join its ranks, lived before the downturn in the regional economy took place. However, we do not attempt to associate these indicators directly to changes in the earnings distribution because we cannot use inferential statistics. Our conclusions must be considered to be suggestive and somewhat speculative.

Our attention will now focus on Cuyahoga County, not the PMSA. This is because the poverty and housing data on which we rely do not exist for the other three counties in the PMSA.

Distribution of Skills in 1980

We have hypothesized that the way in which regional economic change is transmitted to neighborhoods is through the residences of those who benefit or lose from restructuring. Neighborhoods in American metropolitan areas tend to be fairly homogeneous as to the social and economic status of their residents. In the case of Cleveland's change from 1979 to 1987 we expect to see that those census tracts with high concentrations of operators, fabricators, and laborers, occupations that depend on a manufacturing-based economy for work, will be hard hit. Those tracts that will benefit from new-order Cleveland will be those housing managers and other professionals. We also expect that those areas contiguous to the tracts housing the managerial class will also benefit, as new housing is constructed to satisfy the demands of this expanding class. The abutting tracts will be on the periphery of the county.

In this section of the article we describe the residential pattern of old-order Cleveland, as depicted in the 1980 Census. This serves as the baseline against which we will measure changes in our two indicator variables of neighborhood health: the growth in poverty and changes in housing values. Map 10.1 depicts the percentage of workers by census tract employed in managerial and professional occupa-

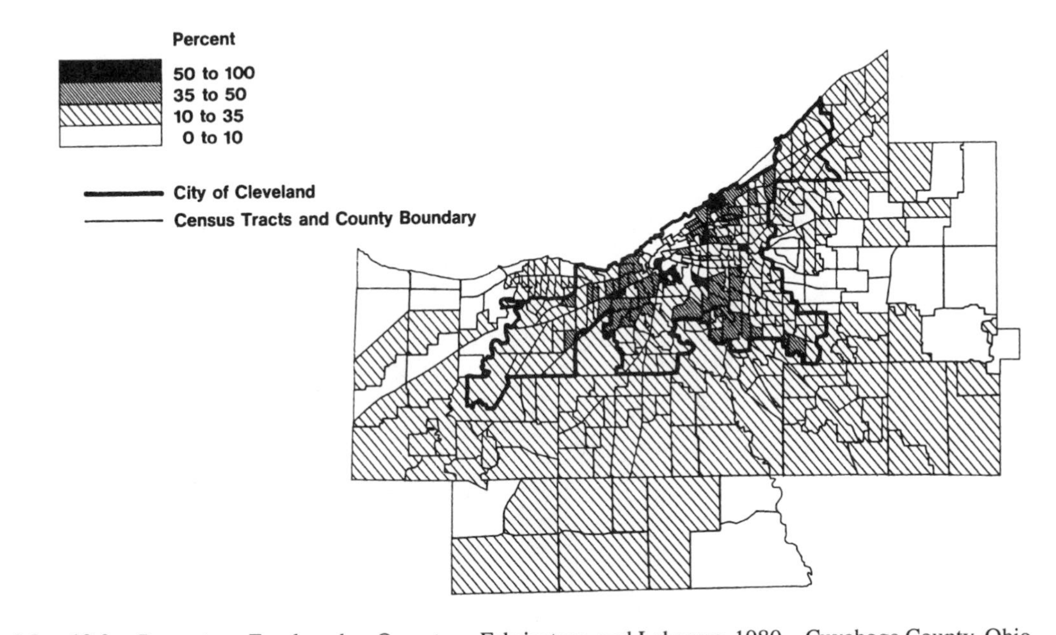

Map 10.2. Percentage Employed as Operators, Fabricators, and Laborers, 1980—Cuyahoga County, Ohio
SOURCE: 1980 Census of Population and Housing. Prepared by Northern Ohio Data and Information Service, College of Urban Affairs, Cleveland State University.

tions in 1980, and Map 10.2 shows the percentage employed as operators, fabricators, and laborers.[12] The neighborhoods that should benefit from the changing structure of the economy are highlighted in Map 10.1, and those that will benefit the least, along with areas with large poverty populations in 1980, are concentrated in the tracts emphasized in Map 10.2. Cleveland's city boundary is outlined with a heavy black line on each map.

The city of Cleveland is nearly devoid of census tracts with large concentrations of managerial or professional workers. The highest concentrations are along the western suburban lake shore, in the cities of Rocky River and Westlake, and in the high-status eastern suburbs, Shaker Heights and those areas that are in a direct eastern corridor from Shaker Heights. Lower, but still significant, concentrations existed in the outer suburban ring. If our hypothesis is correct, we expect that these census tracts will comprise those areas that will benefit from the transition.

Operators, fabricators, and laborers have a completely different residential pattern. The highest concentrations are in the city of Cleveland. No census tract outside of the city had more than 35% of its residential work force in these occupations in 1980. It is expected that residents in these areas will experience the greatest losses from the transition of the regional economy.

These locational expectations are conditioned upon two facts. First, the overall level of economic activity remains stagnant. The new sectors of the economy are not generating so many jobs as to create a temporary shortage in the housing market. More vibrant metropolitan economies were confronted by the twin problems of rapid household formation and job creation that raised the value of land and created demand for upgraded used housing. These two preconditions to gentrification currently do not exist in Cleveland. Second, there are no major geographical constraints to the expansion of the metropolitan area. There is still vacant land available in the southern and western parts of the county with commuting times of less than 30 minutes to the central business district and suburban nodes of employment. This will retard the adaptation of older neighborhoods to meet the demands of increased professional employment.

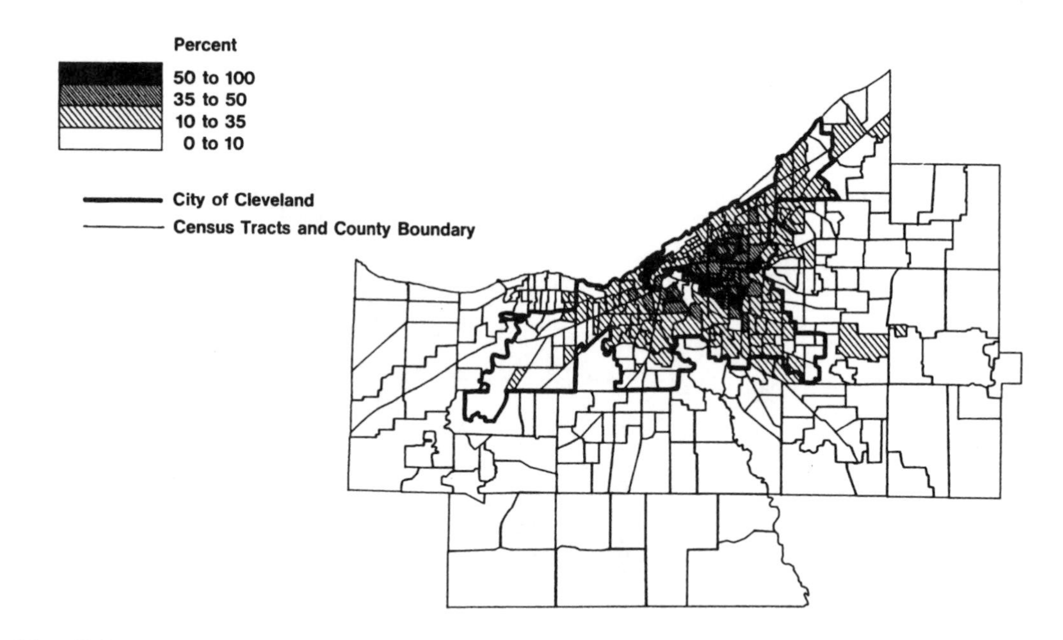

Map 10.3. Percentage of Persons Below the Poverty Level, 1979—Cuyahoga County, Ohio

SOURCE: 1980 Census of Population and Housing. Prepared by Northern Ohio Data and Information Service, College of Urban Affairs, Cleveland State University.

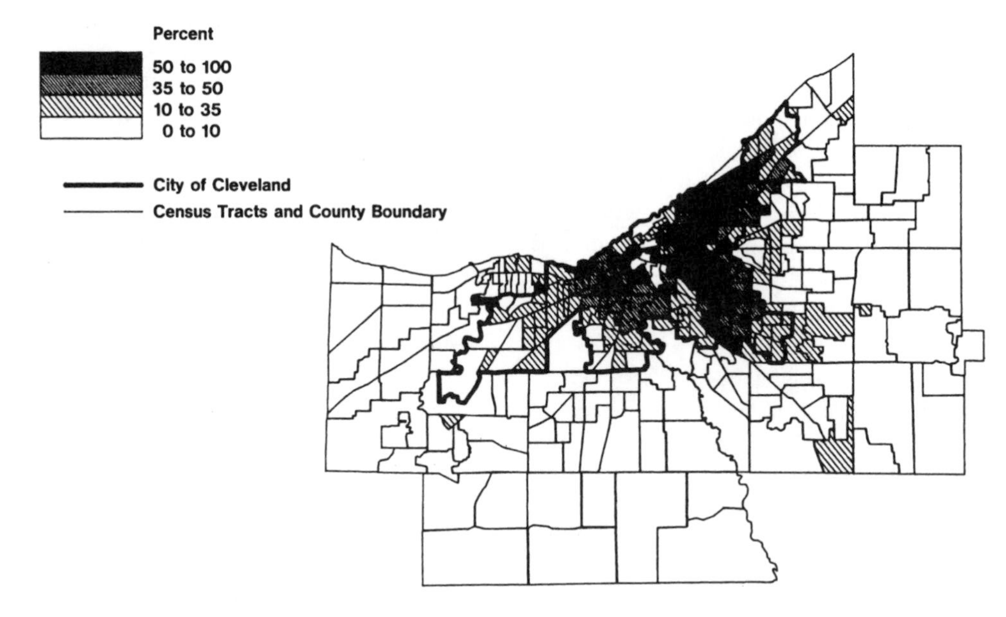

Map 10.4. Percentage of Persons Below the Poverty Level, 1987—Cuyahoga County, Ohio

SOURCE: Council for Economic Opportunities in Greater Cleveland. Prepared by Northern Ohio Data and Information Service, College of Urban Affairs, Cleveland State University.

Impact on Poverty

The poverty population was tightly concentrated around the core of the city of Cleveland when the 1980 Census was taken (Map 10.3). Poverty was present on the west side of the city, but the bulk of the poverty tracts were on its east side. Those tracts with more than half of the population being classified as poor are located

POV87 is the percentage of residents in a census tract with family incomes at or below the poverty line in 1987 (these are data provided by Cleveland's Council for Economic Opportunities);

POV80 is the percentage of the residents in a census tract with family incomes at or below the poverty line as reported in the 1980 Census;

LABOR is the percentage of a tract's residential work force with occupations classified as operators, fabricators, or laborers in the 1980 Census.

	POV87 =	B1	+B2(POV80)	+B3(LABOR)	
COEFFICIENT		−3.625	1.521	0.290	(4)
t-STATISTIC		(3.53)*	(31.84)*	(5.49)*	
R**2 = 87.1					

*Significantly different from 0.0 at the 0.01 significance level.

in either Cleveland or in the predominantly black suburb of East Cleveland.

The poverty population grew rapidly from 1979 to 1987. We now see that nearly half of the population in the city's traditionally black east side neighborhoods is poor, and the poverty rate on the west side reached 25%. What is confusing is that the growth in the numbers of poor do not trend with changes in the unemployment rate. The unemployment rate in Cuyahoga County was 5.6% in 1980 and the poverty rate was 13.6%. Unemployment peaked in 1983 at 12.0%, with a poverty rate of 16.1%. In 1988 the unemployment rate dropped to 5.8%, yet the estimated poverty rate reached 18.6%.[13]

The 1987 poverty estimates derived by the Council of Economic Opportunity for the county indicate that there was a huge expansion in the number of tracts with a majority of the population living in poverty (Map 10.4). The entire east side of the city is dominated by poor residents, as are sections of several suburban municipalities. The growth of poverty on the west side of the city and in one of its western suburbs, Lakewood, is also noticeable.

We used ordinary least squares to correlate the incidence of poverty in the county's census tracts in 1980 with its incidence in 1987 and the occupational characteristics of each tract's residents in 1980. Earlier, we hypothesized that there would be a strong positive association between a tract's poverty rate in 1980 and its rate in 1987. Additionally, we expect a strong positive spatial correspondence between those tracts that housed blue-collar workers in 1980 and 1987 poverty rates. These hypotheses are tested in equation 4. We use the percentage of workers in the tract who claim to be operators, fabricators, or laborers as an approximation of the blue collar work force, where

[SEE BOX ABOVE]

The coefficients in equation 4 conform to our expectations and are significantly different from 0.0 at the .01 level. The absolute impact of poverty in 1980 on poverty levels in 1987 is unexpectedly large. The coefficient indicates that the poverty growth rate from 1980 to 1987 was 50%. By this we mean that if 10.0% of a tract's 1980 population was poor, we expect that 15.2% would be poor in 1987. The coefficient of the variable that approximated the impact of a blue-collar residential work force in 1980 was positive. If a third of the workers living in a tract in 1980 were operators, fabricators, or laborers in 1980 we would expect to see the 1987 poverty rates add 9.6 percentage points to the county's average poverty rate.

We tested the first hypothesis with a third, in equation 5. We expect that the incidence of poverty in a tract in 1987 will be inversely associated with the percentage of white-collar workers living in a tract in 1980. The percentage claiming managerial or professional occupations in the 1980 Census is used as a measure of white-collar employment. The three hypotheses could not be tested in a single regression equa-

MGR is the percentage of a tract's residential work force with occupations classified as managerial or professional in the 1980 Census.

	POV87 =	B1	+B2 (POV80)	+B3(MGR)	
COEFFICIENT		5.278	1.594	–0.161	(5)
t-STATISTIC		(4.04)*	(35.94)*	(4.02)*	
$R^{**}2 = 87.1$					

*Significantly different from 0.0 at the 0.01 significance level.

tion because the combination of the two occupational variables with the poverty variable showed signs of multicollinearity, where

[SEE BOX ABOVE]

The signs of the variables were as expected in equation 5. The proxy variable for a white-collar work force was significantly negative. Every percentage-point increase in the portion of a tract's work force that holds managerial or professional occupations is associated with a decrease of 0.16 percentage points from the average 1987 poverty rate.

Poverty increased in tracts dominated by the poor in 1980. Those areas with large portions of their work force employed in lower-skilled occupations associated with factory work were the spawning grounds for poverty at the end of the 1980s. This strongly suggests that change in the structure of the regional economy had a direct impact on the neighborhoods of the city and on sections of its close-in suburbs.

Impact on Neighborhood Housing

Between 1950 and 1986, 1.5 housing units were built (most of them in the suburbs) for every additional household living in the region. Expansion of the suburban supply at a rate greater than household growth enabled a steady stream of population to move outward. As the population shifted into more preferred housing and locations, the least preferred were abandoned. In the 1970s the city of Cleveland lost 30,000 households; the suburban portion of the metropolitan area gained 74,000. In Cleveland, a lost household resulted in an abandoned unit; Cleveland lost 21% of its households and 30% of its housing units.

Figure 10.6 displays the 1986 price distribution of single-family housing stock of Cuyahoga County by its value.[14] The price of Cleveland homes is low compared with its suburbs. Cleveland's median value is in the $30,000 range, whereas the suburbs are valued in the $60,000 range. Also, the county contains a very large supply of homes priced between $20,000 and $80,000—about 75% of all units fall into this range. Finally, there are relatively high-priced units—12% of the stock is valued at more than $100,000.

This profile of homes, nearly all built before 1979, is largely a product of the Cleveland area economy as it existed before the recession of 1979. It reflects the interaction of supply and demand in an economy dominated by blue-collar earnings and the relatively small number of higher-income households that lived in the region in the past. The distribution of housing values complements the tight distribution of earnings that existed in old-order Cleveland.

The housing industry built products that responded to the expressed demands and incomes of an economy dominated by manufacturing workers. With the economic changes that began in the 1970s and accelerated in 1979, a significant portion of demand for working-class housing was effectively removed. The departing jobs left behind not only empty factories and plants, but also midpriced homes with fewer potential buyers. At the same time economic change created demand for higher-status homes. The old-order housing profile is not compatible with the new-order economy.

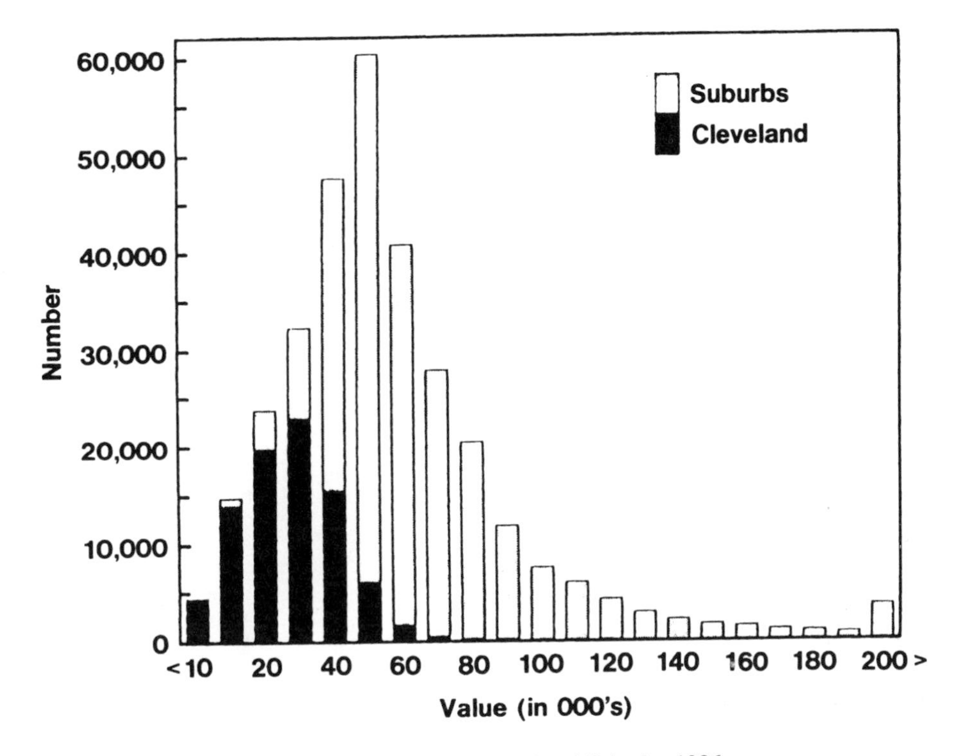

Figure 10.6. Number of Single-Family Homes—Cleveland and Suburbs, 1986

Change: 1979-1987

The impact on housing values stemming from economic change was intense. The data in Figure 10.7 demonstrates the impact of the post-1979 changes in the regional economy on the value of housing. The graph represents median single-family home sales prices in the city of Cleveland and the suburban portion of Cuyahoga County relative to changes in the Consumer Price Index (CPI); 1967 is the base year. If sales price appreciation kept pace with the CPI, the percentage difference will be 0.0; if home prices rise faster than inflation there will be positive appreciation; if they rise less than inflation the gains will be negative.

Between 1967 and 1978 sales prices in the suburbs greatly outpaced inflation. Initially, prices in Cleveland declined relative to inflation but began to recover in 1970; they nearly regained their 1967 real values in 1978. The Gini coefficient gives us a possible explanation of the divergence in appreciation. The earnings distribution was widening at an accelerating pace from 1975 to 1978. Those who were at the upper end of the distribution, or gaining access to the upper end, may have been moving to the suburbs. The partial recovery of the inner-city portion of the market could be attributable to first-time home buyers and those with declining real incomes, searching for relative housing bargains.

In 1979 both the city and suburban markets went into precipitous decline, those who owned homes in 1978 have taken substantial capital losses on their investments.[15] Prices in the city of Cleveland hit bottom in 1982; the suburbs did not bottom out until 1985, when the recovery had been underway for nearly two years. Between 1984 and 1985 the city's housing values gained marginally relative to both the CPI and suburban values. However, in 1986 and 1987 the suburbs began to gain relative to the city.

Losses in median sales prices over time differ among the census tracts. We have to sort out the impact of decline resulting from economic restructuring from the loss in value resulting

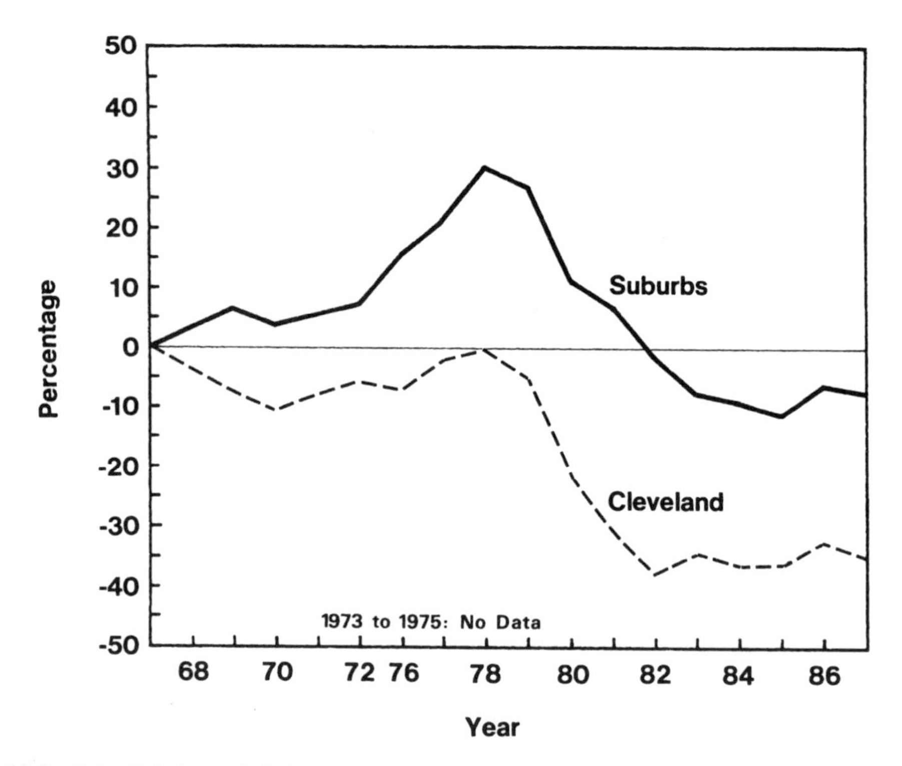

Figure 10.7. Price Relative to Inflation—Single Family Sales

from the overall decline that Cleveland has experienced since 1979. We attempt to do this by estimating three regression equations, each with the same general form but estimated for the 1979 business cycle peak, the 1983 trough, and 1987. The median sales price of single-family housing is estimated as a function of the portion of the residential labor force in the 1980 Census that is managerial and professional and the percentage that is either operators,fabricators, or laborers,[16] where

[SEE BOX ON NEXT PAGE]

The intercept of each regression equation is the average price in the county, expressed in real terms, using 1983 dollars. Each of the estimated coefficients is also expressed in terms of 1983 dollars. Differences in the value of the intercept in each equation is a measure of the general decline in the regional housing market. The intercept declined from $63,269 in 1979 to $43.360 in 1987, a $20,000 loss. There was an $808 rebound in the value of the intercept between the 1983 trough and 1987.

The way in which the loss in value was distributed is demonstrated by the estimated coefficients, B2 and B3 in equations 6 to 8. In all cases the signs of these coefficients were in the expected direction and significant at the .01 level. The presence of managers and professionals in a census tract is associated with a relative increase in housing values, and the blue-collar proxy variable appears to discount sales prices. The value of the coefficients across time is fairly constant. The markup owing to the presence of professionals and managers was $1,402 in 1979, $1,292 in 1983, and $1,438 in 1987, always positive with a slightly higher value in real terms at the end of the time period than at the beginning. The discount for operators, fabricators, and laborers was $1,374 in 1979, $920 in 1983, and $1,014 in 1987. However, the data are biased in terms of minimizing the negative impact of the blue-collar variable. In all, 5% of the tracts could not be used in the 1983 and 1987 estimating equations because they had sales in 1979 and no sales in 1983 or

HV79 is the real median sales price by census tract of single family homes in 1979 in 1983 prices;
HV83 is the real median sales price by census tract of single family homes in 1983 in 1983 prices;
HV87 is the real median sales price by census tract of single family homes in 1987 in 1983 prices;

HV79 =	B1	+B2(MGR)	+B3(LABOR)	
COEFFICIENT	63,269	1,402.50	−1,374.30	(6)
t-STATISTIC	(5.41)*	(5.63)*	(4.49)*	
R** = 52.0				
Coefficient as % of Intercept		2.22%	−2.17%	

HV83 =	B1	+B2(MGR)	+B3(LABOR)	
COEFFICIENT	42,552	1,291.68	−920.29	(7)
t-STATISTIC	(5.23)*	(7.46)*	(4.33)*	
R**2 = 59.7				
Coefficient as % of Intercept		3.04%	−2.16%	

HV87 =	B1	+B2(MGR)	+B3(LABOR)	
COEFFICIENT	43,360	1,437.55	−1,014.21	(8)
t-STATISTIC	(4.58)*	(7.13)*	(4.10)*	
R**2 = 57.3				
Coefficient as % of Intercept		3.32%	−2.34%	

*Significantly different from 0.0 at the 0.01 significance level.

1987. The housing market ceased to exist in these tracts and a large share of their residents were operators, fabricators, or laborers in 1980.

The best way to compare the distributional impact of occupation on housing values in each of the years is to calculate the value of the markup and discount as a percentage of the intercept, or average sales price, for the year the equation was estimated. This figure is reported in the last line following equations 6 to 8. The percentage markup resulting from the presence of managers and professionals increased in each of the years, from 2.22% in 1979 to 3.32% in 1987. This means that a percentage-point difference in the portion of managers living in two census tracts accounts for a 3.32% increase in the value of housing. On the other hand, the discount associated with blue-collar residents has increased over the time period, a sure sign of restructuring and of the loss of economic status among the neighborhoods where the blue-collar labor force lived in 1980. The impact of these differences is quite large and increased over the time period studied. Both results are consistent with our expectations. In 1979 the total differential, adding the blue-collar and white-collar coefficients, between white and blue-collar census tracts was 4.39% of the average house price in the county; it is now 5.66%.

In Cleveland's inner-city neighborhoods, where the lowest priced housing in the county is located (under $20,000 in current dollars) and where the city's poorest residents live, sales price changes have been weakest. By 1987, nearly all of Cleveland's innermost neighborhoods lost one-half, and some lost two-thirds, of their real 1967 sales value. These losses were aggravated by the switch in the economy's base in 1979. The outer neighborhoods of the city, those closest to the suburbs, have fared the best, but even there real prices have fallen 20% to 30%.

Suburbs within the county with median sales prices over $100,000 dominate the group of communities with the highest sales price appreciation. The $100,000-plus communities are almost the only ones where the current value of housing exceeds the inflation adjusted value in 1967. And those high-priced communities are where most of the new residential construction is occurring. Buyers and residents in those communities are linked by their occupations to the economic activity that represents the growing, and best-paying, component of the changed Cleveland economy.

Substantial gains were experienced in the outermost suburbs and those sections of the county where professionals and managers lived before the 1979 recession took hold. The other areas of the county showed losses or little appreciation. These observations are consistent with changes in the earnings distribution. Those areas of the county that specialized in housing the blue-collar middle class have felt the effects of earnings losses; those that specialized in housing the professional and managerial classes have done somewhat better. There have not been large enough gains made in service sector employment or income to produce strength in the portion of the housing market that once depended on the wages of the middle of the earnings distribution.

Conclusion

The lesson to be learned from Cleveland is that the effects of economic restructuring are widespread but uneven. The link between the regional economy and the structure of neighborhoods is direct but difficult to model, owing to the scarcity of time series data. We have demonstrated that the distributional impact of Cleveland's restructuring is evident through the marked growth in poverty and uneven changes in home sales prices. These changes take place in specific parts of the metropolis. The link between the world of work and the residential world is the occupation of workers and the socioeconomic status of their neighborhoods.

Our research also indicates that the existing stock of housing will not be upgraded by market mechanisms without tremendous economic expansion, accompanied by high rates of household formation. These twin forces create pressure to upgrade the incumbent housing stock owing to increases in land values. However, if the expansion in the regional economy is unbalanced there will be a pattern of winning and losing neighborhoods and residents. In the case of regions that have seen erosion in the middle of their earnings distribution, the filtering process will be accelerated. In fact, the poor may become better off as the value of low to moderate quality housing becomes discounted. But they must move to take advantage of the depreciated stock. The city will experience accelerated population losses and those who remain behind will be the poorest of the poor—those who cannot move.

In Cleveland's case, where the recovery has been uneven, the real losers will be those who own middle-valued housing and the neighborhoods in which they are located. Their property values have been affected and their capital losses will make it difficult to move. We suspect that this conclusion holds for many metropolitan regions in the heartland of the United States. Strong demand for older neighborhoods with less than elite status, therefore, requires either reduced suburban construction or area-wide job growth that outpaces suburban development.

The true impact of economic restructuring is felt by more than just those who either lose or change jobs. Because restructuring does not affect a random group of workers spread evenly throughout a metropolis, it generates spillover effects that reach into neighborhoods. Economic restructuring affects classes of workers and their families based on the occupation of the worker. In the case of Cleveland the spatial effects of restructuring resulted in changes in the vitality and prospects of neighborhoods and their residents.

Notes

1. All employment and earnings data are quarterly aggregates of monthly ES-202 reports provided by the Bureau of Labor Statistics. ES-202 data are collected as a by-product of the unemployment compensation insurance program.

Total employment in this data set includes government and private employment.

2. All earnings data are deflated using the Consumer Price Index for all urban consumers (CPI-U) from 1983 to 1997. From 1975 to 1982, CPI-U-X1, an experimental precursor to the current CPI-U series, was used. These two series adjust housing costs using the rental equivalence method rather than the purchase price method. Quarterly earnings are expressed in real dollars for a base period from 1982 to 1984. See *Using the Consumer Price Index for Escalation* (Washington, DC: Department of Labor, Bureau of Labor Statistics, Report 732, October 1986); *Questions and Answers on Homeownership Costs* (Washington, DC: Department of Labor, Bureau of Labor Statistics, January 1983); Janet L. Norwood, "Two Consumer Price Index Issues: Weighting and Homeownership," *Monthly Labor Review,* March 1981, pp. 58-59; and "The Effect of Rental Equivalence on the Consumer Price Index, 1967-82," *Monthly Labor Review,* February 1985, pp. 53-55. CPI data for the Cleveland region were not used owning to the large error component in these estimates.

3. Mark Hoffman et al., *Population and Household Projections: Cleveland Metropolitan Area, 1985-2020* (Cleveland: Housing Policy Research Program, The Urban Center, College of Urban Affairs, Cleveland State University, May 1988); and Bureau of the Census, *Estimates of the Population of Ohio Counties and Metropolitan Areas: July 1, 1981 to 1985,* Current Population Reports, Series P-26, No. 85-OH-C (Washington, DC: Government Printing Office).

4. The data on poverty in Cleveland and in Cuyahoga County have been developed by George C. Zeller and published in *Poverty Indicators, Trends: 1970-1988, Cuyahoga County, Ohio* (Cleveland: Council for Economic Opportunities, 1998). The city of Cleveland is part of Cuyahoga County. The county contains 80% of the population of the Cleveland PMSA. Zeller estimates the number of poor from 1980 to 1988, as follows:

Poverty in Cuyahoga County and Cleveland

	Cuyahoga County			Cleveland		
Year	Number	Growth Rate %	Poverty Rate %	Number	Growth Rate %	Poverty Rate %
1980	203,000		13.6	155,000		27.1
1981	230,000	13.3	15.5	172,000	11.0	30.4
1982	209,000	–9.1	14.1	161,000	–6.4	28.7
1983	236,000	12.9	16.1	181,000	12.4	32.8
1994	251,000	6.4	17.2	193,000	6.6	35.4
1985	258,000	2.8	17.7	199,000	3.1	36.9
1986	271,000	5.0	18.6	210,000	5.5	38.9
1987	273,000	0.7	18.7	211,000	0.5	39.1
1988	270,000	–1.1	18.6	209,000	–0.9	38.7

5. Harry Margulis uses logit analysis to classify neighborhoods and examine home buyer behavior in the city of Cleveland in "Homebuyer Neighborhood Choice in a Fragmented City" (Cleveland State University, College of Urban Affairs [draft]). His work builds on the work of Brian Berry and John Kasarda, *Contemporary Urban Ecology* (New York: Macmillan, 1977) and R. J. Johnston, "Residential Area Characteristics: Research Methods for Identifying Urban Sub-Areas—Social Area Analysis and Factorial Ecology" in *Spatial Process and Form,* eds. D. T. Herbert and R. J. Johnston (New York: John Wiley, 1976). Margulis's paper is an interesting complement to our work because it is clearly based on the literature of factoral ecology.

6. John F. Kain, "Black Suburbanization in the Eighties: A New Beginning or a False Hope?" in *American Domestic Priorities,* ed. John M. Quigley and Daniel L. Rubinfeld (Berkeley: University of California Press, 1985); Douglas S. Massey and Nancy A. Denton, "Trends in the Residential Segregation of Blacks, Hispanics, and Asians: 1970-1980," *American Sociological Review* 52 (1987): 802-25; Karl E. Taeuber, "Racial Residential Segregation, 28 Cities, 1970-1980," Working Paper 83-12 (Madison: University of Wisconsin, Center for Demography and Ecology, 1983).

7. The literature on this subject is growing, but it combines empirical work on both individual earnings and incomes as well as family incomes. See, for example, Barry Bluestone and Bennett Harrison, *The Deindustrialization of America* (New York: Basic Books, 1982); Bennett Harrison and Barry Bluestone, *The Great U-Turn* (New York: Basic Books, 1988); Katherine L. Bradbury, "The Shrinking Middle Class," *New England Economic Review,* September/October 1986, pp. 41-55; Michael W. Horrigan and Steven E. Haugen, "The Declining Middle-Class Thesis: A Sensitivity Analysis," *Monthly Labor Review* May 1988, pp. 3-13. Notes in Horrigan and Haugen provide an extensive set of references on this topic.

8. Average earnings were imputed by dividing each industry's total reported earnings by quarterly average employment.

9. The synthetic earnings distribution will understate inequality because it assumes that earnings are equal within each two-digit industry. One example is in SIC 80, health care. We are forced to assume that all health care workers from janitors to residents and chief executives are paid an equal wage. This is clearly false. If we assume that the dispersion of earnings remains roughly constant within each industry over time, then the movement in the summary measure, the Gini coefficient, is an accurate reflection of the trend in inequality in the region. The same logic holds for median quarterly earnings reported in the article. They may be overstated because they are the industry average earnings for the industry that employs the median worker. These are imperfect measures of the regional earnings distribution, but they are the only ones available in noncensus years.

10. The Gini coefficient is a summary measure of inequality of income or earnings distributions. The coefficient varies from 0.0 to 1.0, where 0.0 means complete income equality and 1.0 complete inequality. Therefore, greater inequality is associated with higher Gini scores. See William C. Apgar and H. James Brown, *Microeconomics*

and Public Policy (Glenview, IL: Scott, Foresman, 1987), pp. 211-13.

11. Bennett Harrison and Barry Bluestone, *The Great U-Turn*; Richard S. Belows, Linda H. LeGrande, and Brian W. Cashell, *Middle Class Erosion and Growing Income Inequality: Fact or Fiction?* (Congressional Research Service, Library of Congress, Report No. 85-203E, November 1985); and Joseph Perkins, "The Poor are Trickling Upward," *Wall Street Journal,* November 3, 1988, p. A 14.

12. Maps 10.1 to 10.4 are adapted from Northern Ohio Data and Information Service, *Cleveland Area Atlas* (Cleveland: Cleveland State University, The Urban Center, College of Urban Affairs, 1988).

13. George Zeller, *Poverty Indicators, Trends: 1970-1988, Cuyahoga County, Ohio,* p. 68.

14. All data on housing sales prices were obtained from deeds and mortgages recorded with the Cuyahoga County Auditor. The data are maintained by the Housing Policy Research Program of The Urban Center at Cleveland State University.

15. The volume of sales activity in the county was also affected by the business cycle. In 1979 16,040 single-family homes were sold. This number declined by 26.3% in 1983 to 12,019. Activity accelerated to 14,464 in 1987, but was still 9.8% below the 1979 level. Another indicator of the impact of restructuring and of the expansion of the metropolis is the increase in the number of tracts without sales over time. There were no home sales in 1979, 1983, and 1987 in 27 tracts; 16 tracts had sales in 1979, but not in the other two years; 28 tracts had more than five sales in 1979 but between one and four sales in 1983 and 1987. The single-family home sales market is weak to nonexistent in nearly 20% of the county's census tracts.

16. Problems of multicollinearity and serial correlation, which we experienced in the earlier regressions, were not encountered in equations 6, 7, and 8. In equations 2 and 3 the problem was autocorrelation between PCTMGR and PCTSERV. This difficulty does not exist in equations 6 to 8 because they are cross-sectional. Equations 4 and 5 could not be combined because of the affect of POV80 on LABOR and MGR. The POV80 variable is not included in equations 6 to 8.

Chapter 11

Neighborhood Initiative and the Regional Economy

Jeremy Nowak

❋❋❋

Neighborhood Development and Poverty Alleviation

During the past decade, the community development financial institution that I manage (the Delaware Valley Community Reinvestment Fund) has invested millions of dollars in the low-income neighborhoods of metropolitan Philadelphia. Since its inception, the fund has operated under the explicit assumption that our financial and technical assistance programs were in support of *neighborhood or community development.* Our mission followed the logic of a generation of similar efforts. Having witnessed an outflow of jobs, capital, and people from the inner city, we conceived the reinvestment fund as a targeted development credit institution that would help reverse that trend by investing money in places that were increasingly poor and isolated from the mainstream economy. We also assumed that the organizational mediation for revitalization would itself be local: geographically delineated development corporations, local business associations and entrepreneurs, local companies, nonprofit institutions, and congregations.

The strategies pursued by our investments and technical assistance are both financial and civic. Above all, they are grounded in place. The questions we ask when we work in West Philadelphia or North Camden are defined by a community geography. How does a particular marketplace get restored? How do the actions of civic associations affect declining real estate values and abandoned housing stock? What financial, technical, and organizational supports are needed in a particular area to affect the economic well-being of its residents? These questions elevate local association and civic entrepreneurship alongside the profit and mobility motives of individuals, entrepreneurs, and investors. They assume that the preconditions to neighborhood change are linked as much to local social organization as to the workings of urban and regional economies.

The neighborhood revitalization model seeks to rebuild low-income neighborhoods and improve the lives of their residents through civic participation and the creation of economic activity that is locally situated. Its ancestry can be traced to federal public policy, ethnic identity politics, international development proj-

Reprinted from *Economic Development Quarterly* Vol. 11, No. 1, February 1997.

ects, the civil rights movement, the policies of national foundations, and 30 years of community organizing efforts. This myriad of influences has created a *community-based development* field that has found more and more public- and private-sector acceptance as a way to combat inner-city decline.

One of the striking features of the tradition—and this may help to account for its persistence—is the contradictory way it is interpreted and claimed. It is a tradition that was defined by the activism of government more than 25 years ago and, alternatively, by the absence of governmental intervention during the past 15 years. It is a tradition filled with the community control language of 1960s Black nationalism, yet is ideologically malleable enough to appeal to *bootstrap capitalist* Republicans. Neighborhood development seemingly offers something for everyone, whether you are a real estate investor interested in internal rates of return or a social scientist concerned with citizen participation and the decline of civil society.

The 30-year tradition of community-based development has had significant success, particularly in low-income housing production and related social service delivery.[1] Every major city in the country contains dozens of community development corporations; local and federal public policy increasingly looks to community-based institutions to implement housing and community development activities; and an emerging infrastructure of community development financial institutions, technical assistance agencies, and bank-sponsored development programs offers support to neighborhood groups. In a period of public policy devolution, community development's emphasis on volunteerism and locality should continue to play well, although with less federal money.

At the same time, the persistence and acceleration of poverty, in the very areas where so much community development activity takes place, reveal the limitations of the approach: limitations of *scale and perspective.* The scale problem is self-explanatory. Just take a drive through the most blighted sections of Philadelphia, Baltimore, or Detroit, and it becomes clear that massive development intervention is required to restore the ordinary mechanisms of the marketplace and make the area a place in which anyone with choice will want to remain or locate.

Although there are good examples of community-based development activity reaching scale in very depressed areas—the section of East Brooklyn rebuilt by the Industrial Areas Foundation-sponsored East Brooklyn Congregations, the work of several development groups in the South Bronx, or the work done in Newark's Central Ward by the New Communities Corporation—examples are few and far between. In most of America's low-income urban neighborhoods, even the best community-based development efforts function as *managers of decline* as much as catalysts of significant renewal. This is not to criticize their hard work, competence, or even their possibilities but, rather, to point out that, in the absence of other economic strategies, community-based initiatives cannot reverse the downward spiral.

But the problem of quantity is compounded by the question of the very appropriateness of the approach. The neighborhood development model, organized around place and community, has tended to consider neighborhoods in terms of constituent service rather than in economic terms. Moreover, the transactions of community development are largely real estate based and thus not in and of themselves organized around requirements of social mobility. Far too little attention has been paid to the core issues of household poverty defined by access to good jobs and the accumulation of wealth. Making a neighborhood real estate market the focus of antipoverty intervention says more about American politics, racial segregation, and the marginality of low-income people than it does about a theory of social mobility. The places of low-income urban residents, increasingly homogeneous by race and class and increasingly detached from the centers of job generation, have less ability to regenerate themselves "as communities" at the very time that community development has become prominent as a self-help antipoverty strategy.[2] As the predicament of inner-city residents has to do with household income, job location, workforce preparation, and the capacity of families and social networks to link to nonneighborhood sources of opportunity, rebuilding communities is as much a *product* of poverty alleviation as its mediation.

When the problem of poverty is interpreted as a problem of place, public policy and popular thought confuse the linkages between the revitalization of a neighborhood and the alleviation of poverty. This lack of clarity has become all too characteristic of community development. Neighborhood revitalization may not affect household income and assets—which, after all, are how we define poverty and affluence. Revitalizing a neighborhood involves rebuilding the physical and civic assets of a locality, the residential and commercial real estate, and the mediating institutions that hold places together—schools, congregations, and voluntary associations. The alleviation of poverty comes from and results in an increase in income security, the accumulation of personal and family assets, and the creation of reliable connections to economic opportunities.

The connections between neighborhood revitalization and poverty alleviation are easy to identify; they exist when the former creates the conditions for the latter. Affordable housing developments can provide both social and economic stability and savings. Inner-city retail and small-business developments offer local services and some employment opportunity. Renewed economic investment in a local real estate market creates construction jobs and new business opportunities as well as increasing real estate values, all of which can have an impact on incomes and assets. Social services that support household stability increase the likelihood that adults will be able to work and children will be able to learn. The renewed civic involvement of local mediating institutions clearly broadens opportunities for residents by entering into association with public and private institutions.

But the disconnection is also apparent. Neighborhood development strategies can reinforce the segregation of the poor by building housing in the worst employment markets. In low-income neighborhoods, the attention of civic associations and politicians to highly symbolic commercial and residential restoration projects often far outweighs the benefit of the projects to residents. The community control ideology of neighborhood development often regards locality in strategic isolation from the rest of the economy. And, as Robert Halpern notes in his recent book, *Rebuilding the Inner City,*[3] neighborhood initiatives are consistently faced with the dilemma of asking the most marginal and powerless persons in society to solve some of society's most difficult problems.

The Challenges of the Regional Economy

The inability of neighborhood development activities to link residents to work and increase the mobility of households is a function not only of emphasis but of social/geographical positioning. Rapid changes in many metropolitan economies, particularly in cities that are no longer the economic engines of their region, place neighborhood-centered approaches at a disadvantage. Philadelphia is a case in point: The past quarter century has witnessed a decline in the city's share of regional jobs and population; moreover, the concentration of poverty within the city has accelerated as a percentage of the regional poor. The numbers are striking. During the past 25 years, the city lost 225,000 jobs and 500,000 residents; during that same period, the percentage of the city's population defined as living below the poverty line rose from 15% to 24%. Median household income in the city between 1970 and 1990 declined by $7,000, after it was adjusted for inflation, and the city's share of the regional tax base declined from 27% to 18% during that same period.[4]

Regional job and population growth has occurred increasingly in the suburbs, particularly in "edge city" clusters.[5] Research by Mark Alan Hughes on Philadelphia and 11 other metropolitan "settlement patterns" shows the extent to which the suburbs are the new economic engines: While the city was losing employment in the 1980s, the surrounding counties were adding a half-million jobs. Some of the trends are particularly worrisome for low-income residents: significant declines in city manufacturing employment, not made up for by new city service jobs, and regional retail employment gains that were almost all suburban.[6]

The deconcentration of people is uneven with respect to low- and moderate-income households that might qualify for new suburban manufacturing and retail employment. Housing trends followed by the Delaware Valley Regional Planning Commission document the problem of regional housing affordability and its location to employment centers. Again, there are some dramatic trends: Philadelphia contains 82% of the region's public housing units and 54% of all units that have any public financing or rent subsidy assistance.[7] In almost 40% of the municipalities in the region, the median-income renter cannot afford to rent a unit; in 80% of all municipalities, the median-income household cannot afford to buy a house.[8] The affordability problem is particularly problematic in areas near the new regional employment centers.

When planners track automobile ownership rates among rental households—which are the poorest households in the region, with median incomes of $23,100—the problem of job access becomes even more clear: 34% of all renter-occupied households are without a car, and 43% have only one car. In the absence of city-to-suburb and suburb-to-suburb public transportation, low-income renters have limited geographical mobility.[9]

Finally, data on educational attainment and workforce preparedness are similarly stacked against city residents. All indices, from SAT scores to high-school dropout rates, point to lower urban workforce capacity. To provide a sense of dimension to the problem, Philadelphia's Private Industry Council estimates that, although as many as 300,000 adult city residents are income eligible for federal job training programs, only 20,000 function above an eighth grade level, and more than half (180,000) function below a fifth grade level.[10]

Regional segmentation by income, skills, race, and job location shed light on the prospects and challenges of inner-city residents. If the focus is shifted from neighborhood renewal to poverty alleviation, the policy questions that have to be addressed revolve around housing deconcentration, transportation policy, and workforce readiness:

- Rather than build low-income housing in city neighborhoods that are increasingly isolated from work, should more money and policy effort be spent on suburban affordable housing development, including the deconcentration of existing public housing stock?
- Should the demonstration efforts around the country that promote reverse transportation strategies be expanded and subsidized at a scale that would systematically connect suburban job clusters with the inner-city work force?
- Should a priority for neighborhood development advocates be the design and implementation of the highest-quality job training programs to prepare workers for family-wage jobs, thus increasing their mobility?

Although these questions do not exclude the importance of much of what passes as community development, the regional context forces us to expand the focus of local activity. What is most daunting about these issues is that serious attention to them requires massive public-sector effort—both legal (in the case of housing location) and financial. They thus pose a paradox that the present atmosphere in Washington, D.C. (and most state capitals) is unable to confront. If we want low-income, inner-city residents to enter the ordinary path of social mobility and leave public housing and income subsidy behind, it may first necessitate renewed public investment in housing, transportation, and job training, based on an explicit regional development strategy. If we do not do this, it is likely that regional class segmentation will accelerate and poverty alleviation strategies aimed at the poorest residents will be limited.

Community Development at the Intersection of Neighborhood and Region

For now, no matter what kind of logical case can be made for residential dispersal strategies, the deconcentration of low-income residents is unlikely to occur. As one who has financed affordable housing in affluent suburban counties near Philadelphia, I know how the cards are stacked against significant progress. Land values, subsidy scarcity, zoning requirements, and public opposition make even incremental steps costly and time-consuming. Although the middle class (including African Americans) and those who make incomes even marginally above the regional median will continue to move toward the urban edge and the inner suburban belt, there will be few opportunities for low-income households to follow. The rental units—publicly subsidized or otherwise—are just not there. Similarly, significant increases in the public subsidy of transportation will not happen overnight, although increased subsidy of public transportation is likely to have a more sympathetic public hearing than is the redistribution of populations through residential location—particularly public housing tenants.[11] Better public transportation will most likely occur as the concentration of jobs in various edge cities reaches critical destination mass.

Community development will remain an important policy alternative, in part because it lacks controversy and a claim on significant public intervention and in part because there are few competing alternatives. The reason for continuing to try to rebuild locally is related not just to the policy and activity vacuum but also to the continuing possibilities of cities and the potential of refocusing local activity in a decidedly more regional and economically relevant manner. Most central cities (certainly, Philadelphia) still have significant public- and private-sector job creation possibilities linked to the service and retail establishments of the central business district, government, the construction and transportation industry, the tourist and hospitality sectors, financial and information services, and the existing health care/educational infrastructure. Community development strategies can, and often do, take advantage of the existing growth potential. And some neighborhoods—although not most—are situated within the urban ecology so that their proximity to ports, highways, and vacant land make them potential warehouse and light manufacturing centers, even in an era of regional and global change. Well-conceived community development strategies can help position those places to maximize their advantages through real estate development, marketing, community safety activity, and job training. Finally, the labor of low-income neighborhoods is an important and underused resource that community development activity should view as central to success. Focusing on labor forces community development corporations and other neighborhood associations to think differently about activity priorities and mission.

The future of a more effective community development requires an explicit emphasis on poverty alleviation, which in turn requires linking the possibilities of the inner city to the regional economy. Working in the intersection of neighborhoods and the region requires five perspectives:

1. *Understanding the needs of local labor.* Community development organizations have to shift their view of the neighborhoods they work in, seeing them primarily not as real estate and social service markets but as labor markets. There is a need to gain familiarity with where people work, what their skills are, what the barriers to employment are, and which training programs or schools best link them to employers. Inner-city neighborhoods require "workforce intermediaries,"[12] who can position themselves between residents and employers, even more than they need real estate developers and social workers.
2. *Understanding the regional economy.* The most relevant economic category for neighborhood activity is not the neighborhood itself but regional retail, housing, and employment markets. Neighborhood intervention strategies—even those that focus explicitly on real estate markets—must have some regional context and rationale. Community development organizations must be aware of sectoral growth dynamics, regional retail trends that will affect neighborhood shopping, residential housing trends that will affect different classes of renters and homeowners, and the evolving work preparation gaps experienced by employers.
3. *Residential housing scale and heterogeneity.* In cities with significant declines in population, some real estate markets will not revive in the near future, others will require permanent density changes, and some should be marketed to new populations and uses. In many city neighborhoods, housing activity that has as its goal the rebuilding and growth of the residential marketplace and the attraction of mixed-income populations is critical to neighborhood

survival. Rebuilding residential markets in these areas means pursuing enough scale that long-term value will act as both stabilizer and attractor.

4. *Local businesses as regional actors.* In many inner-city neighborhoods, there are limited opportunities for creating the kind of commercial and industrial activity that will lead to real local job growth. Where locational, consumer, and entrepreneurial opportunities exist, commercial and industrial development should be pursued for three reasons: (1) as part of the local community development emphasis on rebuilding physical assets and maximizing local tax ratables; (2) as part of an effort to create linkages or associations between the neighborhood and outside companies, franchises, suppliers, and distributors that may be linked to the new businesses; and (3) to create a pipeline of local jobs that can be used to develop job skills and relationships for later employment. Local commercial and industrial establishments should be viewed as part of a chain leading to external employment and business opportunities.
5. *Household service programs and asset accumulation strategies.* Although low-income households need traditional social work and health care, an emphasis should be placed on family service programs that lead to income security. The self-employment credit and individual development account experiments throughout the country are important to follow and (when they are successful) to replicate. The former provide credit for small-scale businesses that supplement other forms of household income, and the latter provide savings for educational, housing, and business investment opportunities. Other service interventions of this sort include household budgeting and mortgage and credit counseling. The goal in all of these efforts is to create relationships between households and the mainstream economy.

One example of our effort to work at the intersection of region and neighborhood involves a partnership between several business associations, a community development corporation (the Ogontz Avenue Revitalization Corporation), and several educational institutions. Working in the northwest section of Philadelphia, in a predominantly African American neighborhood, this partnership is slowly developing a community development framework that, although rooted in place, is self-consciously oriented toward work, business, and the regional economy. The neighborhood's relatively high level of economic heterogeneity, its political leadership, and its location close to the edge of the city's political boundary have made it amenable to development strategies designed to connect the neighborhood to the region.[13]

Along with the usual neighborhood development tools—the rehabilitation of housing units, commercial real estate revitalization, and community organizing and social service activities—the community development corporation has embarked on the planning and implementation of an ambitious job creation and workforce preparation strategy. The strategy includes the following:

- *The systematic assessment and recruitment of local labor.* Through a planning grant from the Annie E. Casey Foundation,[14] the neighborhood association and the Delaware Valley Community Reinvestment Fund are analyzing work skills and social barriers to employment in a target area of about 100,000 persons. The result of the planning project will be an ongoing recruitment and assessment program to refer workers to regional work and job training opportunities, including reverse commuter demonstration programs.
- *The identification of regional economic growth opportunities appropriate to local workers.* As part of the Casey planning grant, the neighborhood association, the reinvestment fund, and the Greater Philadelphia First Corporation (a regional business association) are identifying the most promising growth sectors and employment opportunities for the target labor pool. This identification of growth sectors builds on a prior report, issued by the Greater Philadelphia First Corporation,[15] which details elements of a regional economic development strategy but involves a more systematic look at opportunities appropriate to the neighborhood's workforce.
- *Strengthening the local small-business feeder system.* Based on a study carried out by the Pennsylvania Economy League[16] that ana-

lyzed the state of small businesses in the area, including small-business use and satisfaction with local labor, a workplace training program designed to increase the number and quality of workers able to work locally as a first employment step is being created. A neighborhood small-business loan fund has also been created to assist small businesses with a variety of fixed asset and working capital credit needs.

- *The development of a regional employment and training center.* Over and above the workforce needs of local businesses, the community association has moved to position itself as a major supporter of regional job training opportunities. It has done this in two ways: First, through the development of a successful Philadelphia Community College annex that emphasizes work skills appropriate to regional job growth opportunities and is able to customize training curricula for employers; second, through the sponsorship of a regional training center that will function as a customized training and product development center for specific industry clusters. Two engineering departments from local universities, several major corporations located in the suburbs, and the Delaware Valley Industrial Resource Center (a technical assistance agency that works to improve the competitiveness of manufacturing firms) have signed on to assist and contract with the center. The center will also rent space to a business incubator that will work in close coordination with a local cluster of hospitals and schools.

In the planning and implementation of these strategies, it is the regional perspective of the community institution that has made it possible to construct new business and job training opportunities. This is a critical lesson for urban community development, if it is to respond to the new challenges of the economy. Healthy neighborhoods are first and foremost dynamic places, to which people want to move and where residents and businesses maximize their advantages vis-à-vis other places, relationships, and markets. Neighborhood economic self-sufficiency has no meaning in this context. Strong neighborhoods are destination places and incubators; they are healthy, not because they are self-contained or self-sufficient but because their residents are appropriately linked to nonneighborhood opportunities. The current interest in developing comprehensive community development programs must consider this issue carefully, for, to the extent to which comprehensive programs are service based and neighborhood introverted, the essential linkage between residents and work opportunities will be minimized.

Finally, it should be noted that it is not possible for a regionally defined community development practice to exist unless there are regional institutions that can accommodate the planning, information, and program implementation needs of these communities. There are substantial and vital roles for business associations, industry trade associations, labor unions, community colleges, universities, and corporations in preparing and linking neighborhood residents and associations to the economy. The effectiveness of these mediating institutions will increasingly be critical to successful community development efforts.

Notes

1. Probably the most comprehensive review of the efforts of local community development corporations can be found in Avis C. Vidal, *Rebuilding Communities: A National Study of Urban Community Development Corporations* (New York: Community Development Research Center, 1992).

2. The issue of the spatial mismatch between residence and opportunity involves an extensive literature in economics, demography, sociology, and geography. Some of the issues related to low-income housing location and the "geography of opportunity" can be found in a recent issue of the journal *Housing Policy Debate* (vol. 6, no. 1 [1995]). A particularly good analysis of the problem can be found in Douglas S. Massey, "American Apartheid: Segregation and Making of the Underclass," *American Journal of Sociology* 96 (1990): 329-57.

3. Robert Halpern, *Rebuilding the Inner City* (New York: Columbia University Press, 1995).

4. Craig R. McCoy, Lea Sitton, and Thomas Ferrick, Jr., "Vital Signs," *The Philadelphia Inquirer,* September 24-28, 1995.

5. Joel Garreau, *Edge City* (New York: Doubleday, 1991).

6. Mark Alan Hughes, *Over the Horizon: Jobs in the Suburbs of Major Metropolitan Areas* (Philadelphia: Public/Private Ventures, 1993).

7. *Public and Assisted Housing in the Delaware Valley* (Philadelphia: Delaware Valley Regional Planning Commission, November 1995).

8. *Solutions for Affordable Rental Housing in the Delaware Valley* (Philadelphia: Delaware Valley Regional Planning Commission, August 1994).

9. *Delaware Valley Rental Housing Assessment* (Philadelphia: Delaware Valley Regional Planning Commission, 1993).

10. Private Industry Council data, reported in Theodore Hersberg, *Regional Labor Force* (Philadelphia: Center for Greater Philadelphia, 1990).

11. The issue of regionalizing public housing location is getting renewed legal and policy attention, particularly in Texas and Maryland. For the results of the Chicago demonstration program (Gautreaux), see James Rosenbaum, "Changing the Geography of Opportunity by Expanding Residential Choice: Lessons from the Gautreaux Program," *Housing Policy Debate* 6, no. 1 (1995): 231-69.

12. The term *workforce intermediary* was coined in the Austin Neighborhood Initiative sponsored by ShoreBank, a Chicago-based community development bank.

13. Planning data on the neighborhood can be found in *The Neighborhood Investment Action Plan,* developed by the Ogontz Avenue Revitalization Corporation and the Community Development Institute (Philadelphia: Ogontz Avenue Revitalization Corporation, 1995).

14. The Annie E. Casey Jobs Initiative is a six-city, seven-year project designed to link targeted inner-city residents with regional job opportunities. In the Philadelphia site, the Delaware Valley Community Reinvestment Fund is the regional manager for the initiative, and the Ogontz Avenue Revitalization Corporation is the community-based association responsible for recruiting and assessing local labor. The project is sponsored by the Annie E. Casey Foundation, located in Baltimore.

15. Greater Philadelphia First Corporation, *An Economic Development Strategy for the Greater Philadelphia Region, Philadelphia* (Philadelphia: Greater Philadelphia First Corporation, 1995).

16. Pennsylvania Economy League, *Creating a Learning Community: Improving the Growth of Small Business and Employers in West Oak Lane/Ogontz* (Philadelphia: Pennsylvania Economy League, 1996).

Chapter 12

Hidden Economic Development Assets

John P. Blair and Carole R. Endres

❋❋❋

The informal, or underground, economy is often ignored in analyses of economic development. Both academics and practitioners tend to focus on processes that have measurable outcomes. Unfortunately, informal activities are not tracked by any regular statistical data series in spite of their importance. The purposes of this article are to show that the informal economy can be an important part of economic development strategies and to suggest some ways that policymakers might exploit this potential asset.

A definition of the informal economy is a necessary prologue. The informal economy represents both extralegal and illegal-criminal transactions that are not properly reported to appropriate governmental agencies such as the Internal Revenue Service.[1] However, the focus of this article is primarily extralegal activities—activities that would be legal except for nonreporting or minor regulatory violations. We may think of these activities as the productive informal economy.[2] Examples of productive informal activity include home repair, child care, transportation services, elder care, personal services, unreported retail sales, and so forth.

Size and Growth of Informal Activity

To understand the potential developmental impact of the informal economy, it is helpful to consider the size and growth of unobserved activity in both the legal and illegal-criminal sectors. Simon and White indicated that more than half of informal activity is legal but for minor violations or nonreporting.[3] Estimating the size of a sector of the economy that is unobserved by statistical reporting systems is like studying a black hole. Substantial inference is required. Nevertheless, there have been sufficient studies using a variety of methodologies to establish a reasonable size estimate.

The consensus of opinion among experts on the informal economy is that it comprises between 16% and 20% of national income.[4] Al-

AUTHORS' NOTE: We thank an anonymous referee for several insightful comments.

Reprinted from *Economic Development Quarterly* Vol. 8, No. 3, August 1994.

though size estimates of the informal sector depend on the methodology used, any reasonable size estimate suggests that the informal sector is large enough to be a potential development resource.

The informal economy also appears to be growing more rapidly than the economy as a whole. Carson replicated the methods used by other analysts to examine unobserved sector growth.[5] In almost all cases, the estimated growth rate of the unobserved sector was substantially higher than the growth rate of GNP. Between 1974 and 1980 Carson's average estimate of unobserved sector growth was 14.5%, compared to GNP growth of 11%. Feige reached a similar conclusion, suggesting that the informal sector began to significantly outpace the mainstream economy in the mid-1970s and has continued to grow relative to GNP throughout the late 1980s.[6] Future growth is likely to continue to outpace GNP.[7]

The Inner City

The case for including the informal sector in an economic development strategy is particularly strong in inner-city areas, where effective economic development policies may be needed most. There are several reasons for believing that the informal sector is more active in poor inner-city neighborhoods than in the economy as a whole. First, the opportunity cost of participating in informal activities is lower for individuals without mainstream employment opportunities. The decline in the real wage rate among low-skilled workers has also lowered opportunity costs. The continued mismatch between available jobs and the skills possessed by central city residents will aggravate the problem of inner-city job opportunities.[8] Accordingly, the number of persons available for work in the inner-city informal economy will increase.

Second, informal activities allow individuals to circumvent means tests for certain government programs.[9] Hidden income will appeal to the high portion of inner-city residents who receive government support that decreases as earned income increases. Some families may find that combining government income support and informal earnings is a feasible response to the declining real value of many government transfer programs.

Third, a dense net of social and institutional relationships characteristic of some urban neighborhoods has been shown to be useful in supporting the informal economy. Case studies by Dow and Tokatli illustrated how close-knit ethnic neighborhoods provide a network of labor, skills, information, contacts, and codes of loyalty that support informal activities.[10]

Next, the cover function is also provided in an inner-city environment. Many productive informal activities operate with one foot in the informal and one foot in the mainstream economy. Mainstream activities sometimes shelter the informal economy. Many inner-city commercial operations are small proprietorships in which tax avoidance is relatively easy and work rules and other regulations can be skirted. In contrast, suburban businesses tend to be more formal, often franchised, and have more rigorous accounting and inventory control systems. In such organizations, informal practices are more difficult to conceal.

Finally, the demographic composition of central cities will contribute to the strength of the informal economy. Inner cities have a disproportionate share of minority groups, female-headed households, and immigrants. Case studies and consumer services indicate that these groups tend to participate in informal activities both as suppliers and customers.[11] Lower incomes among inner-city residents may make it worthwhile for them to exert the extra effort necessary to find informal suppliers who may provide goods and services at a lower cost.

Policy Implications

The size and growth of the informal sector raise important policy issues. By focusing only on potential loss of tax revenue, many policymakers have failed to recognize four potentially significant implications of the productive informal sector for urban development: (1) the meaning and interpretation of urban data, (2) the safety net function, (3) the human resource and training function, and (4) the potential for entrepreneurial growth and job creation.

Interpreting Local Data

Houston showed that under certain circumstances the informal sector may move in opposite directions from the mainstream economy.[12] For instance, increases in unemployment in the mainstream economy may contribute to increased informal employment. Similarly, price changes in the informal sector could moderate price increases in the observed sector. As prices increase in the mainstream economy, individuals may seek lower-priced substitutes in the informal sector. Thus the amplitude of the statistical business cycle may exaggerate fluctuations in the real (formal and informal) economy. To the extent that the informal sector plays a stabilizing role, observers who focus only on official statistics could misinterpret real unemployment, price, and income trends in their areas.

Urban policy makers may find it useful to systematically monitor productive informal activities. Such an approach may be critical to understanding changes in inner-city neighborhoods. Toward this end, observation techniques of anthropology will be as important as traditional survey research, because informal sector participants are naturally reluctant to answer questionnaires honestly.

Development officials might monitor informal activities by observing and initiating casual conversations with street vendors, gypsy cab operators, and other small business proprietors, who may operate partly or entirely in the informal sector. Discussions with other inner-city and neighborhood leaders may also be useful. As rapport with individuals in the informal sector grows, a better understanding of inner-city economies will emerge, and public officials may gain insights into policies that might help informal producers.

The Safety Net Function

The role of the informal sector in providing a source of support for unemployed workers or individuals receiving public assistance is an important function of the unobserved sector. The largest percentage of informal workers tends to be from disadvantaged groups living in central cities.[13] Many individuals combine welfare and other transfer payments with earnings from informal work. To the extent that informal activity enables welfare recipients and the working poor to maintain higher consumption levels, the extra spending will contribute to the support of neighborhood retail businesses. Furthermore, earnings from informal activities supplement transfer payments and may moderate pressure to increase welfare programs.

Informal work opportunities also provide a safety net for some workers who have income and wealth above the level to qualify for many public assistance programs but who are experiencing transitory unemployment. Income from informal activities may be a critical part of a financial package (including draw-down of savings, unemployment compensation, the earnings of additional family members, and so forth) that helps maintain at least a near-comparable living standard for some time after a layoff. Transitory assistance from informal activity may be particularly important in light of the fact that most unemployment spells are for less than 20 weeks.

The Human Resource and Training Function

The human resource and training function refers to the role of informal employment in upgrading work skills. For many new job entrants, the informal sector represents a "sandbox" in which they learn skills and make contacts that can lead to mainstream jobs. Individuals with low productivity may develop both social and technical job skills that often help them obtain mainstream jobs. Minimum-wage laws, employer payroll taxes, and some work rules add to the cost of labor. Under these conditions, employers may find it unprofitable to hire low-productivity workers in the mainstream economy. Workers can be hired at lower cost in the informal sector. In the best cases, employees will use these experiences to increase their productivity and obtain better, mainstream jobs. Administrators of job-training programs may wish to build on informal experiences by encouraging individuals to participate in training programs while continuing their informal work. Unfortunately, some casual labor conditions may actually discourage job-skill enhancement by not rewarding improvements and not punishing sloppy work habits.

A strong case can be made for benign neglect in the enforcement of welfare work rules. It is reasonable to suppose that individuals who received public assistance while working off-the-books would have a better chance of becoming self-supporting than someone who received only public assistance. Informal work can often provide opportunities to develop contacts, skills, and work habits that enhance employment prospects in the formal sector, thus reducing public assistance dependency. Of course, the argument for benign neglect is not absolute. Considerable judgment will be required to determine when the level and type of off-the-books income earned by a public assistance recipient becomes dysfunctional or an abuse of the system.

The developmental function of the informal sector has been widely recognized in less-developed nations.[14] It has smoothed the transition of individuals migrating from rural to urban areas. The United States has recognized the significance of the informal sector in less-developed countries by funding the PISCES Project.[15] Because informal microenterprises are the largest employers in most less-developed countries, the PISCES Project was undertaken to identify these enterprises and to assist them without requiring them to meet the obligations of formal businesses. Despite their support of this project in less-developed countries, U.S. policymakers have not fully appreciated the importance of informal activities in the United States. Consequently, urban policymakers have failed to adopt a domestic-policy agenda that exploits the economic development potential of the informal sector.

Entrepreneurship and Job Creation

Urban development planners are increasingly focusing on entrepreneurial development as a means of stimulating local growth. At the same time, many observers have suggested that economic development efforts should be oriented toward supporting existing local businesses. Because of their small size, informal activities tend to be very entrepreneurially oriented—and therefore good targets for development focus.[16] For some informal businesses, movement into the mainstream economy can stimulate growth by providing greater access to capital, encouraging advertising, and presenting opportunities for linkages with other mainstream firms. A challenge for policymakers is to stimulate informal activities and in some cases to help move them into the mainstream economy. The previously mentioned PISCES Project, for example, has encouraged entrepreneurship and has resulted in job creation in less-developed countries.

The job creation process in the urban United States can be illustrated in the case of a small, informal business. A woman who provides off-the-books day-care services in her home may find it advantageous to report only a fraction of her income. As her business and day-care skills improve, she may recognize advantages from starting a formal, licenced facility with a few employees. Consequently, some or all of her services may shift from the informal to the formal sector. Development officials could play several roles in encouraging this process, including helping the business owner visualize the growth potential, advising the proprietor regarding various aspects of starting a business, minimizing complicated regulations, and assisting with start-up capital.

Balkin suggested that movement into the formal economy might be expedited by a graduated regulatory process.[17] He suggested that small businesses that operate partly or completely in the informal sector should be subject to reduced inspection, compliance, and reporting requirements. As businesses expand they would become subject to stiffer requirements. Thus informal activities could be legitimized without having to meet all the requirements of established businesses. Earlier recognition by public officials could enhance future growth prospects.

Some development groups have recognized the potential economic benefits of informal activity. There have been well-publicized cases of community banks, economic development agencies, and savings associations making loans to very small enterprises that probably had their origins in informal activities. Revolving loan funds, repayable weekly, have been suggested to assist homeless men in starting small "microenterprises" that would probably operate in the informal sector.[18] Gerry suggested that with supportive public policies, the informal economy "could be transformed from a stagnating and inward-looking complex of coping mechanisms of the urban poor into an authentic engine of growth."[19]

Community development bank officials should be sensitive to the tax-avoidance behavior of many small businesses. When entrepreneurs operate with one foot in the informal economy, their official records will reflect neither their true financial position nor their ability to repay a loan. Since community development banks are designed to encourage growth among small area businesses (which may operate at least partly in the informal sector), loan officers may want to consider the possibility that unreported income may contribute to a firm's ability to repay loans. However, informal income tends to be less stable than formal income, in part because legal actions can shut down informal activities. Consequently, decisions regarding unreported income probably require sound, experience-based judgment.

Development professionals may have difficulties identifying promising informal endeavors that constitute all or part of a business activity. Many businesses may be unwilling to deal with governmental agencies because some of their activities may be illegal. Consequently, informal entrepreneurs may best be approached incrementally. Business development specialists have reported that owners of small proprietorships are often reluctant to reveal financial and accounting data. Therefore, small business advisors may wish to emphasize marketing and organizational development services during initial contacts. As the firm grows, activities may move into the mainstream economy to exploit mainstream opportunities and through necessity as informal activity becomes harder to mask.

Conclusion

To exploit the safety net, developmental, and job creation functions, policymakers should distinguish among the variety of informal activities and treat some activities with benign neglect or even encouragement. This policy requires a distinction between destructive informal activities and activities that contribute to the safety net, developmental, or job creation functions. More careful analysis of the potentially positive functions of the informal sector and formulation of relevant policies appropriately implemented could contribute to the economic development prospects of many inner-city areas.

Jane Jacobs described the process of economic development as "adding new work to old."[20] As development planners work with informal enterprises to add new lines of business in the formal sector, they may pioneer a useful twist on this well-known strategy.

Notes

1. Given our definition, some businesses may operate simultaneously in both the formal and underground economies.

2. Some observers have argued that many illegal activities are productive. For instance, prostitution and drug dealing are activities that provide services for which individuals are willing to pay. In some instances illegal-criminal activities may be productive. However, most of these types of activities are illegal because their social costs outweigh other benefits, hence our generalization that illegal-criminal activities tend to be unproductive.

3. Carl P. Simon and Ann D. Witte, *Beating the System: The Underground Economy* (Boston: Auburn House, 1982).

4. C. S. Carson, "The Underground Economy: An Introduction," *Survey of Current Business,* U.S. Department of Commerce, Bureau of Economic Analysis, 64 (May 1984): 21-37; and Edgar L. Feige, "The Meaning and Measurement of the Underground Economy," in *The Underground Economy,* ed. Edgar L. Feige (New York: Cambridge University Press, 1989), pp. 51-3.

5. Carson, "The Underground Economy: An Introduction," 106-17; Edgar L. Feige, "A New Perspective on Macroeconomic Phenomena—Theory and Measurement of the Unobserved Sector of the United States Economy: Causes, Consequences, and Implications" (Paper presented at the 93rd Annual Meeting of the American Economic Association, September 6, 1980); Peter M. Gutmann, "The Subterranean Economy," *Financial Analysts Journal,* November/December 1977, pp. 26-7, 34; Carl P. Simon and Ann D. Witte, *Beating the System: The Underground Economy*; Vito Tanzi, "The Underground Economy in the United States: Annual Estimates 1930-1980," *International Monetary Fund Staff Papers* 30 (June 1983): 283-305; and U.S. Internal Revenue Service, "Noncompliance with U.S. Tax Laws—Evidence on Size, Growth and Composition," in *Income Tax Compliance* (American Bar Association, 1983): 15-111.

6. Feige, "The Meaning and Measurement of the Underground Economy."

7. John P. Blair and Carole R. Endres, "Prospects for the Informal Economy," in *The Future of Urban Environments,* ed. Gappert and Knight (Newbury Park, CA: Sage, forthcoming).

8. J. D. Kasarda, "Urban Industrial Transition and the Underclass," *Annals of the American Academy of Political and Social Science* 501 (1989): 26-47.

9. Carson, "The Underground Economy: An Introduction," 21-37.

10. Leslie M. Dow, "High Weeds in Detroit: The Irregular Economy Among a Network of Appalachian Migrants," *Urban Anthropology* 6 (1977): 111-28; Nebahat S. Tokatli, "Immigrants in a Restructuring Economy: The Case of Turkish Immigrants in Paterson, New Jersey and Brooklyn, New York" (Paper presented at the Joint International Congress of ACSP and AESOP, July 9, 1991); and Tokatli, "Imported, Informalized and Place-Bound Labor: A Turkish Immigrant Community in Paterson, New Jersey" (Paper presented at the Annual Meetings of the Association of Collegiate Schools of Planning, November 3, 1990).

11. Louis A. Ferman, Louise Berndt, and Elaine Selo, *Analysis of the Irregular Economy: Cash Flow in the Informal Sector,* Institute of Labor and Industrial Relations (University of Michigan—Wayne State University, March 1978).

12. Joe F. Houston, "The Policy Implications of the Underground Economy," *Journal of Economics and Business* 42 (February 1990): 27-37.

13. Phillip Mattera, *Off the Books: The Rise of the Underground Economy* (New York: St. Martin's, 1985).

14. A. Portes, M. Castells, and L. Benton, eds., *The Informal Economy: Studies in Advanced and Less Developed Countries* (Baltimore: Johns Hopkins University Press, 1989); and J. J. Thomas, "Synthesis of Comments and Discussion: Methodology and Theory," *The Informal Sector Revisited,* ed. D. Turnhan, B. Salome, and A. Schwartz (Paris OECD, 1990): 88-92.

15. Jeffrey Ash, *The PISCES II Experience: Local Efforts in Micro-Enterprise Development* (vol. 1) (Washington, DC: Agency for International Development, 1985).

16. We use the term *entrepreneur* to mean someone who organizes resources and takes risks in the production process. One-person operations, such as gypsy cabs, which characterize many informal activities, clearly have a high ratio of entrepreneurs to total employees.

17. Steven Balkin, "A More Favorable Playing Field: Regulatory Policy for Micro-Enterprises of the Poor" (Paper presented at the Self Employment Strategy—Building the New Economy conference, October 1989).

18. S. Balkin, "Entrepreneurial Activities of Homeless Men," *Journal of Sociology and Social Welfare* 19 (December 1992): 129-50.

19. Chris Gerry, "Developing Economies and the Informal Sector in Historical Perspective," *Annals of the American Academy of Political and Social Science* 493 (September 1987): 9.

20. Jane Jacobs, *The Economy of Cities* (New York: Random House, 1969).

PART IV

State and Regional Issues

Chapter 13

The Next Wave

Postfederal Local Economic Development Strategies

Susan E. Clarke and Gary L. Gaile

❋❋❋

Recent changes in economic conditions and national policies mark out a new local terrain. In the past, American communities were able to rely on some measure of aid from state and national government. Although this external aid was substantially less than in other industrialized countries, there is strong evidence that it allowed local governments to offset the effects of structural economic changes.[1] As such aid wanes, local governments are seeking new ways to generate sufficient revenues. This article assesses the policy choices American local officials are making in response to resource changes, particularly the withdrawal of federal funds for local economic development activities in the 1980s. It draws on findings from a national survey of local economic development officials in 1989 to describe the postfederal policy context, local policy responses to changing resources, and the possible economic effects of current strategies.

The New Centrality of Locality

A volatile local resource base is a fundamental, enigmatic feature of this new setting. There is little reason to hope local resource dilemmas will improve in the near future. Independent of who controls the White House, Congress, or state capital, the effects of global competition, recessions, increasing poverty and crime, changes in military priorities, and environmental concerns are apt to dominate political agendas in the next decade. The potential for a community-oriented policy response is weaker than in the past, thanks to ideological interpretations of economic restructuring processes that

AUTHORS' NOTE: This research was supported in part by the Center for Public Policy Research, University of Colorado, and the Economic Development Administration, U.S. Department of Commerce, Grant 99-07-13709. The statements, findings, conclusions, and interpretations are those of the authors and do not necessarily reflect the views of the Economic Development Administration.

Reprinted from *Economic Development Quarterly,* Vol. 6, No. 2, May 1992.

trivialize the role of cities in the national economy and undercut the rationale for national urban policies.[2] Neither the perceived crises nor the political constituency concerns that drove urban policy over 2 decades ago[3] are as salient now as international crisis and competition issues. In the absence of alternative interpretations of economic trends, national and state policymakers will be reluctant to seemingly jeopardize economic productivity goals by directing resources to communities. Thus there is every sign that the chilly climate for cities will persist in the near future.[4]

Ironically, these broader trends may enhance local politics in the coming years. If contemporary economic change processes are uneven and imperfect rather than single, uniform, monolithic global processes, there may be greater potential for local political discretion and choice.[5] And, as Mayer points out, many of the conditions necessary for the attraction and expansion of new production processes and complexes cannot be organized centrally by either the national, state, or the multinational corporation.[6] In these settings, local governments play critical roles in coordinating the economic and political resources necessary for national economic growth and development. This claim for the new centrality of locality is not necessarily a more optimistic view nor one promising more local autonomy. But it does emphasize that place-specific differences can be competitive advantages[7]—that is, cities matter. By default, if not design, current global trends and national policies promote a new economic localism and the potential resurgence of local politics.[8]

Why anticipate that this new localism will engender significantly different economic development activities? The direction of local policy change is problematic, given the withdrawal of federal resources previously supporting local efforts to reduce investment uncertainty and risk. At a minimum, there is a possible reversion to caretaker roles and more risk-aversive, less active local agendas. And even if global economic trends do create a need for greater local policy coordination, local officials now must resort to their own revenues to provide such incentives and to entertain greater risk in joint projects.[9]

The Contemporary Policy Context

In a national study of local economic development efforts, we analyzed patterns of local economic development activities from 1978 to 1989. We examined whether cuts in federal resources affected the level of local economic development efforts, and whether local efforts relying on own-source revenues, rather than federal resources, appeared to be relatively risk aversive or risk taking.[10] Here we report on the new local policy context—local strategies undertaken in the absence of federal resources. The study includes all American cities with populations above 100,000 in 1975; these were the cities primarily targeted for federal economic development aid, and these are the cities now coping with its withdrawal. Data on current and historical local economic development activities were collected for the 178 cities above the population threshold. Officials in all cities above the 1975 population threshold were contacted to collect information on current economic development strategies; the findings here reflect responses from 101 cities.[11]

Resource Change and Local Policy Efforts

Federal program funds prompted the acceleration of local economic development activity in the late 1970s and 1980s. As Table 13.1 shows, by the mid 1980s federal program funds were the most important revenue sources for local economic development. All together, 49% of the revenue sources for local economic development important before 1984 involved federal program funds. Community Development Block Grant (CDBG) funds received most mentions (20%) although Urban Development Action Grant (UDAG; 10%) and Economic Development Administration (EDA; 10%) funds were also important. But even during this period of federal attention, local General Funds constituted 16% of the sources mentioned.

In the late 1980s, the volatility of federal resources, as well as absolute cuts in the level of resources, forced many local officials to turn to nonfederal resources to support local economic development activities. How did this change in

TABLE 13.1 Revenue Sources Most Important for Economic Development Activities (percentage of sources mentioned)

	Before 1984	*Since 1984*
Community Development Block Grant	20	18
General fund	16	25
Urban Development Action Grant	10	—
Economic Development Administration	10	—
General federal programs	9	7
Tax increment financing	8	10
Program income	—	7
Private/corporate contributions	—	8
Number of sources mentioned	174	226

resource availability affect local economic development efforts? Of the 237 problems mentioned by local officials as constraints on local economic development efforts, resource issues stand out: Lack of funds, tight budgets, and lack of staff were mentioned most frequently (24%) as a major problem.[12] But city officials also report a tremendous increase in public and private cooperation. For 75% of the cities, this is a dramatic increase in public/private activity compared to 5 years ago.

With the loss of federal program funds, local officials turn to three other revenue sources to support economic development activities: relying increasingly on local general funds, shifting CDBG funds to economic development purposes, and using revenues from successful redevelopment projects (Table 13.1). The first two suggest reallocation of funds across functional areas; in the absence of federal funds previously used to subsidize local economic development efforts, cities turn to general funds or the declining pot of CDBG funds. Cities increasingly devote CDBG funds to economic development: the median use of 10% in fiscal year 1988 is double the median reported by these cities for 5 years ago. Greater reliance on tax increment financing and program income from previous redevelopment projects signal attempts to make economic development a self-financing local enterprise.

Overall, local economic development efforts are expanding in the postfederal era. The local focus has clearly shifted to the state level although there is no perception that state programs have compensated for the federal absence or offset federal program cuts.[13] The findings suggest a more politicized, more fragmented local setting in the wake of federal withdrawal; increased competition over general fund revenues, CDBG funds and tax increments makes consensus more difficult, a situation only compounded by basic problems of aging infrastructure and land shortages.

Shifts in Local Economic Development Policy Orientations

The new local policy strategies reflect a shift away from conventional economic development orientations toward market-based, or entrepreneurial, approaches.[14] Two features distinguish market-based economic development strategies from more conventional approaches: (1) the focus on facilitating value-creating processes by private investors and (2) the investment and risk-taking role adopted by local officials.[15] Variously labeled "generative development," "enterprise development," and "entre preneurial" strategies, these new approaches center on public policies that encourage wealth creation rather than subsidize locational decisions or employment strategies.[16] As such, they demand a new range of local roles and responsibilities.

Because this new generation of complex, entrepreneurial local policy initiatives operates at the grass-roots, local strategies are less visible and less easily characterized than nationally designed local economic development programs. These initiatives are not embodied in specific programs or agencies but, rather, in particular policy tools or instruments sharing certain common features that distinguish them from previous approaches.[17] Greater tolerance for *risk* is one of these features; others include differences in *purpose* (stimulating new enterprise rather than stabilizing or protecting);[18] in *focus* (using government authority to shape market structure and opportunity rather than influencing the functions of individual businesses);[19] in *criteria* (using market criteria, such as maximizing rates of return, rather than political criteria in setting priorities for allocation and investment of public funds); in *finance* (leveraging public and private resources rather than relying on one or the other); in *public roles* (relying on joint public/private ventures for implementation of economic development projects rather than bureaucratic approaches); in *administrative ease* (administering through quasi-public agencies rather than line agencies); in *decision processes* (involving negotiated decisions on a case-by-case basis rather than juridical, standardized decision processes); and in *linkages* (establishing contractual, contingent relations with those affected rather than linkages based on rights or entitlement).

This new generation of policy approaches is rooted in a legacy of diverse federal programs encouraging local discretion, investment orientations, and greater risk taking. The early National Development Council traveling seminars on packaging "deals," the UDAG leverage ratios and kicker features, and the introduction of CDBG float loans and Section 108 guarantees were particularly important in stimulating new entrepreneurial practices. Over time, these entrepreneurial strategies began to diffuse across cities and over programs. The federal urban legacy, however, was ambivalent: It promoted certain redevelopment goals, encouraged large commercial and industrial projects, initiated new local development institutions, and endorsed innovative financing techniques. When cities were cut loose from federal ties, they retained some of the entrepreneurial aspects of past federal programs but geared them to local ends. But local entrepreneurial approaches remain crucially dependent on state enabling legislation and fiscal discretion authorities. Many of the more entrepreneurial tools require positive state action, sometimes including changes in state Constitutional limitations on local fiscal powers.[20] Ironically, although the extensive local use of these tools is often attributed to the withdrawal of federal oversight, federal tax code changes, and the deregulation climate for private investment, their ultimate range may be limited by state governments.

In 1989, local officials reported notable policy shifts prompted by cuts in federal economic development programs. These shifts included changes in local objectives, smaller projects, more diverse projects, and smaller public shares of development costs but little change in sectoral, spatial, or minority targeting efforts in most cities. Nearly 40% of the cities responding claimed that federal cuts resulted in more diverse economic development objectives, in many instances including more housing and social programs responsive to local priorities. About one-third (29%) of the cities saw these cuts as negative, leading to restricted, more narrow local objectives. Most cities are now involved in smaller projects (55%) and many reported the nature of local projects is now more diverse (38%), often because the choices are market-driven but more often because they are now more open to local political and organizational pressures. Public shares of development costs have decreased in over 46% of the cities responding. Targeting efforts appear to be unchanged in a majority of cities, although the percentage of cities claiming to do more minority (17%) and low income (20%) targeting is similar to the percentage claiming to do less (15% and 21% respectively).

Local officials believe their current policy orientations differ in important ways when compared to the past: current policies are oriented more toward risk-taking, job growth, downtown development, job creation, local

TABLE 13.2 Most Frequently Used Economic Development Strategies (percentage of cities that ever used strategy)

Strategy	*Percentage*
Comprehensive planning	93
Capital improvement budgeting	91
Marketing and promotion	86
Infrastructure as in-kind development contribution	83
Land acquisition and demolition	80
Revenue bonds	79
Strategic planning	74
Revolving loan fund	73
Streamlining permits	73
Selling land	69
Industrial parks	68
Below market rate loans	67
General obligation bonds	65
Local development corporations	63
Annexation	62
Historical tax credits	60
More metropolitan and regional cooperation	58
Tax increment financing	56
Industrial development authorities	55
Enterprise zones	55
Use program income for economic development	55
Special assessment districts	54
Community development corporations	52
Land leases	52
Trade missions abroad	50

NOTE: Italics indicate majority of cities first used strategy after 1980.

concerns, market feasibility, and aiding local firms when compared to policies 5 years ago. They also are more likely to see the city's role as a public developer engaged in contractual relations rather than responding to entitlement obligations. This brief profile intimates a more entrepreneurial orientation, one emphasizing job creation and indigenous growth and accommodating more risk-taking by local officials.

Local Policy Choices When Using Nonfederal Resources

These impressions of shifting local policy orientations are borne out by comparing the clusters of tools adopted recently with those reported in use before 1980. Local officials reported whether and when they had first used 47 strategies using nonfederal resources. At least 50% of the cities reported using 25 core strategies; of those 25, 7 were first used after 1980 by at least half the cities. As listed in Table 13.2, planning, management, and marketing strategies are prominent and pervasive before 1980. Land acquisition, building demolition and the use of infrastructure to support development are also classic local development tools. Direct federal influence is evident in the importance of tax-exempt bonds, historical tax credits, and program income. Federal programs also spread or diffused more widely the use of revolving loan funds, local development corporations, enterprise zones, strategic planning, below market rate loans, and community development corporations. Thus the current menu of local economic development strategies is shaped by past federal initiatives and residual local powers.

TABLE 13.3 Patterns of Use of Local Economic Development Strategies (percentage of cities that used strategy)

	Before 1980	*After 1980*	*Under Study*
Comprehensive planning	84		
Capital improvement budgeting	77		
Revenue bonds	69		
Land acquisition and demolition	67		
Infrastructure as in-kind development contribution	64		
Selling land	59		
General obligation bonds	59		
Industrial parks	57		
Annexation	53		
Marketing and promotion	47	39	
Strategic planning	46		10
Industrial development authorities	40		
Local development corporations	38		5
Special assessment districts	37		11
Land leases	37		5
Land banks	34		10
Below market rate loans	33	34	
Enterprise funds for public services	33		
Historical tax credits	31		
Streamlining permits		59	7
Enterprise zones		52	
Revolving loan funds		45	
Business incubators		40	21
Trade missions abroad		39	
More metropolitan and regional cooperation		36	14
Use program income for economic development		33	
Tax increment financing		32	12
Foreign trade zones			10
Export and promotion			10
Equity participation			10
Taxable bonds			9
Tax abatements—targeted at new business			7
Equity pools: public-private consortiums			7
Venture capital funds			5
Linked deposits			5
Sale-leasebacks			5

Patterns of Policy Change

But some of the most heavily used local strategies are relatively recent innovations. Table 13.3 sorts out strategies used before 1980 from those first used in the postfederal era. The first wave of policies—those reported as first used before 1980 by at least 30% of the cities responding—rely heavily on cities' ability to regulate and facilitate development by land use controls, public services, and provision of infrastructure (Table 13.3). Only limited, traditional financial tools entailing debt or cheap loans are noted; there is scant evidence of the revenue-generating financial tools or higher-risk city roles that characterize recent entrepreneurial approaches.

In the second wave—tools adopted since 1980—cities are characterized by a stronger investment and entrepreneurial approach; the use of revolving loan funds, below market loans, and program income indicates generation of revenue streams independent of federal programs and tax revenues. Further, both enterprise zones and tax increment financing districts let cities reorganize their local fiscal structure to channel future revenues into allocation procedures outside the usual budgetary processes. In this sense, revenues come into the city on the basis of public investment decisions, rather than tax policies or federal programs, and future resources may be allocated with little public notice or accountability. Finally, attention to markets and business startups—two hallmarks of entrepreneurial approaches—also characterize this new postfederal era.

Looking ahead, entrepreneurial orientations predominate.[21] Although a wide range of programs are reported as "under study," those mentioned most often share a number of traits. The importance of business incubators as a recent and future strategy underscores the salience of recent approaches encouraging new business start-ups rather than conventional "smokestack chasing" strategies. There is also a clear interest in spatial reorganization of the local tax and resource base. The interest in greater metropolitan and regional cooperation on development issues suggests that economic regionalization issues may be more manageable than political consolidation efforts. Reports of the extensive use of regional economic development associations further attest to the viability of a regional approach despite assumptions of interjurisdictional competition.

This is complemented by an interest in carving the local tax base into foreign trade zones, special assessment districts, tax increment financing districts, and enterprise zones. In each instance, these spatial arrangements tend to reduce the revenue base allocated by normal budgetary processes; tax revenues from these arrangements often are dedicated to debt service and further areal redevelopment rather than enhancing the city tax base. Finally, it appears that local officials are adopting a more business-oriented approach toward use of local assets; potential land use control strategies emphasize the flexible management of land, and its exchange value, rather than permanent transfers or sales. Similarly, there is a growing inclination to view public capital in terms of its investment potential, with the greater public risk taking that implies.

Changes in Institutional Arrangements

In most cities, mayors take the lead in promoting economic development. Many cities also indicate that deputy mayors have special responsibilities in this area. Economic development line agencies reporting to the mayor and the mayor/city manager or city council have lead responsibility for policy formulation in most cities; over 50% place lead responsibility in economic development line agencies that are separate from community development or planning agencies. Thus, the design of economic development policy rests with elected officials or those appointed by them. In contrast, economic development line agencies often share implementation responsibilities with special authorities or quasi-public organizations. Cities often turn to organizations such as citywide development corporations or special authorities such as redevelopment agencies and port authorities to carry out economic development projects. Many of these organizations have special financing authorities or resources not available to line agencies; further, they are able to bring expertise and resources from the private sector that are especially germane to economic development projects.

In the absence of federal programs, additional, place-specific means of coordinating private and public sector commitment will be critical. In this sample, 63% of the cities report using local development corporations. Cleveland credits its recent turnaround, for example, to institutional coordination of the business community, neighborhood groups, and local government agencies. The institutional arrangements include Cleveland Tomorrow and its offshoot, the Neighborhood Program Inc., a citywide development corporation, the Cleveland

Foundation, the city's Neighborhood Partnerships Program, and a network of burgeoning neighborhood local development corporations.

Some hope for disadvantaged groups in American cities rests in community development corporations (CDCs) and local development corporations (LDCs). Some cities, such as Miami, view these organizations as an alternative delivery system for carrying out economic development activities in the postfederal era. They may become pivotal as cities reorganize their tax base into discrete areas; LDCs and CDCs in those areas may claim some voice in the development promoted in those areas as well as in the allocation of revenues generated by these new arrangements. Yet this presumes an autonomous capacity and technical knowledge base not yet apparent in most such groups. At this stage, most LDCs and CDCs are capitalized by public funds and staffed by city agencies and private organizations; many have displaced older voluntary organizations or share power with these groups in an uneasy alliance.

Analyzing the Effects of Local Choices

Attempting to assess the effects of local economic development strategies is a quagmire of good intentions and bad measures. An interest in finding out whether policies have the desired effect is laudable, but there is little consensus on appropriate measures of success or impact.[22] For many city officials, project completion is unique enough to be a salient and sufficient success measure.[23] Few have the means or interest in monitoring project-specific revenue and job impacts so there is little systematic data on project effects, especially those without federal funds or monitoring requirements.

Indeed, most local officials adopt a portfolio mentality in which they assume some projects, such as a convention center, are "loss leaders" but that the overall local policy effort contributes to local economic growth.[24] Nevertheless, analysts persist in measuring the success of local policy efforts in terms of changes in local per capita income and local employment per resident. Not surprisingly, there is little evidence that local economic development policies affect either of these measures. As any local official will point out, the former is a measure of wealth beyond the control of local efforts and often an artifact of jurisdictional boundaries. There is little evidence that changes in residential wealth, as reflected in income measures, are associated with changes in a city's fiscal well-being.[25]

Jobs are another matter. Jobs for city residents are important goals for city officials for many reasons. Most local economic development strategies emphasize job creation and retention as their primary justification. Trying to shape a city's employment structure seems more tangible than influencing the distribution of wealth; it involves negotiations with developers and merchants in which the city can bring something to the table. Cities can negotiate jobs for city residents and set-asides for local workers and contractors.[26] These concessions will not necessarily alter the overall employment structure but they do display city initiative and generate local revenues. Unfortunately, employment growth may not be associated with income growth due to commuting patterns, wage differences, and overlying jurisdictional boundaries. Thus cities with healthy job growth may also have impoverished populations.[27] Further, increases in employment may be mixed blessings because they entail higher service costs that may not be borne by commuting workers.[28] But for both political and fiscal reasons, job-related measures are the primary concern of local officials and is the success indicator we use in the following preliminary assessments.

Due to the newness of these postfederal strategies, it is too early to determine conclusively if they have brought about net changes in local jobs and revenue. Given this necessary disclaimer, the last two tables compare local growth features of cities adopting specific tools and those that have not. For Table 13.4, local policy activity is measured with an additive index of specific entrepreneurial tools: whether a city first used revolving loan funds, venture

TABLE 13.4 New Economic Development Tool Usage Index and City Growth Features

Job growth rate, 1987	.20
Fast-growing companies as a percentage of new firms, 1987	.19
City tax per capita, 1985	–.18
Property tax per capita, 1985	–.18
City government employees per 10,000 population, 1985	–.18
City government general expenditures per capita, 1985	–.16

SOURCES: New Economic Development Tool Usage Index—An additive index measuring whether a city had used the following tools for first time in 1980 or later: revolving loan fund, venture capital, net cash flow participation, interest subsidies, equity participation, equity pools established by private/public consortiums, or use of program income for economic development purposes. Usage measures are based on a national survey of local economic development officials. Job Growth Rate—Change in private employment between January 1983 and March 1987. Bureau of Census and Bureau of Labor Statistics, as reported in *Inc.*, March 1988, p. 76. Fast-Growing Companies—The percentage of all companies founded between January 1983 and July 1987 by its percentage of employment growth during the same period. Companies with an index of 20 or higher were classified as high growth. As reported in *Inc.*, March 1988. City Fiscal Data—County and City Data Book.

NOTE: All coefficients significant at .05 level.

capital, net cash flow participation, interest subsidies, equity participation, equity pools established by private/public consortiums, or used program income for economic development purposes in 1980 or later.[29] Table 13.4 suggests that, controlling for region, cities using entrepreneurial strategies have higher job growth rates and higher proportions of fast growing new firms than cities without these tools. Further, they are operating with significantly lower taxes, lower expenditures, and lower levels of city government employment than nonadopting cities.

In addition, there is some modest support for associating use of specific policy tools with these city growth features. Statistical analyses comparing group means (*t* tests) of cities who had ever used a particular tool with those who had never used a particular tool (Table 13.5) show several instances of tools whose use appears to effectively distinguish among communities with fast economic growth and those with slow growth. Not all of the tools associated with distinctive growth differences fit a strict entrepreneurial definition; interestingly, they also represent a cross-section of capital, land, and institutional strategies. Although neither Table 13.4 nor Table 13.5 should be interpreted as attributing these growth features to specify policy choices, the findings do make a case for further exploration of these new local strategies.

Conclusions

Most American cities in the postfederal period are making increased use of relatively entrepreneurial strategies that entail risk of own source revenues and substantial opportunity costs. There are concomitant shifts in policy orientation: the ascendance of market feasibility over social criteria, the resurgence of downtown as the locus for redevelopment, a trend toward reliance on nonprofit organizations rather than line agencies for implementation, and the redefinition of city responsibilities as a public developer. We find some evidence that use of these entrepreneurial strategies distinguishes communities with job growth, new firm formation, and fast-growing firms from those less fortunate. This is not necessarily a causal relationship but it indicates a clear difference in the public investment climate in communities adopting market-oriented strategies and those with more conventional orientations.

Although this analysis indicates that entrepreneurial approaches are becoming more characteristic of city economic development strategies, several factors guard against anticipating a wholesale conversion to this orientation. Most significantly, an entrepreneurial policy orientation capable of comprehensively dealing with the economic transformation is-

TABLE 13.5 Economic Development Tool Usage and City Growth Features; Means of Standardized Residuals from Trend Surface Analysis

Policy Tools	*City Growth Features*	*Ever Used Tool*	*Never Used Tool*
Marketing and promotion	Increasing Job Growth Rate	.22	–.58
Cash flow participation		.69	–.09
Selling land		.24	–.31
Equity pools funded by public-private consortium	Fast growing companies as a percentage of new firms	.21	–.07
Cash flow participation		.51	–.17
Marketing and promotion		.13	–.53
Community development corporations		.58	–.20
Enterprise zones	Increasing relative business growth rate, 1980-1986	–.15	.01
Interest subsidies		1.23	–.07
Land acquisition and demolition		1.15	–.06
Land leases		1.07	–.05
Loan guarantees		.83	–.03
Infrastructure as in-kind contribution	Increasing business birthrate	–.06	.68

SOURCES: Job growth rate—Change in private employment between January 1983 and March 1987. Bureau of Census and Bureau of Labor Statistics, as reported in *Inc.*, March 1988, p. 76. Fast-growing companies—Young companies enjoying high growth rates. The percentage of all companies founded between January 1979 and July 1987 by its percentage of employment growth during the same period. Companies with an index of 20 or higher were classified as high growth. As reported in *Inc.*, March 1988. Relative business growth rate—Change in business growth rate, 1980-1986. As reported in *Inc.*, March 1988, p. 76. Business birthrate—This is derived by dividing the total number of business enterprises in an area by the number founded since January 1983 that had 10 or more employees by July 1987. As reported in *Inc.*, March 1988.

NOTE: Difference of means per *t* test all significant at .05 or greater.

sues impacting local governments is beyond the organizational and institutional capacities of American local government. Entrepreneurial policies present cross-cutting distributional and redistributional policy issues that are not easily dealt with through extant policy-making processes. The benefits and costs of these approaches are less visible and possibly more long-term than most economic development issues; this feature, in addition to the obvious fiscal and political advantages, may partially explain why these policies are so often the province of joint public/private or off-budget institutions. In the European context, this need for reducing certainty and increasing consensus on complex economic policies often results in the establishment of corporatist procedures and institutions, but in the American setting the result is the institutionalization of off-budget arrangements. Few current American political institutions are geared to the consensual, cooperative decision processes demanded by these approaches.

Even if the institutional and organizational capacity could be established, there is rarely a compelling consensus on entrepreneurial goals or effective strategies for achieving them. The entrepreneurial emphasis on wealth generation quickly breaks down into distributional issues. Curiously enough, the interest group politics characteristic of distributional politics are less evident in the emergence and spread of this new approach. There is no coherent voice from the business community nor any clear dissent from other groups. Older manufacturing interests clearly benefit more from traditional subsidy orientations; newer growth sectors and smaller, newer businesses are more likely to gain from the market facilitation slant of entrepreneurial strategies. But there are few instances where entrepreneurial initiatives have been thwarted by the actions of other interests. To some extent, cloaking entrepreneurial approaches in the symbolic language of "growth," "wealth," and "jobs" dissuades informed discussion or active opposition to these risk-taking activities.

Despite this lack of organized opposition to entrepreneurial approaches, interviews with local officials revealed a great deal of conflict over these policies. There appear to be two dimensions to this local conflict: the appropriate city role in economic development and the balance of executive and legislative voice in the formation and implementation of economic development policy in the postfederal era. Federal economic development resources obscured ideological issues over the appropriate degree of local activism on economic development issues; because outside external resources were at risk, it was possible to move ahead without resolving disagreements over local roles. With the withdrawal of federal resources, local economic development issues once again center on whether cities should return to "caretaker" roles and a more risk-aversive stance or pursue more activist, risk-taking roles.

Presumably, this traditional debate became a moot point in the face of federal revenue losses and increasing service demands, but it is a surprisingly salient issue in many cities. Local officials explain it in terms of divergent political constituencies: Mayors and administrators appreciate the "politics of announcement" and the career-enhancing, credit-taking opportunities offered by economic development activities. Council members are concerned about immediate constituency needs and are less willing to risk own source revenues on ventures with real opportunity costs and, at best, long-term payoffs. These conflicts in institutional incentives, coupled with lack of consensus on goals and tools, may contribute to a stalemate on further initiatives and to increased local political conflicts over future economic development strategies.

Notes

1. See analysis in Helen F. Ladd and John Yinger, *America's Ailing Cities* (Baltimore, MD: Johns Hopkins Press, 1989).

2. These implications are drawn out in Timothy Barnekov, Robin Boyle, and Daniel Rich, *Privatism and Urban Policy in Britain and the United States* (Oxford: Oxford University Press, 1989).

3. Michael Brintnall, "Future Directions in Federal Urban Policy," *Journal of Urban Affairs*, Winter 1989, pp. 1-19.

4. See Jonathan Walters, "The Urban Crisis, Act II," *Governing*, April 1991, pp. 26-32. Like Brintnall, Walters stresses the shifting political constituencies and coalitions that work against future efforts at formulating *urban* policy. He takes particular note of the salience of urban versus suburban hostilities for both national and state policymakers.

5. John Logan and Todd Swanstrom, eds., *Beyond the City Limits: Urban Policy and Economic Restructuring in Comparative Perspective* (Philadelphia, PA: Temple University Press, 1990), p. 30.

6. Margit Mayer, "Local Politics: From Administration to Management" (Paper presented at the Cardiff Symposium on Regulation, Innovation and Spatial Development, University of Wales, September 1989), p. 1.

7. John Logan and Harvey Molotch, *Urban Fortunes: The Political Economy of Place* (Berkeley: University of California Press, 1988).

8. Susan E. Clarke and Andrew M. Kirby, "The Mysterious Case of the Local Corpse," *Urban Affairs Quarterly*, March 1990, pp. 389-412.

9. An insightful series of case studies on how local officials in different cities respond to these dilemmas is available in Bernard J. Friedan and Lynne Sagalyn, *Downtown, Inc.* (Cambridge MA: MIT Press, 1989).

10. Findings from the larger study are reported in Susan E. Clarke and Gary L. Gaile, *Assessing the Characteristics and Effectiveness of Market-Based Urban Economic Development Strategies*, Grant #99-07-13709 (Washington, DC: U.S. Department of Commerce, Economic Development Administration, 1990). The study includes comparisons of city economic development strategies when using federal economic development funds with tools used in the absence of federal resources. The analysis indicates that needy cities were more likely to use risk-taking entrepreneurial tools in federal programs whereas better-off cities are more likely to use such tools when relying on nonfederal resources.

11. Field visits to 15 cities with distinctive federal program participation profiles used unstructured open-ended interviews with local officials to identify current approaches and those under study. Based on these discussions, we developed a mail survey instrument to systematically collect information from the 178 cities meeting the 1975 population criterion. In each city, a government official responsible for economic development was identified and asked to provide this information. The inquiries focused on use of 47 economic development tools that do not require federal resources, the timing of policy adoption, and the assessment of policy effectiveness. Of the 178 cities in this population, usable responses were gathered from 101 cities, for a response rate of 57%. A statistical comparison of this subpopulation of respondents to the overall population indicated that the respondent subpopulation was not significantly different. For more detail, see Susan E. Clarke and Gary L. Gaile, *Assessing the Characteristics.*

12. State laws restricting local fiscal powers and fiscal authority are also perceived as serious barriers; they constitute over 9% of the problems noted. Fragmented public voice is also a concern; 9% of the problems cited involved the lack of local political consensus and leadership, and the absence of focus for economic development efforts. Cities also are limited by land availability problems; these issues constituted over 7% of all problems mentioned. Aging infrastructure and the lack of amenities or perceived image problems limit many cities. Finally, federal restrictions on program eligibility, fund uses, environmental impacts, and other guidelines as well as the loss of federal funds are seen as constraints (6%) but rank behind lack of resources, state laws, and land availability issues.

13. State aid is problematic. With the exception of state enterprise zones, traditional state programs addressing infrastructure needs (11%) and providing financing options (10%) are seen as most useful of the 169 programs nominated. State enterprise zone programs were mentioned more often (17%) than any other state program. Although other innovative programs such as business incubators and high-tech programs were noted (5%), they are not perceived to be as useful as conventional programs such as infrastructure assistance, revolving loan funds, linked deposit schemes, job training programs, and marketing and promotion efforts, including international trade promotion. The diffuse endorsement of state programs, and the 9% specifically mentioning that state programs are not useful, indicates that state governments have yet to fill the gap left by the absence of federal funds.

14. A cogent description of these orientations is set out by Peter Eisinger, *The Rise of the Entrepreneurial State* (Madison, WI: University of Wisconsin Press, 1988). For an empirical analysis of state economic development policy shifts, see Susan E. Clarke and Gary L. Gaile, "Moving Towards Entrepreneurial State and Local Economic Development Strategies: Opportunities and Barriers," *Policy Studies Journal* 17 (1989): 574-98.

15. Roger J. Vaughn and Robert Pollard, "Small Business and Economic Development," in *Financing Economic Development in the 1980s*, ed. Norman Walzer and David L. Chicoine (New York: Praeger, 1986).

16. These distinctions are traced by several observers, including the Committee for Economic Development, *Leadership for Dynamic State Economies* (New York: CED, 1986); Ann O'M. Bowman, *Tools and Targets: The Mechanics of City Economic Development* and *The Visible Hand* (Washington DC: National League of Cities, 1987); Peter Eisinger, *The Rise of the Entrepreneurial State* (Madison, WI: University of Wisconsin Press, 1988); and Richard D. Bingham, Edward W. Hill, and Sammis B. White, eds., *Financing Economic Development* (Newbury Park, CA: Sage, 1990).

17. We discuss these attributes in more detail in Susan E. Clarke and Gary L. Gaile, "Moving Towards Entrepreneurial State and Local Economic Development Strategies."

18. As noted by Melvin J. Dubnick and Barbara A. Bardes, *Thinking About Public Policy* (New York: Wiley, 1983).

19. One of the many useful distinctions made by Ernest Sternberg, "A Practitioner's Classification of Economic Development Policy Instruments, with Some Inspiration from Political Economy," *Economic Development Quarterly* 1 (1987): 149-61.

20. For example, the strategies reported as least used by over 75% of the cities responding include those that would interfere with private investor decision making, compete with private investment capital, or require significant state actions loosening local fiscal constraints. This includes mechanisms allowing cities to pool public and private capital or to influence the use of private investment funds, some of which face state or local legal restrictions. Mechanisms for pooling capital receive high ratings as future strategies currently "under study" but it is currently something cities may not do (in many cities, pension funds are controlled by state personnel systems), or lack the capacity or private sector cooperation to do (zero coupon bonds, linked deposits, venture capital funds, public/private consortium equity pools).

21. Since writing this article, a recent publication also using the wave analogy to describe innovative local development strategies has come to our attention. See R. Scott Fosler, ed., *Local Economic Strategy* (Washington, DC: International City Management Association, 1991).

22. See Harry Hatry, Mark Fall, Thomas O. Singer, and E. Blaine Liner, eds., *Monitoring the Outcomes of Economic Development Programs.* (Washington, DC: The Urban Institute, 1990).

23. Bernard J. Friedan and Lynne Sagalyn's interviews with local officials in *Downtown, Inc.* underscore this perspective.

24. Ibid., p. 270.

25. As documented in Ladd and Yinger, *America's Ailing Cities.*

26. For a national study analyzing the conditions under which such local linkage policies are adopted, see Edward Goetz, "Type II Linkage Policies," *Urban Affairs Quarterly* 26 (1990): 170-90.

27. Ladd and Yinger, *America's Ailing Cities,* p. 17.

28. Ibid., p. 290.

29. We explicitly incorporated and controlled for the expected spatial variation in local entrepreneurial activity and its outcomes through use of trend surface analysis techniques. Analysts and local officials are well aware that a given policy effort in a region characterized by distress will not have the same impact on outcomes as a similar level of effort in a region characterized by economic prosperity. That is, the same strategy applied in Buffalo and San Diego can be expected to have differential impacts. They both may be positive relative to their region, but this success may be overwhelmed by the statistical noise in regional variations in the overall economy. In analyses ignoring locational differences or attempting to control for them with dummy variables for regions, significant explanatory power is lost.

The use of spatial statistical techniques to identify and control for these regional variations is discussed in Gary L. Gaile and Cort J. Willmott, *Spatial Statistics and Models* (Dordrecht: D. Reidel, 1984). Here, we use trend surface analysis to measure the strength of regional variables and to create residuals to be used as independent nonautocorrelated variables in further analyses (see Fritz Agterberg, "Trend Surface Analysis," in Gaile and Willmott, *Spatial Statistics and Models*). More specifically, before analyzing the relationship of the new policy index and city growth features, it was necessary to develop residuals through trend surface analysis for each variable with strong regional patterns. The principal methodological argument is that regional trends exist in most of the variables typically used to identify policy effectiveness. The hypothesized strength of these regional trends is such that they would likely mask or misrepresent significant relationships between new policy efforts and city growth measures.

Essentially, this is a straightforward modification of standard regression techniques; it uses regression to analyze a variable's locational attributes by disaggregating the variable into a *regional component* that is described by the regression "surface" and a *local component* that is described by the residuals from that surface. The validity of trend surface results can be judged using the same appropriateness measures as in simple regression analyses. Thus F-tests are performed, and significance levels of the analyses are readily interpretable. Our trend surface analyses of the policy index showed no statistically significant regional trends. This suggests that these entrepreneurial, market-based policies have largely become employed on a national basis without strong regional emphases.

Chapter 14

State Economic Development in the 1990s

Politics and Policy Learning

Peter Eisinger

❀❀❀

State and local economic development policy is clearly in a state of ferment at mid-decade. This policy domain has always been marked by a high degree of volatility, particularly in the last 20 years. Thus the presence of shifts and cross-currents in the approaches by which subnational governments seek to stimulate and direct private investment within their borders is not surprising. What is interesting is that the current changes have certain markings of a watershed, although at first glance they are surprisingly difficult to read. It is not readily apparent whether some transformational shift is taking place in the reigning assumptions or whether these are simply adjustments of a less fundamental sort.

Establishing some sort of compass reading is important. From the perspective of the policy practitioner, the ability to discern leading developments at the frontier of this policy domain provides crucial information—not only about the efforts of competitor states but also, potentially, about promising new approaches in a field in which many existing programs do not seem to have yielded clear results. To track changes in this policy domain, then, is to plot the location of the pioneer states in the economic development field.

From a more analytical point of view, the effort to sort through these changes promises to illuminate the process of "policy learning"—that is, the set of evaluative experiences that is presumed to lead to adjustments in policy. Scholars often speak of the life course of public policy in terms of a policy cycle in which information derived after program implementation is used to initiate policy modifications. By determining the nature of the changes in state and

AUTHOR'S NOTE: An earlier version of this article was delivered at the Conference on New Perspectives on State Government, Policy and Economic Development, sponsored by the Institute of Government and Public Affairs, University of Illinois, Chicago, May 13-15, 1993. I am grateful to Beverlee Garb, Paul Passavant, Ted Reuter, and Greg White, who conducted the interviews of economic development officials on which much of this article is based.

Reprinted from *Economic Development Quarterly*, Vol. 9, No. 2, May 1995.

local economic development, we gain an understanding of the policy learning taking place. The character of these changes may be related to the sorts of information on which states base policy modifications: formal program evaluations, anecdotal evidence, or political signals, among other sources. Specifically, I am interested in the extent to which states actually modify policy approaches as the result of learning. Or are there other outcomes of the learning process besides policy modification, such as withdrawal from or abandonment of the field?

Exploring the character of these changes in economic development policy is also important analytically for an understanding of the American political economy. Many observers believe that most American states adopted little "industrial policies" in the 1980s, pursuing an activist, interventionist, entrepreneurial role in the economy that stood them in sharp contrast to the free-market stance of the national government in the Reagan years.[1] The present ferment may provide clues as to whether some version of subnational industrial policy has taken hold or whether the states are in retreat from economic activism.

New Directions

State and local economic development policy has already undergone one major strategic shift. In the late 1970s and early 1980s, industrial recruitment strategies, born in the Depression and honed to cutthroat sharpness in two decades of bitter postwar interstate competition for footloose industry,[2] gradually gave way to what has been called "state entrepreneurialism."[3] Industrial recruitment (contemptuously called "smokestack chasing") was pursued on the supply side through proffering public subsidies for private firms in the form of tax abatements, investment credits, low-interest loans, land writedowns, and labor-training grants. These subsidies, which reduced the costs of operating a business in particular states, were designed to influence business location choices.

Evidence began to mount not only that such incentives rarely influenced business investment decisions but also that they had no significant job-generating effects. Then, as a series of larger economic transformations emerged that propelled Americans into the postindustrial international marketplace, which raised concerns about the competitiveness of American business, states turned increasingly to strategies designed to foster indigenous firms and local entrepreneurial capabilities. Guided by strategic plans, states sought to identify and develop market opportunities for selected home-based businesses and new industries with high growth potential. High-technology programs (to create the industries of the future), export assistance (to exploit new markets), and venture capital (to nurture new small-business firms) became hallmarks of the entrepreneurial state. Such policies never entirely displaced the old industrial recruitment programs, but they came generally to surpass the latter in effort, spending, and political glamour.[4]

Now there are signs of another directional shift, yet its outlines lack sufficient coherence to discern their course clearly. There is evidence to build cases for at least three different directional scenarios. One is that states are leaving the field—that is, their commitment to active economic intervention, exemplified by the entrepreneurial programs of the 1980s, is beginning to flag in the face of doubts about the efficacy and political viability of their initiatives. A second is that the states are regressing to their old industrial recruitment habits, forsaking the uncertainties of the entrepreneurial state for the instant political gratifications of winning an industrial relocation prize. A third possibility is that the states are embarking on a "Third Wave," a set of new strategies designed to overcome the lack of scale and scope of entrepreneurial programs. Let us examine the evidence for each of these scenarios.

Are States Abandoning Economic Development?

At the height of the much-vaunted "Massachusetts Miracle," commentators pointed to the revival of the city of Lowell as a stellar example of what state government could do through an enlightened economic development program in partnership with private business. Lowell was, in David Osborne's words, "the flagship of Dukakis' urban development policy."[5] But the collapse of the Massachusetts economy, just as Dukakis was running for the presidency, and Lowell's regression from a museum piece of the industrial revolution to the struggling New England mill town it had been before took some of the bloom off the rose of state economic development claims. If Massachusetts, with its pioneering venture capital and technology development programs, its centers of excellence and innovative capital leveraging programs, could not stave off decline, what hope was there for other, less activist states?

State economic development officials have had to contend not only with evidence of this apparent futility in some larger in economic sense but also with the extraordinary difficulty of proving the efficacy of particular interventions. Evaluation has always been the central difficulty in economic development policy; establishing a causal link between state development incentives and job-creating investment is rarely simple.[6] As a Maryland development official pointed out in the survey conducted for this study,[7] if it were not for the governor's strong support of the state's economic development department, "We would be a vulnerable department because . . . much of our performance cannot be accurately measured."[8] Thus, given both the failure of policy initiatives in Massachusetts and other states to counteract the recession of the late 1980s and the struggle to demonstrate convincingly with hard numbers the return on investment of development dollars spent, it would not be surprising if state dedication to economic development began to diminish.

Signs of withdrawal from the economic development commitment forged in the 1980s appeared in programs as diverse as science and technology partnerships and state venture capital funds. As to the former, once "the cynosure of state economic development strategies" most are now reported to be in a defensive posture.[9] According to Goldie Blumenstyk, "Governors and legislators are demanding more evidence that university research designed to develop new technologies and products has a demonstrable impact on their states' economies."[10]

In the 1991-1992 fiscal year, Pennsylvania reduced the financing for its model Ben Franklin program by nearly 20%. Similar programs were cut, after years of steady growth, in Massachusetts, Iowa, Illinois, Virginia, and Nebraska. In Ohio, a gubernatorial task force recommended that the state's development agency eliminate altogether its seed development fund for university research, as well as seven small-business technical innovation centers.[11] In general, there is an apparently widespread and growing sense that the time from university research to product development, and from commercialization to job generation, is both too long and too uncertain to justify the price of science and technology partnerships.

In the realm of public venture capital programs, several states—Michigan, Florida, and Massachusetts—have actually terminated their initiatives; others have refused to fund or reauthorize programs already on the books. Two surveys tell the story. The first, conducted in 1989-1990, found that 23 states were administering a total of 30 different programs involving the use of public funds for venture investments.[12] The second survey, a follow-up conducted in 1991, found that only 17 states were still running venture capital programs, and the number of the latter had declined to 23.[13]

The dead and dying programs foundered on a combination of problems, including their extremely modest job-generating effects, the clash between the long-term nature of venture investments and the short-term political needs of state officials driven by the electoral cycle, and the political unacceptability of the high failure rate inherent in venture capital investing.

Second thoughts about economic development initiatives apply to more than just science partnerships and venture capital programs. For example, at least a dozen states have, closed or reduced trade offices in Japan.[14] More generally, William Nothdurft reports that state international trade programs "are under the gun almost everywhere," in part because states cannot demonstrate convincingly that their overseas trade offices and export promotion efforts have measurable effects.[15] "No one," Nothdurft observes, "has a clue as to what all this activity has produced."[16]

States are clearly engaged in a process of winnowing extending beyond science and technology, venture capital, and foreign trade programs. Respondents in 19 states reported in the national survey that their state had, in the past couple of years, terminated economic development programs or allowed a program to die by denying it funding, accounting for a reported total of 34 program deaths.[17] These included not only some of the initiatives noted above but also programs that marketed the state, provided public financing for businesses, and encouraged technology transfer.

Since Herbert Kaufman's observation that government organizations have "an impressive ability to stay alive," could, in many circumstances, just as easily apply to particular programs,[18] this rate of winnowing may indicate a powerful sense of disaffection or disappointment about the economic development enterprise as a state government function. But these program terminations must be put in proper perspective; that states have eliminated programs is in itself insufficient evidence to conclude that they are losing interest in economic development generally. For one thing, no systematic rationale emerges to explain the different program deaths; for another, there were many more new program births than deaths in this period. Respondents in 38 states reported the passage and implementation of at least 85 new programs.[19] Nevertheless, the data on program deaths—in a field characterized during the 1980s by such intense, competition-driven program innovation—suggest a certain novel state ambivalence about economic development.

This sense of ambivalence is reinforced, to some extent, by information on trends in state economic development budgets. Respondents were asked a series of questions about the magnitude of state spending for the state's economic development agency and for all economic development spending, including programs outside the agency. Economic development budgets cover not only entrepreneurial programs but also the more traditional industrial recruitment efforts. Thus the examination of budget trends, useful for gauging a state's general priorities, is a relatively blunt tool when it comes to ascertaining the strength of the state's commitment to a particular development strategy.

Many states have cut the budgets of their economic development departments. Table 14.1 shows budget comparisons for 1993 compared to the previous fiscal year. Half the states in the survey ($n = 24$) report lower agency budgets in the current year; another 25% report a flat budget trend. Officials were also asked to characterize the budget trend for their agency over the last five years. Department budgets have reportedly declined over that period in 27 states (57%) and stayed flat in 8 others (17%).

As Table 14.2 suggests, declining budgets may have as much to do with a state's relative economic performance and prospects as with any reevaluation of its economic development role. States experiencing economic hard times relative to national trends, measured by an "economic momentum" index, were somewhat more likely to cut or maintain rather than increase development agency budgets.[20] Officials in a number of states—Ohio, Missouri, Illinois, Florida, Iowa, Maryland, and Virginia—indicated explicitly that the fortunes of their economic development agencies were tied directly to the state economy.[21] "There is less money this year than last . . . because the financial situation in the state is very bad," said an official with the Alabama Department of Economic and Community Affairs.[22] Yet not all states follow this logic: The nine states in the upper-right-hand box of Table 14.2 demonstrate that, for some at least, economic development spending is countercyclical. Greater spending effort is required especially during hard times.

It is possible to argue that when agency budget fortunes are so dependent on good economic times, then the commitment to that policy area must be somewhat tenuous. But there is some counterevidence to indicate that state interest in economic development policy may be, at base, more durable in a significant number of states. Respondents in 40% of the states reported that all economic development spending—including that for programs often outside agency budgets, such as tax expenditures and business loans—was up in their state, compared to five years ago; only 13% said it

TABLE 14.1 Economic Development Department Budget This Year Compared to Last Year

Budget		*State*		n
Higher	OK	IN	NC	12 (25%)
	KS	ND	NM	
	MS	OR	MA	
	MI	NH	CA	
Same	AR	PA	FL	12 (25%)
	NY	NE	HI	
	ID	MO	WV	
	AK	KY	CO	
Lower	NV	VT	MN	24 (50%)
	WA	GA	WI	
	RI	AL	WY	
	TX	IL	CT	
	MT	UT	MD	
	NJ	AZ	SC	
	LA	IA	TN	
	OH	VA	ME	

NOTE: ($N = 48$).

TABLE 14.2 Department Budget This Year Compared to Last Year, by Economic Momentum Measure[a]

	Momentum Category			
Budget	*Above Average*	*Average*	*Below Average*	
Higher	OR	CA	OK	NH
	NM		KS	NC
			MS	MA
			MI	IN
			ND	
	($n = 2$)	($n = 1$)		($n = 9$)
Same	ID	AR	NY	
	AK	NE	PA	
	CO	KY	MO	
	HI	FL	WV	
	($n = 4$)	($n = 5$)		($n = 3$)
Lower	NV	OH	RI	IA
	WA	MT	NJ	WY
	TX	MN	LA	CT
	UT	WI	VT	GA
	AZ	MD	AL	IL
		SC	TN	ME
			VA	
	($n = 5$)	($n = 6$)		($n = 13$)

SOURCE: The economic momentum index in Jordan's "A State Fiscal Report," *Governing* 5 (1992), 34.

NOTE: $\chi^2 = 6.404851$, sig. < .170.

a. This is an index based on percentage changes in population, employment, and personal income, compared to national average, for 1990-1991.

was down. Furthermore, officials displayed great optimism about the future: Nearly twice as many (*n* = 20), including five states that reported cuts from the prior year, anticipated budget increases in the coming fiscal year as expected cuts (*n* = 11).

To summarize, it would be difficult to make the case that states are abandoning the commitment to economic development that emerged so powerfully in the 1980s, but there are signs of loss of ardor.[23] Program births are offset, to some degree, by a number of program deaths; long-term funding increases for economic development are offset by recent declines. If we cannot maintain that states are engaged in headlong retreat from this policy function, we can conclude that economic development is no longer regarded in the 1990s as the keystone policy area on which all other state functions depend and to which other funding claims must be subordinated.

Are States Reverting to Industrial Recruitment Strategies?

The initial strategic transition from industrial recruitment to entrepreneurial approaches was fueled in part by the belief that luring established firms from out-of-state locations contributed little in the long run to a state's growth. Home-grown firm births and indigenous firm expansions turned out to be much more important sources of job generation. It is difficult to demonstrate empirically that industrial recruitment activities declined during the rise of the entrepreneurial state, but, at least in terms of planning priorities, development rhetoric, and program invention, recruitment appeared to take a back seat. There are signs now, however, that industrial recruitment activities are still central features of state development policy and may even be enjoying a surge. Robert Guskind reports that "the practice of actively recruiting firms from other states has intensified."[24] A North Carolina development official noted in the national survey, "We're basically industry hunters. . . . Smokestack chasing works. Manufacturing in North Carolina has been a success in terms of growth and job creation . . . and we've been successful without giving away the store."[25]

One indicator of the resurgence of industrial recruitment efforts, according to *State Policy Reports* (which proclaims that "smokestack chasing isn't dead"), is the increasing level of bidding for major investments like the United Airlines maintenance base, the McDonnell Douglas facility, and the BMW plant.[26] For the biggest prizes at least, states are willing to commit ever-higher sums. The cost in public incentives per plant employee for the big automobile-plant competitions, for example, has risen sharply in the last decade from the roughly $11,000 per worker paid by Tennessee for Nissan in 1984 to about $68,000 that South Carolina committed for the BMW plant announced in the summer of 1992, to Alabama's pledge of nearly $200,000 per job for the Mercedes plant.[27] The state of Iowa has further increased the ante by offering a Canadian steelmaker nearly a quarter of a million dollars in state incentives per direct job.[28]

Some states have tried to lessen the costs of regional bidding wars by signing antipirating agreements with their neighbors, but these have rarely lasted very long. The temptation to capture a footloose firm from across the border is a powerful one. The *New York Times* reported not long ago, for example, that the agreement signed by Connecticut, New Jersey, New York State, and New York City not to recruit one another's businesses was close to collapse after only one year.[29]

Perhaps the most telling evidence of a resurgence of interest in industrial recruitment is contained in the biennial survey of state development expenditures, conducted by the National Association of State Development Agencies (NASDA). This notoriously unreliable survey of state economic development bureaus has a relatively low response rate; its questions are open to multiple interpretations; and the questionnaire is frequently filled out by low-level officials. Nevertheless, some of the data are suggestive.

For example, NASDA asks its respondents to determine the proportion of their agency's budget spent on "attraction" versus "retention" of firms. The former suggests industrial recruitment; the latter, which focuses on indigenous businesses, is a central concern of the entrepreneurial state. In the 1990 survey, 21 of the 39 states responding reported that they spent a greater proportion of their budgets on attraction than on retention.[30] Not only did a slight majority favor attraction, but a significant number indicated that the percentage of their budget devoted to attraction had *increased* since the previous survey. Only 24 of the 39 states surveyed in 1990 were surveyed successfully in 1988; however, of those 24 states, 11 reported an increase in their attraction budgets, 8 reported no change, and only 5 reported a decrease in favor of retention.

That states continue to devote resources to industrial recruitment is perhaps not surprising: The political payoffs are dramatic, particularly when a big industrial prize is at stake. Writing in South Carolina's in-house development publication of his state's successful bid for the BMW plant, Governor Carroll Campbell observes, "With 4,000 plus jobs and a billion-dollar impact, South Carolina truly has scored a grand slam." In the same issue, a story on the BMW deal notes that landing "a jewel of a company" was "quite a coup. . . . Now it's South Carolina's turn on the international stage."[31]

Not only does industrial recruitment produce satisfying political rewards, but the economic benefits appear to be immediate, at least compared to the slower and riskier entrepreneurial economic development strategies. Whereas funding science and business partnerships as a strategy for developing new commercial products and the firms and jobs to produce them is a long and uncertain enterprise, winning a new plant offers what appears to the public to be a tangible sign of successful economic development. Explaining why he devotes so much time talking "CEO to CEO" in an effort to lure industry, the governor of Mississippi compared the time required for an industrial recruitment strategy versus developing human capital: "It takes 10 or 12 years of concerted effort to make a noticeable difference in an education system. We obviously don't have 10-12 years to wait on jobs."[32]

Data from the 1993 survey of state development agencies also provide support for the proposition that industrial recruitment remains popular among the states. But these same data do not indicate that states regard industrial recruitment as their primary development strategy, or that they are retreating from some other strategy in favor of smokestack chasing. Respondents were asked whether their state focused more on "old-style" economic development, defined as industrial recruitment, or "new-style," defined in terms of small-business programs, high-tech development, or export promotion.[33] Only one state (North Carolina) suggested that its efforts were geared primarily toward industrial recruitment; 25 states said that they pursued both strategies equally, while 22 said they emphasized new-style approaches.

Another question sought to elicit more detail about each state's development strategy. Respondents were asked to list the main programmatic approaches in their state. Table 14.3 breaks down their answers according to whether the approaches mentioned were industrial recruitment, entrepreneurial, or "Third Wave"—a category discussed in the next section. The data show that entrepreneurial programs account for most mentions—although, again, industrial recruitment is still a significant preoccupation.

Neither the survey questions on style nor those on programmatic emphasis provide any clue as to whether states have shifted their strategies. Another question offers some perspective on the time dimension: Respondents were asked whether, "in the last couple of years," their state had passed and implemented any major new economic development programs. Officials in 38 states responded positively, enumerating 85 new programs. Table 14.4 arranges the new programs by strategic approach, using the same categories as Table 14.3.

If there were a rush to return to the recruiting wars, it is unlikely that entrepreneurial program innovation would dominate the pattern of new adoptions. Although Table 14.4 confirms that industrial recruitment and entrepreneurial strategies coexist, there is clearly a strong continuing interest in risking state resources in ven-

TABLE 14.3 Answers to Question B2a: What Is the Emphasis of Your State's Economic Development Strategy?

Industrial recruitment[a]		25 (24%)
Recruiting new business	19	
Lowering taxes and regulatory barriers	3	
Tourism	3	
Entrepreneurial state programs[b]		59 (57%)
High-technology promotion	11	
Small-business development	12	
Technology transfer	3	
Export assistance and promotion	11	
Enterprise zones	1	
Existing business development	13	
Minority and female-owned business development	2	
Job retention	6	
Third Wave capacity-building[c]		20 (19%)
Education and job training	8	
Industrial modernization	2	
Building local government capacity	7	
Industry clusters, networking	3	
Other[d]		
Rural development	3	

a. Pertains to all activities designed to attract firms from outside the state, including recruiting efforts, marketing the state, promoting the business climate, and reducing factor costs, including taxes and regulatory burdens.

b. Pertains to activities in which the state risks resources in the effort to help indigenous businesses and entrepreneurs identify and capitalize on new markets and business opportunities.

c. Pertains to activities whose main goal is to build general institutional or individual capacity.

d. These are difficult to classify without knowing specific details. They may involve capacity-building; they may also involve development through tax and other incentives in the old style; they may involve efforts to help firms or entrepreneurs identify or exploit new markets. These are not included in the totals or percentages.

ture investments and export-market development to foster indigenous business. Obviously, the political temptations inherent in industrial recruiting maintain a certain hold on states, but their continuing efforts to lure established industry do not appear to come completely at the expense of the newer entrepreneurial programs.

Third Wave Strategies

Some analysts have suggested that the signs of shift in state economic development policy signal neither withdrawal nor reversion to older patterns but rather the emergence of a "Third Wave." "State and local economic developers," Doug Ross and Robert Friedman write, "are in the process of inventing a new generation of development strategies."[34]

In the view of these analysts, Third Wave programs are primarily, though not exclusively, responses to the lack of scale and accountability of the entrepreneurial programs. By the late 1980s, Dan Pilcher argues, state economic development officials were "increasingly troubled not only by sticky problems of how to measure the results of economic development programs but the inability of these efforts to make a real difference in the state's economy.[35] The new-generation strategy seeks to reduce the role of state government as direct supplier of programs in favor of a role as "wholesaler," in which the emphasis is on seeding, leveraging, and *general* capacity-building functions. The notion is that government proves most effective when it enables individuals, institutions, firms, and communities to develop the means to help themselves.

TABLE 14.4 New Economic Development Programs Adopted in the Last Couple of Years

Industrial recruitment		17 (20%)
Targeted industrial recruitment	4	
Industrial financing	8	
Tourism promotion	3	
Other	2	
Entrepreneurial programs		46 (54%)
International trade promotion	9	
Enterprise zones	4	
Small-business development	12	
Venture capital	5	
High-tech development	8	
Targeted industries assistance	2	
Minority- and female-owned business assistance	3	
Other	3	
Third Wave initiatives		18 (21%)
Job training	8	
Local government assistance	3	
Community development corporations	4	
Industrial modernization	2	
Other	1	
Other		4 (5%)
Total		85

Analytically, it is not always clear how to distinguish Third Wave programs from entrepreneurial programs; many of the latter operate by the same principles, although they seem designed to provide firms with more specific programmatic direction than the general capacity-building of the Third Wave. Those who believe that a Third Wave is emerging typically point to a common set of programs that exemplify the new strategy. These include investment in job training and education, industrial modernization initiatives, support for community-level economic development planning, and encouragement of industrial clusters of firms for the purpose of pooling resources to achieve higher levels of international competitiveness than each firm could manage on its own.

Using these guidelines, we may examine the data from the national survey to see whether there is evidence of an emerging Third Wave. Although a handful of respondents explicitly mentioned the term in the interviews, Table 14.3 shows that the programmatic emphases most often cited were solidly identified with the entrepreneurial state. Third Wave programs were even less likely to be mentioned than industrial recruitment efforts.

If there was significant dissatisfaction with the programs of the entrepreneurial state, in favor of the capacity-building efforts that exemplify the Third Wave, we should find this shift reflected in Table 14.4. But programs passed in the last several years primarily fall under the entrepreneurial category. Third Wave initiatives were virtually tied for a distant second place with new programs for industrial recruitment. Thus, although some states are beginning to explore Third Wave alternatives to the entrepreneurial programs put in place during the 1980s, there is no indication of a sea change in the making.[36]

Policy Learning and Policy Consolidation

The frenetic pace of policy innovation that characterized the entrepreneurial state in the 1980s has clearly come to an end. There are strong indications that states are reassessing their economic development strategies, but no single pattern has emerged. Program terminations and budget reductions—virtually unknown during the period of intense interstate competition a decade ago—are now relatively common. Industrial recruitment, scorned as wasteful and unproductive in the heyday of the entrepreneurial state (although never abandoned entirely), seems to have regained some of its luster. Finally, acknowledgment of the problem of scale inherent in state development programs has encouraged a few states to seek better ways to use their resources to build capacity and leverage the assets and capabilities of other governments, institutions, and individuals. But as none of these patterns is dominant, how might we interpret the picture?

The ferment of the 1990s is, probably, a product of policy learning in which feedback on economic development programs and strategic emphasis comes not primarily from the experience of program administration but rather in the form of various political signals. The lessons of this process have not led mainly to program modifications in response to implementation experiences but rather to what may best be characterized as *consolidation*—a defensive posture having more to do with political survival than with effective economic development. Ideally, the policy-learning feedback loop is driven by experience with programs in the field.[37] As imperfections of program design or unanticipated problems come to light, policymakers have an opportunity to modify or adjust the program at a later stage. Such policy learning may come about as the result of an informal process in which anecdotal evidence and the experiences of program administrators in the field provide the basis for adjustment. But much policy learning is also the product of formal evaluations and audits.

Unfortunately, the survey of state development agencies provides no estimate of the importance of anecdotal evidence in policy learning, but it is possible to examine the incidence among the states of more systematic scrutiny. Survey respondents were asked whether there had been formal audits or evaluations of their economic development programs by a legislative audit bureau, auditor general, or some outside contractor.[38] Officials interviewed in 34 of the 48 states responded in the affirmative. Thus, while audits or evaluations were relatively common, they are by no means universal practices. Fourteen states had pursued no systematic examination of their economic development programs at all, and thus have little rigorous analytical basis for making program modifications.

What did the 34 states with program evaluations learn from their experience? Of the 34 evaluative exercises, 12 resulted in no recommendations at all, essentially giving economic development agencies and their programs an uncritical vote of confidence.[39] Another 12 evaluations recommended various administrative reorganizations and adjustments, but they said nothing about program modifications, terminations, or budget cuts. Only eight audits or evaluations recommended program modifications. In short, the audit and evaluation process, a critical element for systematic policy learning, produced recommendations for program adjustments or changes in only 8 of the 48 states in the survey. We may conclude that the policy-learning feedback loop in most states contains little information about program operations derived from formal program scrutiny.

Yet all our evidence about program terminations, loss of ardor, revived interest in industrial recruitment, and the search for Third Wave strategies suggests changes are occurring in state economic development. If program assessments do not seem to be driving these changes, what are?

Possibly, this conjunction of changes is the result of less formal internal evaluations by program managers, but the changes occurring at present in economic development also seem to stem from a strategic recalibration of political costs and benefits that accrue to the pursuit of different economic development strategies. That policy learning and the resultant policy change are as much driven by political as by technical considerations is widely acknowledged. Still, the literature on the policy cycle seems to focus principally on the partisan and ideological components of this political calculus, rather than strategic factors.[40] I am suggesting that policy change in state economic development is occurring primarily because of a change in strategic calculations by state-elected officials—much more than as the result of program evaluations, changes in partisan control, or ideological transformations.[41]

This is not an easily testable proposition,[42] but at least the changes occurring in economic development are consistent with this interpretation. The economic recession of the later Bush years was clearly a powerful force shaping state policy. Indeed, in the national survey, respondents repeatedly and spontaneously mentioned economic stringency as a constraint on policy.[43] Not only did the recession impose severe limits on government action, but it put a high premium on the creation of jobs—higher perhaps even than that during the formative period of the entrepreneurial state in the early Reagan years, when international and interstate competitiveness, rather than recession, seemed a more important spur to state development policy.

Under pressure to produce jobs during these years of economic hardship, state officials began to back off development initiatives with long gestation periods. As a Maryland respondent noted in the national survey, "International trade takes perhaps 6 to 10 years of sustained effort before one realizes sizable returns directly attributed to program effort. Legislators have (at best) a 24-month time horizon."[44] Hallmark programs of the entrepreneurial state, such as state venture capital initiatives and high-tech research partnerships, whose development benefits lie in the uncertain future, lost much of their attraction. These were initiatives of a more optimistic period: the growth years of the mid-1980s, when it was easier to contemplate a more leisurely approach to development. With the recession late in the decade, however, state officials began to divest themselves of programs that did not bear early fruit. But they were not in retreat from economic development in general.

The pressure to show quick results, generated by economic hardship, also accounts for the persistence of—and perhaps renewed interest in—industrial recruiting. Dennis Grady has shown that the worse the economic conditions in a state, the greater the efforts of its governor to recruit industry.[45] Given its apparent immediate gratification and the possibilities for dramatic credit claiming, capturing a footloose firm is far more satisfying politically than waiting for slower and riskier entrepreneurial programs to mature.

Finally, the pattern of budget reductions for economic development derives from a strategic calculation that in these hard times, other responsibilities of state government command more attention. Economic development must defer in most states to more fundamental needs like education and health care.[46] As budgets are cut, however, interest tends to grow in leveraging scarce state resources. Thus a few states have been impelled to explore Third Wave programs, whose emphasis on seed resources to encourage capacity-building, rather than "retailing" programs, fits well in an environment of fiscal stringency.

The patterns of change suggest a retreat along what had appeared the road to serious, relatively coherent state-level industrial policies. By the late 1980s, with their embrace of mid-range strategic planning and targeted development assistance, many states had crafted an economic role that resembled the industrial policies common in Europe and developed Asia. But, in contrast to the long-term strategic thinking central to industrial policy, the changes of the 1990s emphasize short-run advantages and a more detached state role. Thus, although many features of the new American political economy can still be discerned, its foundations are apparently shallower than had once been believed.[47]

In sum, the shifts and currents in state economic development policy form a pattern of recession-induced consolidation, a process not of program modification or adjustment in pursuit of greater effectiveness, but of paring down and focusing on strategic political considerations. Consolidation discourages risk taking and innovation. It puts a premium on affordability. But it does not suggest withdrawal. Political survival during fiscal hard times dictates, not that state officials abandon the obligation to foster their economies through development initiatives, but that they do so with maximum visible effect using the least resources. That maxim has produced the mix of program terminations, budget cuts, recruitment wars, and flirtation with Third Wave programs that characterizes state economic development in the mid-1990s.

Notes

1. See Peter Eisinger, *The Rise of the Entrepreneurial State* (Madison: University of Wisconsin Press, 1988); Virginia Gray and David Lowery, "The Corporatist Foundation of State Industrial Policy," *Social Science Quarterly* 71 (March 1990): 3-24; Otis Graham *Losing Time: The Industrial Policy Debate* (Cambridge, MA: Harvard University Press, 1993); Susan Hansen, "Industrial Policy and Corporatism in the American States," *Governance* 2 (April 1989): 172-97.

2. James Cobb, *The Selling of the South* (Baton Rouge: Louisiana State University Press, 1982).

3. Eisinger, *Rise of the Entrepreneurial State.*

4. Graham, *Losing Time,* p. 198.

5. David Osborne, *Laboratories of Democracy* (Cambridge, MA: Harvard Business Press, 1988), p. 23.

6. Chris Evans and Linda Triplett, "Measuring the Immeasurable," *State Legislatures* 15 (March 1989): 19-21.

7. A telephone survey of state economic development agencies was conducted for this study in February and March 1993. Officials in 48 of the 50 states were interviewed about their states' economic development budgets, strategic priorities, program innovations and terminations, and audits and evaluations. In all, a total of 54 officials in every state but South Dakota and Delaware (which declined to participate) were interviewed. The officials interviewed filled a variety of roles in their respective agencies: economic development specialists, division and agency directors, research managers and analysts, and public information officers.

8. Interview conducted February 24, 1993.

9. Irwin Feller, "The Effectiveness of State Investments in Science and Innovation" in *Industrial Policy for Agriculture in the Global Economy,* ed. Stanley Johnson and Sheila Martin (Ames: Iowa State University Press, 1993), p. 131.

10. Goldie Blumenstyk, "States Re-Evaluate Industrial Collaborations Built around Research Grants to Universities," *The Chronicle of Higher Education,* February 26, 1992, p. 1.

11. *State Policy Reports,* 10 (April 1992), p. 22.

12. Peter Eisinger, "The State of State Venture Capitalism," *Economic Development Quarterly* 5 (1991): 64-76.

13. Peter Eisinger, "State Venture Capitalism, State Politics, and the World of High Risk Investment," *Economic Development Quarterly* 7 (1993): 131-9.

14. *State Policy Reports,* 10 (October 1992), p. 5.

15. William Nothdurft, "The Export Game," *Governing* 5 (August 1992): 57-61.

16. Ibid., p. 58.

17. The wording of the questions was as follows: "In the last couple of years or so, have there been any major economic development program terminations?" This was followed by "Have any programs simply died out from lack of funding or attention?" Respondents named 19 separate program terminations and cited a total of 15 that had died from lack of funding. The total number is probably conservative, since respondents often could not provide a complete account.

18. Herbert Kaufman, *Are Government Organizations Immortal?* (Washington, DC: Brookings Institution, 1976), p. 64.

19. Again, the figure is probably conservative, since many respondents provided examples rather than exhaustive tallies, despite interviewer prompting. It should be acknowledged that counting programs as an indicator of economic development activity is highly imperfect. Not only is it difficult to accomplish a complete census, but counting tells us nothing about budgetary commitments and priorities. But program counts continue, in the absence of any other quick measure, to serve both as a rough guide to comparative state involvement in this policy area and as an indicator of growing or declining policy commitment. See, for example, Paul Brace, *State Government and Economic Performance* (Baltimore: Johns Hopkins, 1993), pp. 91-92; and Keon Chi, "State Business Incentives" *State Trends and Forecasts* 3 (June 1994), pp. 4-6. 20. "The economic momentum index" is the creation of Anne Jordan, "A State Fiscal Report Card," *Governing* 5 (January, 1992), p. 34. Note that economic status seems to be a better predictor of budget cutting than region. Between 45% and 56% of all states in each of the four major census regions—Midwest, Northeast, South, and West—were in the category of budget cutters.

21. A survey by the National Conference of State Legislatures indicates that "the budget will be the issue in 1992. Other pressing issues—like economic development workers' compensation, and solid waste management—must be tabled in many cases while [states] struggle to bridge widening gaps between revenues and projected expenditures." National Conference of State Legislatures, *State Issues: A Survey of Priority Issues for State Legislatures* (Denver: NCSL, 1991), p. 1.

22. Interview conducted January 25, 1993.

23. Eisinger, *Rise of the Entrepreneurial State,* pp. 16-21; Eric Herzik, "Policy Agendas and Gubernatorial Leadership," in *Gubernatorial Leadership and State Policy,* ed. Eric Herzik and Brent Brown (Westpoint, CT: Greenwood, 1991), pp. 25-37.

24. Robert Guskind, "The New Civil War," *National Journal* 14 (April 3, 1993): 818.

25. Interview conducted February 22, 1993.

26. *State Policy Reports,* 10 (April 1992), p. 21; Keon Chi, "State Business Incentives," *State Trends and Forecasts* 3 (June 1994): 11.

27. H. Brinton Milward and Heidi Newman, "State Incentive Packages and the Industrial Location Decision," *Economic Development Quarterly* 3 (1989): 203-22; South Carolina State Development Board, "A Corporate Gem Comes to South Carolina," *Economic Developments* 2 (summer 1992): 4-8; *New York Times,* October 4, 1993.

28. *Wall Street Journal,* February 3, 1994.

29. *New York Times,* November 30, 1992.

30. National Association of State Development Agencies, *State Economic Development Expenditure Survey*

(Washington, DC: National Association of State Development Agencies, 1990).

31. South Carolina State Development Board, "A Corporate Gem," pp. 1, 4.

32. *State Policy Report,* 10 (April 1992), p. 22.

33. The question read as follows: "A lot of people make a distinction between old-style economic development (by which they mean smokestack chasing or industrial recruitment) and new-style development, which focuses on encouraging small business growth, fostering high-tech industries, encouraging export, and supporting human resource development. Would you say that your state focuses on one sort of development more than the other?"

34. Doug Ross and Robert Friedman, "The Emerging Third Wave," *The Entrepreneurial Economy* 9 (autumn 1990): 3-10.

35. Dan Pilcher, "The Third Wave of Economic Development," *State Legislatures* 17 (November 1991): 34-7.

36. There is an ironic supporting footnote to this assessment: At the urging of an administration unsympathetic to activist state economic development policy, the Michigan legislature recently ended the earmarking of a percentage of gas and oil lease revenue for the Michigan Strategic Fund (MSF). The MSF has always been regarded as the prototypical Third Wave program; without a substitute revenue stream, however, it is expected to shrivel away. In Michigan, at least, the Third Wave has dissipated on the outgoing tide.

37. Paul Sabatier, "Policy Change over a Decade or More," in *Policy Change and Learning,* ed. Paul Sabatier and Hank Jenkins-Smith (Boulder, CO: Westview, 1993), p. 19.

38. The term "audit" refers here to program audits, not fiscal audits. The latter is an examination of accounting practices; the former is an assessment of program effectiveness.

39. Thirteen were carried out by legislative audit bureaus, nine by executive branch auditors, four by evaluation units internal to the economic development bureau, and four by outside contractors. Two evaluations were conducted by ad hoc commissions. There is no information on the remaining two.

40. Guy Peters, *American Public Policy: Promise and Performance* (Chatham NJ: Chatham House, 1986); see also, for example, Rita Mae Kelly and Bruce Frankel, "The Federal Decision to Fund Local Programs: Utilizing Evaluation Research," in *The Policy Cycle,* ed. Judith May and Aaron Wildavsky (Beverly Hills, CA: Sage, 1978), pp. 237-58.

41. This argument is not unrelated to the point made by in Sabatier (see "Policy Change," pp. 19-20) that policy change is often less a product of formal policy analysis than of noncognitive forces, such as changes in macroeconomic conditions or the rise of new governing coalitions. But the argument differs from Sabatier's in its emphasis on the strategic considerations that political actors make when faced with these noncognitive changes or the possibility of them.

42. Future research might, for example, explore whether terminations of programs with long gestation periods track the electoral cycle, peaking in the year prior to elections. We may also hypothesize that big industrial recruitment efforts are more likely to precede elections, while controlling for changes in partisan control of the statehouse.

43. Since we did not ask about the impact of economic hard times on policy, we cannot estimate the percentage of states that adjusted their policy to take account of diminished resources. But a number of respondents spontaneously mentioned the state of the economy as a limiting factor. Some examples: in Ohio, "the recent decline in the economy has led to a drop in our [development] budget"; in Alabama, "the financial situation in the state is very bad"; in Illinois, "the budget crisis [produced] a sharp cut [in economic development spending]"; in Florida, "the state is hurting"; in Iowa, the development budget "has risen and fallen with the state's general fund, which hasn't been great lately"; and in Virgina, "we were hurt by defense cuts, so we're in a period of sluggish growth." Other states, such as Colorado and Indiana, suggested that the priority ranking of economic development on the governor's agenda rose and fell with the state of the economy.

44. Interview conducted February 24, 1993.

45. Dennis Grady, "Governors and Markets: Corporate Recruitment from the Gubernatorial Perspective," in *Market-Based Public Policy,* ed. Richard Hula (London: Macmillan, 1988), pp. 36-51.

46. In the national survey, respondents ranked economic development among the top three state policy priorities in 31 states. In all but two cases, education and/or health care ranked higher. In eight states, economic development was ranked as the number one policy priority; in four states, it did not make the top three.

47. Paul Brace writes that as their economics seem "caught in boom and bust cycles," states' efforts in economic development have come to be seen as lacking both sufficiency and durability. Expensive economic development efforts could not be sustained in periods of economic constraint (Brace, *State Government and Economic Performance,* p. 121).

Chapter 15

Competition and Cooperation in Economic Development

A Study of the Twin Cities Metropolitan Area

Edward G. Goetz and Terrence Kayser

❋❋❋

Cities compete with each other for economic development.[1] Local government officials attempt to attract private sector investment to maintain or enhance local economic and fiscal conditions. Most analysts argue that this is the result of the tandem influences of federalism and capitalism.[2] Local governments' authority is limited geographically but private capital is mobile and can locate wherever it wishes. The fact that local governments rely heavily on property taxes to provide municipal services means they are interested in enhancing that tax base. The attraction and retention of economic enterprises is one major way to support and enhance the tax revenues. The increased mobility of private capital, due to technological advances, the decreasing imperative of agglomeration, and the globalization of markets results in a kind of competitive game in which local governments are forced into offering subsidies and incentives to private enterprises in the hopes of attracting their investment. Although much of the extant literature on public sector competition for economic development is set at the state level[3] or examines the competition between major cities,[4] the fiercest competition for private investment is often between neighboring cities or cities within the same region. This article examines the dynamics and the spatial patterns of economic development competition within the Twin Cities region. In addition, we investigate the potential for a cooperative approach to regional economic development among municipalities in the Twin Cities region.

Competition for Economic Development

The competition between local governments for economic development creates both inefficiencies and inequities. Inefficiencies result

Reprinted from *Economic Development Quarterly,* Vol. 7, No. 1, February 1993.

from the escalating cost of public subsidies provided to private capital resulting from the imbalance in information between public and private actors in the negotiations over development and investment. This premise is based on an understanding of land development as a negotiated settlement between developers on the one hand and local officials on the other. Developers wish to invest (or announce the potential for investment) and hope to receive maximum public support for that investment. Local officials desire to have investment occur but also wish to keep public subsidies to a minimum. In some respects, then, the local government is "buying" the private investment of capital. Under circumstances in which local governments compete with each other for development, a "sellers' market" occurs in which the price is driven up by competing bids. Further, in any given instance of development negotiations, the private enterprise knows the precise level of public subsidy needed to induce investment and this minimum amount can only be guessed at by the public sector. Local officials are driven to offer more subsidy if they are in competition with other municipalities, and private capital is in the position of playing one locality against another. This produces what Jones and Bachelor call "the corporate surplus."[5] The corporate surplus is the amount of subsidy provided to the private sector in excess of what would have been minimally sufficient to make them invest. Thus competition between cities tends to increase the amount of the corporate surplus.

A second type of inefficiency relates to the provision of public incentives and subsidies to firms that relocate from one place to another. In this instance, public subsidies are used to simply move jobs and investment around rather than creating a net increase in employment and capital activity. From a regional or national standpoint, public resources have been expended without a net economic benefit.

Inequities arise from the differential treatment that new or relocating firms receive from the public sector. The heavy subsidization of their operations puts them in an advantageous position with their already established competition.[6] Indeed, this realization has even led some private enterprises to oppose subsidy packages targeted to other competing businesses.[7]

Data

In May and June 1991, the Twin Cities' Metropolitan Council's Economic Research Program and University of Minnesota collaborated on a survey of 140 metropolitan area municipalities to determine the extent of economic development effort being undertaken in the region. A total of 109 questionnaires were returned after an original mailing and one follow-up. Fifteen questionnaires were returned with notes indicating that for one reason or another those cities did not have any formal effort in economic development.[8] This left a total of 94 usable questionnaires for analysis. An examination of the nonrespondents revealed that the majority of them (81%) were smaller cities with populations under 10,000 scattered throughout the region with no apparent concentration in any of the seven counties. Because of their small size we have anticipated that they did not have active economic development programs. Thus we have nearly complete data on the municipalities within the region that are interested and involved in economic development.

In the following analyses municipalities are characterized by population size and growth, by a classification of localities developed by the Metropolitan Council (called the Metropolitan Development Investment Framework, MDIF) that distinguishes between (1) fully developed cities, (2) developing cities, (3) free-standing growth centers, and (4) agricultural/rural centers,[9] and finally, by whether or not the municipality has a formal economic development plan. The first two of these classificatory schemes are aimed at illuminating the role of city size, growth rate, and development status on the competitive and cooperative orientation of localities. The final distinction is a direct measure of the amount of planning that municipalities have put into economic development and it, too, will be examined for its impact on competition and cooperation.

The Twin Cities metropolitan region consists of the seven counties including and immediately surrounding the Twin Cities of Minneapolis and St. Paul. The region has 140 municipal governments, covers 3,000 square miles and is home to 2.2 million people. The Twin Cities region is the economic center of both the state of Minnesota and the entire upper midwest region from Wisconsin to Montana.[10]

Dimensions of Competition

There is much consensus among the economic development officials contacted that there is competition for development within the Twin Cities region. Eighty-five percent of the respondents either agreed or strongly agreed with that premise. Further, this attitude was shown by cities of all types and in all areas of the region. Officials in large and small cities and quickly growing and fully developed cities all agreed that competition exists.

Only about one-half of the cities (49.5%) feel they are doing well in that competition, however. This is, in fact, about what would be expected in a zero-sum competitive game. For every municipality that considers itself doing well in competition, there is another that feels it is not doing well. Further analysis shows, logically enough, that faster growing cities feel better about their competitive status ($r = .20$, $p < .05$).[11] Using the MDIF classification, we find that agricultural and rural centers feel significantly worse about their competitive status than do the other categories of cities. Finally, cities that have a formal economic development plan tend to believe they are doing better in the competition for development than cities without plans (mean response of 3.0 compared to 2.24, $p < .001$).

Well over half of the respondents indicated that they need to be more aggressive in competing for economic development. This indicates, we believe, that there is little chance that competition will lessen in the future. Consistent with the previous findings, agricultural and rural centers and cities without economic development plans feel they need to become more aggressive in their pursuit of development. These are the types of cities that felt they were not doing well in the competition for development. Indeed the correlation between the rating of their competitive status and the need for more aggressive efforts was strong and statistically significant ($r = -.44$, $p < .001$). This relationship reveals a ready acceptance of competition as the normal mode of development policymaking. Those who are not doing well do not question the premise of interjurisdictional competition, but instead they feel they must simply work harder to become more aggressive to fare better in that competition.

As described earlier, the competition between cities leads to ever greater levels of public subsidy. From this premise, we hypothesize that the greater the level of competition, the less negotiating power felt by local officials. In fact, the perception of competition and officials' assessment of their own negotiating power are highly correlated ($r = .42$, $p < .001$); meaning that those officials who felt more strongly that there is a great deal of competition for development also agreed more strongly with the premise that competition does negatively affect their bargaining position. Over half (53%) of the city officials who felt there is competition for development also felt they were in a disadvantaged negotiating position. On the other hand, none of the city officials who felt that there is not much competition felt disadvantaged.

Further investigation shows that attitudes toward negotiating power with the private sector are negatively correlated ($r = -.33$, $p < .001$) with the respondents' rating of their competitive status. That is, officials who felt their city was doing well in the competition also felt that the competition did not put them at a disadvantage with the private sector. This is shown in Table 15.1. Only 37% of those cities that felt they were doing well in competition with others, felt they were in a disadvantaged negotiating position, compared to 63% of the cities who were not doing well.

TABLE 15.1 The Relationship between Competitive Status and Negotiating Power: Attitudes of Twin Cities Development Officials

		"We are doing well in the competition for economic development."	
"City is at a disadvantage negotiating with private developers."		Agree	Disagree
	Agree	15 (33%)	25 (58%)
	Disagree	31 (67%)	18 (42%)

NOTE: Chi-square = 4.9, $p < .05$.

Summary

This section has analyzed some of the basic attitudes toward competition for economic development reported by the surveyed officials. In addition, some simple relationships have been discovered that seem to confirm some of the intuitive hypotheses presented. For example, almost 9 of 10 officials acknowledged that there is indeed competition for development between municipalities in the region. Cities doing well in the competition are those that have a formal planning document for economic development and cities that experienced a good deal of population growth in the last 20 years. Cities that have not done well in the competition feel that they must be more aggressive in competing with their neighbors, and they feel that they are put at a disadvantage in negotiating with the private sector. The analysis thus further illuminates the relationship between the two dimensions of development policy competition; the competitive position of a city with other municipalities and the negotiating position of a city with the private sector. Officials who feel the competition more than others also tend to feel less powerful in negotiating with capital. Further, officials who feel their cities are not competing well with others also feel that the private sector has the upper hand in development negotiations.

Who Competes with Whom?

Spatial Analysis

Respondents were asked to list the three municipalities they regarded as their chief competitors for economic development. Of the 94 usable questionnaires returned, only 72 contained responses to this question (77%). Twenty of the nonresponding officials indicated at this point in the questionnaire that their city does not compete with others.[12] This means there are, in fact, only two missing values on this questionnaire item.

Figure 15.1 is a display of competition between municipalities in the Twin Cities metropolitan area. The lines connect cities with the competitors that they named on the questionnaire. The arrow denotes the direction of the relationship. For example, the arrow from St. Paul to Minneapolis (the two larger dots in the center of the map) indicates that St. Paul named Minneapolis as one of their major competitors. The names of municipalities have been left off the map in order to improve the clarity of the data presentation and also because the item of interest here is not so much who picked whom, but the overall spatial pattern of competition. Although the map seems a jumble of criss-

Figure 15.1. Intrametropolitan Competition for Economic Development in the Twin Cities Region

crossing lines, further analysis does reveal an underlying logic.

In fact, if we deconstruct Figure 15.1, the spatial analysis of local competition reveals six distinct patterns of competition. First, most cities choose competitors that are nearby. The lines connecting cities with their named competitors are rarely long. Few cities name as competitors cities in other regions of the metro area. There are, in fact, only two exceptions; Vadnais Heights, a northern suburb of St. Paul named Eagan, a southern suburb as a competitor; and Eden Prairie on the southwest side of Minneapolis named Woodbury, located southeast of St. Paul, as one of its competitors. Otherwise, city officials named other cities in roughly the same geographic subregion of the metro area.

Second, some cities were named more often than others and thus represent what we call competitive nodes. The major competitive nodes are Eagan in the south metro area, Plymouth on the west, and Brooklyn Park to the northwest. These are all second-ring suburbs of the Twin Cities. The competitive nodes (municipalities named by five or more cities) are highlighted in Figure 15.2. They run on roughly a double corridor from the north to the west of the Twin Cities and from the southeast to southwest. The major exception to this pattern is the city of St. Paul, which was named as a competitor by seven of its immediate suburbs.

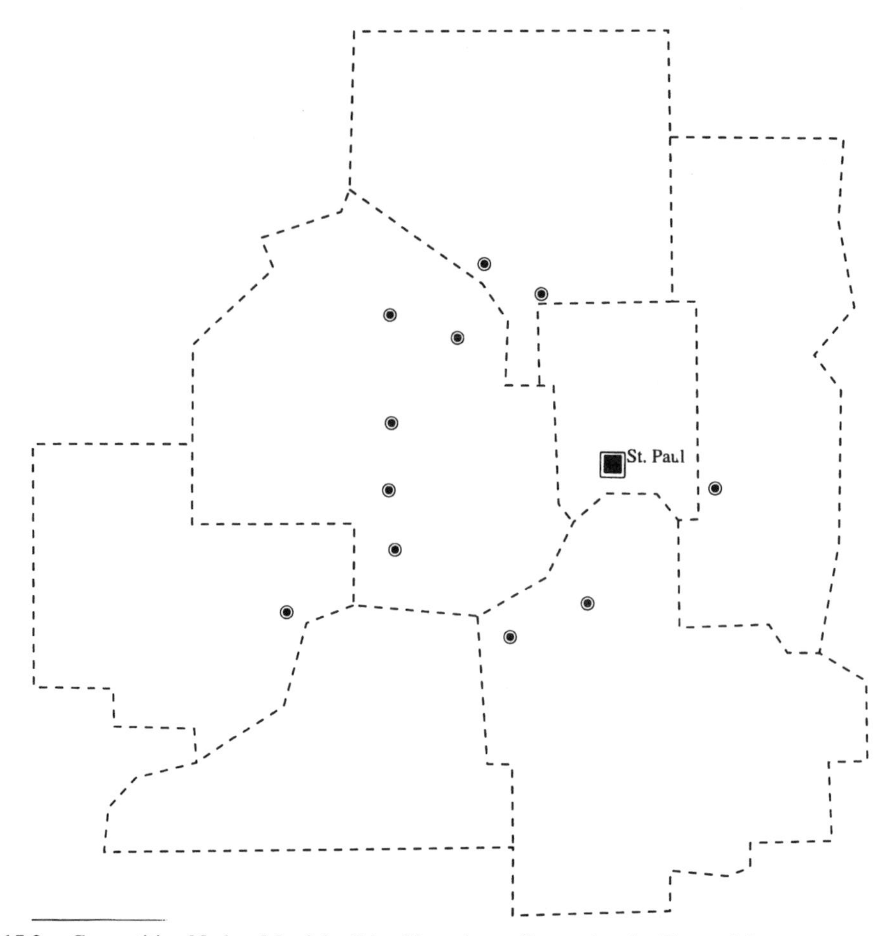

Figure 15.2. Competitive Nodes: Municipalities Named as a Competitor by Five or More

The third emerging pattern depicts smaller cities on the periphery of the region either directing their competition inward to the competitive corridors or targeting each other in a series of submarkets in which they compete with their closest neighbors (see Figure 15.3). This is seen most clearly on the far southwest and far eastern portions of the metro area where the pull of the developing ring is not as strong. In these areas, the municipalities on the periphery of the region form smaller systems of competition that are separate from the growth centers of the region. The rest of the outlying municipalities tend to direct their competition inward.

Fourth, the spatial analysis shows a number of lines from the inner cities and first-ring suburbs outward to the developing suburbs (see Figure 15.4) indicating that the inner cities and the inner ring of suburbs are orienting their development competition toward the competitive corridors in the second ring of suburbs.

Finally, there is a pattern of competition between cities located within the corridors of competition (Figure 15.5). That is, these cities almost exclusively identify each other as competitors, ignoring to a large extent both the inner core and the peripheral areas of the region.[13]

Thus most of the lines tend to converge on the double corridor of intense economic development competition. The final three spatial patterns, in fact, combine to create a single finding; that there is a convergence of named competitors that sandwich the inner cities in a pattern from north and east to southwest. All but one of the major competitive nodes are located within these corridors of competition. Further, the cor-

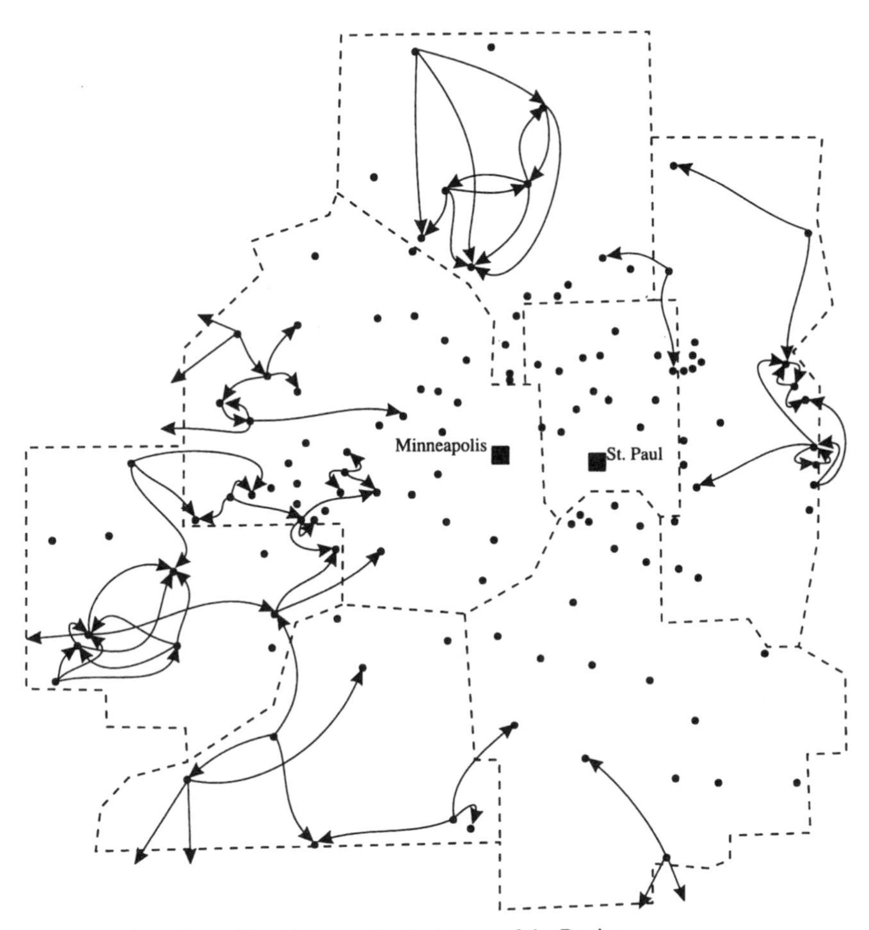

Figure 15.3. Competitors Named by Cities on the Periphery of the Region

ridors, with one exception, do not include the inner cities nor the inner-ring suburbs.

The patterns discovered in the spatial analysis are confirmed by looking at the pattern of competition by county and city classification. Table 15.2 shows the distribution by county of both the respondents and the competitors they named. The clear diagonal pattern shows that most of the named competitors were within the same county, reinforcing the notion that most named competitors were nearby municipalities. For example, for cities from Anoka County, 69% of the competitors named were also located within Anoka County. This pattern is repeated for all the counties.

Table 15.3 shows the breakdown of cities and their competitors by MDIF classification. As the table shows, a similar diagonal pattern emerges in which cities in each category generally choose other cities in that same category. The only exception is that a majority of inner-ring cities choose fully developed suburbs as competitors. Although the fully developed ring constitutes only 48% of responding cities in the metro area, they are named as competitors over 60% of the time. This further reinforces the spatial analysis in Figures 15.3 through 15.5.

Economic Profile Analysis

The following analysis looks at the growth and the revenue profiles of cities to determine whether competitor cities matched each other in demographic or fiscal terms. The working hypotheses are that cities will name competitors

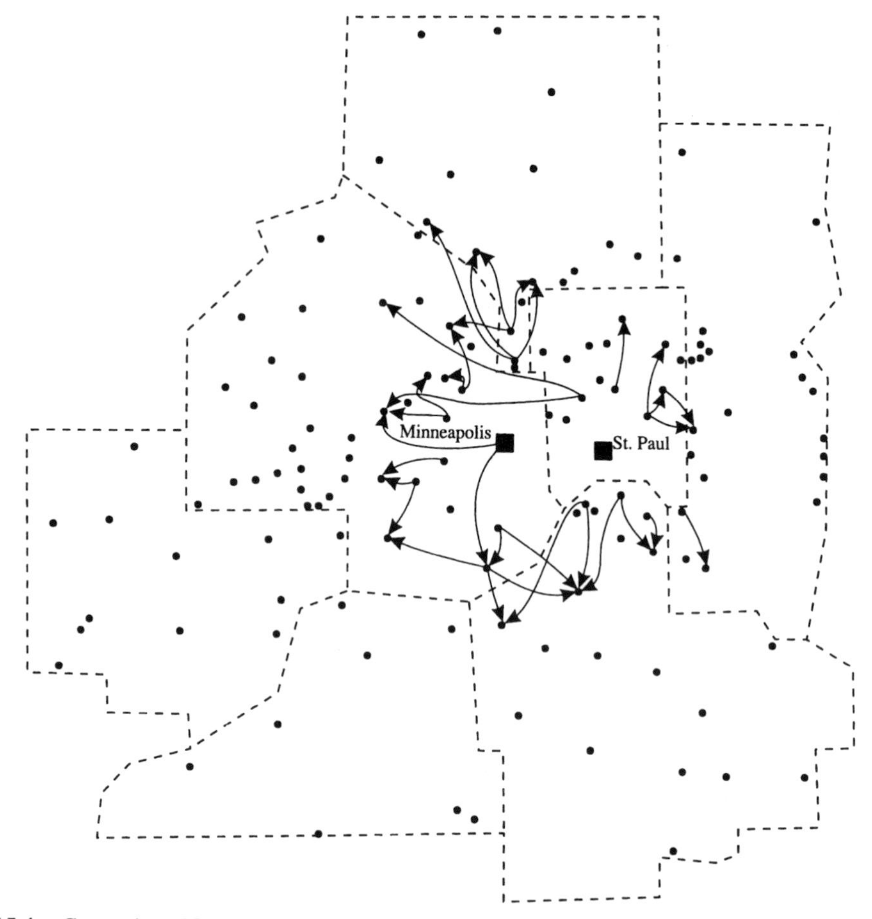

Figure 15.4. Competitors Named by Inner City and First-Ring Suburbs

that are similar in size, revenue, and growth characteristics. Six measures of population size were examined, population in 1980, 1985, and 1990, and the number of households in 1980, 1985, and 1990. In addition, six growth measures were considered, the percentage of change in population between 1970 and 1980, between 1980 and 1990, and between 1985 and 1990; and the percentage of change in the number of households between 1970 and 1980, 1980 to 1990, and 1985 to 1990. Table 15.4 shows the Pearson correlations between the respondent cities and their three named competitors. According to the six measures of population size, the first city named by respondents closely resembles the respondent's city; all of the correlations are very high and statistically significant. This is not, however, true of the second and third cities named. Only 3 of the 12 correlations are significant and even these are moderate in strength.

Interestingly enough, the correlation between competitive cities is much stronger for absolute population size than it is for any of the growth measures. As the bottom half of Table 15.4 shows, only 4 of the 18 possible correlations using growth measures are positive and statistically significant. The second and third competitors match the respondents' cities somewhat in terms of household growth in the 1980s. There is less similarity between competing cities in growth than in absolute population size.

In the economic analysis, more correspondence between cities and their named competitors is established. Correlations between re-

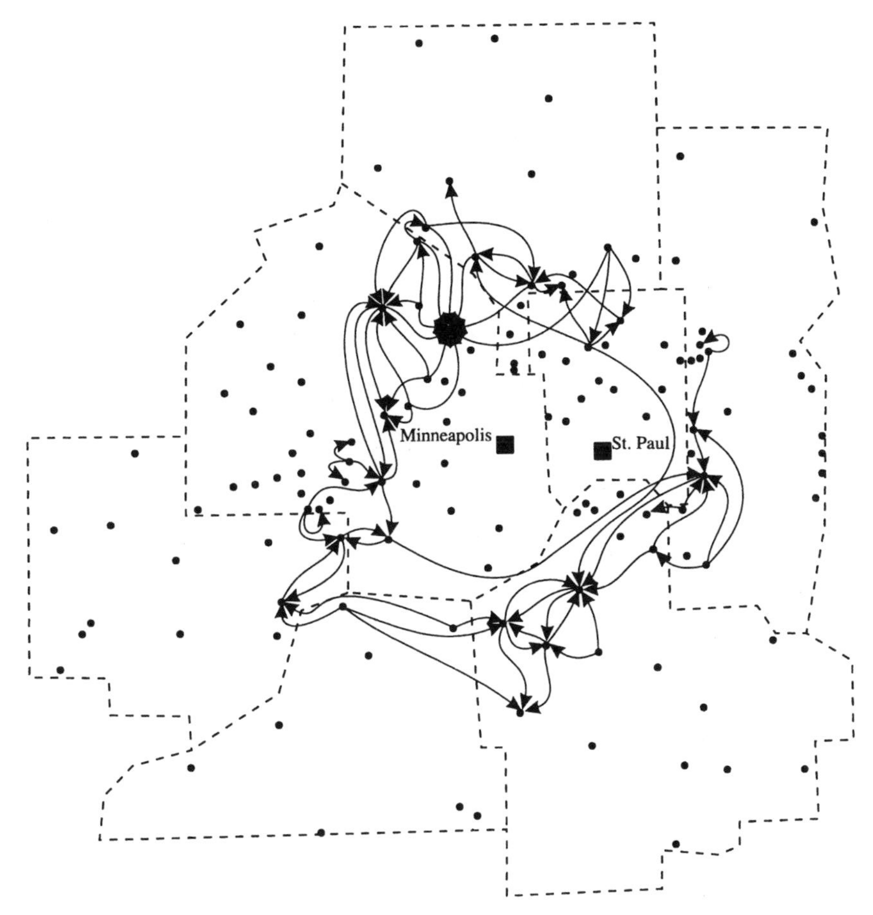

Figure 15.5. Competitors Named by Second-Ring Suburbs

TABLE 15.2 County of Respondent City, by County of Named Competitors

	Respondents' County						
Competitors' County	*Anoka*	*Carver*	*Dakota*	*Hennepin*	*Ramsey*	*Scott*	*Washington*
Anoka	20 (69)	—	—	2 (4)	2 (12)	—	1 (4)
Carver	—	13 (68)	—	—	—	3 (23)	—
Dakota	—	—	21 (82)	4 (8)	1 (6)	4 (39)	3 (12)
Hennepin	6 (21)	4 (21)	1 (4)	43 (81)	4 (22)	—	—
Ramsey	3 (10)	—	3 (12)	3 (6)	9 (50)	—	4 (16)
Scott	—	2 (11)	—	—	—	6 (46)	—
Washington	—	—	1 (4)	1 (2)	2 (12)	—	17 (68)

NOTE: Numbers in parentheses are column percentages.

TABLE 15.3 Relationship between Respondent City and Named Competitors, by Metropolitan Development Investment Framework (MDIF)

	Respondents' MDIF			
Competitors' MDIF	*Inner-Ring Cities*	*Developing Ring*	*Growth Centers*	*Rural/Agricultural Centers*
Inner-ring cities	15 (37)	10 (12)	—	—
Developing ring	25 (62)	66 (80)	2 (20)	11 (27)
Growth centers	—	6 (7)	8 (80)	11 (27)
Rural/agricultural centers	—	1 (1)	—	19 (46)

NOTE: Numbers in parentheses are column percentages.

TABLE 15.4 Population Size and Growth Rate: A Correlational Analysis of Respondent Cities and Their Named Competitors

	Competitor 1	*Competitor 2*	*Competitor 3*
POP80	.71***	.16	.19
POP85	.72***	.17	.22
POP90	.72***	.18	.27*
HH80	.71***	.13	.20
HH85	.71***	.15	.24*
HH90	.72***	.17	.30*
		Percentage change	
POP70-80	.19	.12	.19
POP80-90	.02	–.01	.19
POP85-90	–.01	–.02	.12
HH70-80	.14	.17	.19
HH80-90	.07	.33**	.25*
HH85-90	.10	.29**	.27*

NOTE: Numbers are Pearson product moment correlations. POP = population, HH = number of households. The numbers following these abbreviations indicate the year or time frame that the population or number of households were under consideration.
*$p < .05$; **$p < .01$; ***$p < .001$.

spondent cities and their competitors were run for the following measures: total assessed value in 1985 and in 1990, total assessed value for real property in 1985 and 1990, total assessed value for commercial property in 1985 and 1990, and finally, total assessed value for industrial property in 1985 and 1990. In addition, four variables measuring the change between 1985 and 1990 for each tax category were computed. The correlations are found in Table 15.5. Strong and positive correlations show for the variables measuring the absolute value of the different tax categories. Again, however, the change variables are generally insignificant and sometimes in the negative direction. This indicates that as with population size, officials name competitors with similar revenue levels, not cities with similar growth trajectories.

An additional hypothesis tested is that officials would choose competitors with similar tax structures. That is, primarily residential cities would choose as competitors other cities with a largely residential tax structure. The same would hypothetically be the case for industrial and commercial tax revenues. Thus six variables were created measuring the percentage of total assessed value accounted for by real, commercial, and industrial properties for both 1985 and 1990. Table 15.6 shows the correlations between respondent cities and their named com-

TABLE 15.5 Tax Revenues and Revenue Growth: A Correlational Analysis of Respondent Cities and Their Named Competitors

	Competitor 1	*Competitor 2*	*Competitor 3*
Assessed value 85	.67***	.26*	.33*
Assessed Value 90	.14	.08	–.02
Residential 85	.66***	.26*	.33*
Residential 90	.67***	.32**	.41***
Commercial 85	.61***	.22*	.32**
Commercial 90	.46***	.39***	.38**
Industrial 85	.60***	.32**	.20
Industrial 90	.51***	.46***	.26*
		Percentage change 85-90	
Assessed value	.02	–.21*	.92***
Residential	.12	.16	.07
Commercial	.01	.14	.12
Industrial	.11	.04	–.07

NOTE: Numbers are Pearson product moment correlations.
*$p < .05$; **$p < .01$; ***$p < .001$.

TABLE 15.6 Tax Structure: A Correlational Analysis of Respondent Cities and Their Named Competitors

Tax Structure (in percentages)	*Competitor 1*	*Competitor 2*	*Competitor 3*
Residential 85	.16	.04	.28*
Residential 90	.38***	–.03	–.06
Commercial 85	.28*	.05	.18
Commercial 90	.21*	–.03	–.04
Industrial 85	.28*	.18	.15
Industrial 90	.29*	–.02	–.04

NOTE: Numbers are Pearson product moment correlations.
*$p < .05$; **$p < .01$; ***$p < .001$.

petitors. The correlations are generally low and insignificant. Only 6 of the 18 variables are positive and statistically significant.[14] Thus the tax structure was somewhat important in determining the initial competitor but did not seem to play a role in the other choices.

Summary

The analysis of competition has shed some light on the types of cities that officials regard as competitors. The strongest relationships are the purely spatial ones and those based on tax revenues. Competitors are quite simply those cities that are close at hand. In the aggregate sense, the Twin Cities reveals a pattern of strong economic development competition in the second-ring suburbs ringing the city from the northwest to the southeast. The first competitors named resemble the respondent city on population size. Otherwise, population size and growth are generally not similar between competitive cities. Absolute levels of tax value are, however, similar between cities and their named competitors. Thus officials identify as competitors cities of similar assessed value levels but generally do not name cities with similar tax structures or rates of growth (in both tax or population terms).

The Prospects for Cooperation in Regional Economic Development

In this section we explore the issue of cooperation between municipalities in economic development. The first part of the analysis looks at the attitudinal foundation for cooperation. That is, do local development officials agree that cooperation is beneficial or likely? Also, we explore officials' attitudes about the impact of economic development and whether there exists any objective basis for regional cooperation as a viable economic development strategy.

There is a moderate amount of support for the explicit notion of regional economic development. Thirty nine percent of the respondents agreed with the statement "economic development is best done regionally." Those who agreed with this statement tend to be officials in smaller cities ($r = -.17$, $p < .05$).

The following analysis examines whether there might be any objective basis for regional or even subregional cooperation. One prerequisite for such cooperation would be the feeling that economic development benefits spill over from one community to the next. Respondents were asked the degree to which they agreed with the following statements: "Economic development in my municipality is a benefit to the entire region," and "Economic development in my municipality is a benefit to the entire county." Fully 84% of the respondents agreed with the statement that development benefits redound to the entire region, and 87% agreed with the more modest statement that development was beneficial to the entire county. These findings show that the objective basis for regional cooperation exists. That is, an overwhelming number of respondents felt that the benefits of economic development are not confined by city limits, but do indeed spill over into neighboring communities. In fact, 84% believe that the benefits are regionwide, impacting all seven counties in the metropolitan area. This belief in the wide-ranging nature of development benefits is held to a greater extent by officials in large cities ($r = .30$ and $.25$, $p < .01$),[15] officials in growing cities ($r = .20$, $p < .05$),[16] and cities with a development plan.[17] It is held less strongly in the rural and agricultural centers of the region.[18] Finally, the belief that development has widespread benefits is positively correlated with respondents' belief that their own municipality is doing well in the competition for economic development ($r = .44$ and $.35$, $p < .001$). Even though there are some types of cities in which the belief is stronger than in other cities, it should be emphasized that the spillover of development benefits is almost a universally accepted idea.

If the nature of development benefits cannot be contained by one city, then a logical basis for cooperation exists. The reluctance of 61% of the respondents to accept regional planning must be explained by other factors, and merits further investigation. Some potential explanations are examined in the next section.

Finally, we asked respondents, "What type of economic development is best done at the regional level?" A full 55% of those who responded answered, "None." That is, more than half of those responding to this question feel that there is no type of economic development that is best suited to regional cooperation. Of those development objectives listed, transportation is the most mentioned (27%) followed by various development supports such as planning, marketing, and grantsmanship (16%), and infrastructure (14%). The full range of responses is listed in Table 15.7.

Summary

Some of the findings reported above support the idea of regional cooperation in economic development. First, already one-third of development officials in the Twin Cities metropolitan region agree that development policy is best made at the regional level. Although many more need to be convinced before regional steps can be effectively taken, this is a significant base of support. Further, almost all development officials agree that development has countywide and even regionwide benefits. This seems to be an important starting point for the discussion of regional cooperation. If benefits are shared,

TABLE 15.7 Potential Regional Development Projects: Attitudes of Twin Cities Area Development Officials

"What types of development projects should be done on a regional basis?"	
Type of Development	*Number of Responses*
Transportation	**12 (27)**
Development support, e.g., planning, marketing, grantsmanship	**7 (16)**
Infrastructure	**6 (14)**
Individual issues, e.g., tourism, housing, shopping centers	**5 (11)**
Tax base sharing	**4 (9)**
Sewage	**3 (7)**
Other	**7 (16)**

NOTE: Numbers in parentheses are percentages.

even in the current environment of active competition between municipalities, then the potential for more active cooperation exists. Finally, metro area development officials felt that transportation and infrastructure might be promising areas for a regional approach to economic development. In addition, officials felt that regional provision of development supports such as planning and marketing might also be useful.

Regional Cooperation in Economic Development

This section investigates attitudes toward actual experiences in cooperative economic development policy. We first inquired about a specific example of regional development policy, the Fiscal Disparities program. The Twin Cities metropolitan region is widely known for its tax base sharing program. Given the fact that one of the main reasons local governments pursue economic development is to enhance the local tax base, we asked respondents whether or not they felt the tax base sharing plan was a disincentive to development. Two-thirds of the respondents did not feel that Fiscal Disparities was a disincentive, although officials in larger cities were more likely to see Fiscal Disparities as a disincentive ($r = .26$, $p < .01$). There is also a slight tendency for officials from cities that make larger contributions to the Fiscal Disparities pool to feel that it is a disincentive.[19] Notwithstanding that tendency, most officials in the Twin Cities metropolitan area do not feel that the system of regional tax base sharing is a significant disincentive to economic development. That is, the regional spread of benefits was not seen as an impediment to the individual pursuit of growth.

We also hypothesize that negative attitudes about regional approaches to development policy would be positively correlated with negative attitudes about Fiscal Disparities. In fact, there is little relationship between these attitudes ($r = -.03$, n.s.). Thus, the tax-sharing program and officials' attitudes toward it are not related to a larger ambivalence about regional cooperation.

When asked if they have ever considered a development project in cooperation with another municipality, 52% of the respondents answered "yes." Cities that have entertained the idea of cooperative development are slightly more likely to be larger than cities that have not considered cooperation. There is no variation by growth rate or geographic location. Cities with a formal economic development plan, however, are significantly more likely to have considered cooperative development. Fully 69% of the cities with plans have initiated cooperative development compared to 44% of the cities without formal development strategies.

When those officials who indicated that they had considered cooperative development were asked whether they actually carried through with the idea, 23 of 50 said they had. Thus, just less than one-fourth of all respondents have some actual experience in economic development in cooperation with one or more other municipalities. Cities that have actually engaged in cooperative development are much more likely to be larger than cities that have not cooperated ($r = .39, p < .001$).[20] Again, however, there is no variation by growth rate or geographic location, but there is significant difference between cities with development plans and those without. Forty-four percent of cities with formal plans have carried out cooperative development compared to only 14% of the cities without formal planning documents.

In another attempt to explain the source of ambivalence toward regional development, we analyzed the difference in opinion between officials who had engaged in some form of cooperation and those who had not. We hypothesize that those who have had experience in joint ventures will feel more disposed to a regional development approach. In fact, the relationship is in the other direction. City officials who have ever engaged in a cooperative development venture are slightly more likely to think that regionally based development is inappropriate ($F = 4.38, p < .05$). Thus some of the reluctance to engage in a regional approach to economic development might be based on some (perhaps negative) experiences with cooperative development.

But the cooperative ventures that have been tried by governments in the region have generally been satisfactory; 66% of the respondents felt they had had a positive or very positive experience in cooperating with other municipalities. Another 29% could not make a judgment one way or the other and only 5% felt that the cooperative venture had been negative. Further, reported satisfaction with cooperative development is unrelated to attitudes toward regional development ($r = .11$, n.s.).[21]

Finally, 97% of the respondents said they would consider cooperative economic development with another municipality in the future. Although this may be seen as an easy response to make because officials can answer this question affirmatively without committing themselves to any specific action in the future, it does indicate at the very least that they have not ruled out a more cooperative stance on economic development issues.

Summary

These findings, taken together, suggest that there is a difference in the minds of development officials between regionwide development and the cooperation of municipalities within the region. Although a majority reject a regional approach, there seem to be positive experiences and goodwill toward cooperative development between municipalities. A number of subregional, some countywide, development partnerships exist and one-fourth of all municipalities in the region have engaged in cooperative development, most of it highly satisfactory.

Conclusion

This article has presented the findings of a questionnaire survey of development officials in the Twin Cities region. A number of basic hypotheses about the existence of competition between municipalities for economic development were examined and supported by the data. Competition does exist at the subregional level. Further, at least for those who feel the competition the most, and for those who feel their jurisdiction is not doing so well, the competition negatively affects their negotiating position with the private sector leading, in all likelihood, to a growing corporate surplus.

A spatial and economic analysis of competition indicates that municipalities generally compete with nearby cities of the same size and similar tax revenue levels. Further, there exists a number of cities that can be regarded as competitive nodes in that they are named by a large number of other cities as competitors. These nodes are arranged in a pattern of parallel corridors from the north of the inner cities to the southwest, and from the southeast to the southwest. These corridors do not include the inner cities or the inner-ring suburbs but generally consist of second-ring developing suburbs.

The existence of widespread competition notwithstanding, the attitudes of development officials and the experiences of some cities provide a basis for the prospect of increasingly cooperative development between cities or at the county level. Most officials agreed with the notion that development benefits spill over into neighboring municipalities; indeed, over 80% of the officials felt that development has regional benefits. Further, the attitude of city officials who have engaged in cooperative development shows that it can be a positive experience.

The existence of a smattering of cooperative development projects does not alter the basic fact of widespread and localized competition between municipalities in the region. In fact, cooperative ventures exist in the same environment as strong interjurisdictional competition. The results of this research show that competition and cooperation for economic development are not mutually exclusive. Cooperative ventures or subregional "partnerships" for development in all likelihood also compete for economic development and merely displace the locus of competition from the municipality level to the county or subregional level. If the experience of the Twin Cities region is instructive, competition among governments, with all of its inefficiencies and inequities, continues to characterize the economic development policy of local governments.

Notes

1. The competition between cities for economic development is well documented. For a review, see "Interjurisdictional Tax and Policy Competition: Good or Bad for the Federal System?" (Washington, DC: Advisory Commission on Intergovernmental Relations, April, 1991).

2. See Ann O'M. Bowman, "Competition for Economic Development among Southeastern Cities," *Urban Affairs Quarterly* 23:511-27. See also Dennis R. Judd and David Brian Robertson, "Urban Revitalization in the United States: Prisoner of the Federal System," in *Regenerating the Cities: The UK Crisis and the US Experience,* ed. Michael Parkinson, Bernard Foley, and Dennis R. Judd (Glenview, IL; Scott Foresman, 1989).

3. Roger J. Vaughan, *State Taxation and Economic Development* (Washington, DC: Council of State Planning Agencies, 1979).

4. See, for example, John P. Blair and Barton Wechsler, "A Tale of Two Cities," in *Urban Economic Development,* ed. Richard D. Bingham and John P. Blair (Beverly Hills, CA: Sage, 1984); and Bowman, "Competition for Economic Development."

5. Bryan D. Jones and Lynn W. Bachelor, "Local Policy Discretion and the Corporate Surplus," in Bingham and Blair, *Urban Economic Development.*

6. Vaughan, *State Taxation.*

7. See, for example, Larry Bennett, "Beyond Urban Renewal: Chicago's North Loop Redevelopment Project," *Urban Affairs Quarterly* 22:242-60. A coalition of downtown hotel operators and land owners in Chicago mobilized in the early 1980s against a proposed tax abatement from the city of Chicago to the Hilton Corporation in return for Hilton's participation in the North Loop redevelopment project.

8. For example, three municipalities noted that their jurisdictions are zoned entirely residential and thus have no economic development policy. Other officials returned questionnaires indicating their jurisdiction is too small or is simply not involved in economic development.

9. *Metropolitan Development and Investment Framework,* Publication No. 640-88-122 (St. Paul, MN: Metropolitan Council of the Twin Cities, December, 1988). "Fully developed areas" are communities that were more than 85% developed by 1984 and that are contiguous to one another (p. 20). In these areas the need for maintenance, upgrading, and rehabilitation has surpassed the level of new development occurring. The developing area is defined by the Metropolitan Council as "that portion of the region that is in the path of urban growth" (p. 20). These areas are beyond the fully developed area that constitutes the core of the region. "Freestanding growth centers are the large urban centers located within the rural portion of the seven county" area (p. 21). These are smaller towns with a full range of economic services and an employment base sufficient to provide work for the local population. They are generally surrounded by undeveloped land. Rural and agriculture centers are even smaller towns or villages in undeveloped or agriculture areas.

10. Thomas L. Anding, John S. Adams, William Casey, Sandra de Montille, and Miriam Goldfein, *Trade Centers of the Upper Midwest: Changes from 1960 to 1989* (St. Paul-Minneapolis: Center for Urban and Regional Affairs, University of Minnesota, 1990).

11. The two variables correlated are (1) population change between 1970 and 1990 and (2) response to the questionnaire item, "My municipality is doing well in the competition for economic development" rated on a scale of 1 (*strongly disagree*) to 4 (*strongly agree*).

12. This is not entirely inconsistent with earlier results showing an overwhelming support for the proposition that competition exists between cities in the region. The earlier item was in reference to the general case of all cities in the region. This later questionnaire item refers to the responding jurisdiction specifically. Some municipalities apparently felt that although there is competition in the region, they are not participating in it.

13. Figures 15.3, 15.4, and 15.5 highlight the subregional patterns discussed in the text of the article. In order to more clearly present these patterns visually, anomalous relationships (i.e., lines representing competitive relationships that did not conform to the generalization) were omitted from these maps. Nevertheless, these three maps account

for a full 89% of the competitive relationships named by respondents. The major exceptions are the eight St. Paul suburbs that named St. Paul as a primary competitor. The fact that the inner city of St. Paul attracted the competition of a number of its immediate suburbs does not fit the generalization that the inner ring is orienting its competition toward the developing ring.

14. An alternative method of computing tax structure, computing the ratio of commercial to residential revenue and the ratio of industrial to residential revenue, produced even less correlation between respondent cities and their named competitors.

15. The correlations are for the relationship between population in 1990 and reported attitudes on the two questionnaire items, "Development in my municipality is a benefit to the entire region," and "Development in my municipality is a benefit to the entire county." The questionnaire responses were from 1 (*strongly disagree*) to 4 (*strongly agree*).

16. The correlation reported is the same for two relationships: the relationship between population growth from 1980 to 1990 and the two questionnaire items regarding the regional and countywide nature of development benefits (see note 15).

17. The t test for difference in means significant at $p < .001$.

18. ANOVA, F test significant $p < .05$.

19. The correlation between the per capita amount contributed to Fiscal Disparities in 1985 and the belief that the program is a disincentive to development is .24 ($p < .05$). However, the correlation for the 1990 per capita contribution is neither positive nor statistically significant. Further, cities that receive more from Fiscal Disparities are not significantly more likely to say that it is not a disincentive. Although the correlations for 1985 and 1990 are both in that direction, neither are statistically significant.

20. The correlations are for the relationship between population in 1990 and a dichotomous variable that is coded 1 if the city has ever completed a cooperative development and coded 0 if not.

21. A variety of cooperative development projects were reported by respondents, including marketing and cooperative planning efforts. In addition, there are two examples of ongoing cooperative approaches to economic development in the Twin Cities region. The "Metro East Association" and the Scott County Economic Development Coalition are both multi-city associations that address subregional cooperative development issues.

Chapter 16

Exurban Industrialization

Implications for Economic Development Policy

Arthur C. Nelson
with William J. Drummond and David S. Sawicki

❋❋❋

During the 1970s, researchers detected a movement of population and jobs back to rural areas. This was called the "rural renaissance" and "rural industrialization."[1] The literature generally asserts that during the 1970s, the two-century-long trend of population and manufacturing concentration in urban areas was reversed: Nonmetropolitan areas gained population and manufacturing jobs faster than metropolitan areas did. Much of the rural-renaissance phenomenon disappeared, however, when formerly nonmetropolitan counties were deemed metropolitan by the U.S. Bureau of the Census in 1983.[2] The trend was revealed as nothing more than continued population deconcentration.

But the Census Bureau's scheme for classifying counties as either metropolitan or nonmetropolitan, urban or rural, led researchers to overlook the most interesting aspect of counterurbanization: the emergence of *exurbia.* Exurbia is not discernible in analyses based on the Census Bureau's characterizations. The problem, of course, is that exurbia overlaps both urban-suburban and rural-hinterland areas. Because of the Census Bureau's statistical arrangements, some exurban industrialization is placed in metropolitan areas, the balance in nonmetropolitan areas—resulting in confusion about the nature of contemporary industrialization patterns. This leads, in turn, to overstatement of rural and understatement of exurban development, and thereby to unwise economic development policy and planning.

My colleagues and I have pioneered research in defining exurbia. Generally speaking, exurbia can be characterized as (a) having low population density, (b) beginning at the edge of ur-

AUTHORS' NOTE: Support for this report was provided by the U.S. Department of Commerce, Economic Development Administration, and by the Georgia Institute of Technology. The statements, findings, conclusions, and recommendations are those of the authors and do not necessarily reflect the views of the Economic Development Administration or the Georgia Institute of Technology.

Reprinted from *Economic Development Quarterly,* Vol. 9, No. 2, May 1995.

ban development, (c) ending more than 50 miles beyond that edge, and (d) being more than 100 miles from the center of the largest central cities in the United States. We have since developed more precise definitions and characterizations.[3] We had previously written that exurbia is the fastest growing component of the American landscape; by some methods of reckoning, a fifth of the U.S. population already lives there.[4] By the early 21st century, much territory within the contiguous 48 states may no longer be distinguishable as either urban or rural but may well be characterized by low-density exurban development.

In this article, we assess changes in the geography of manufacturing between the early 1960s and the late 1980s, and we draw implications for economic development in exurban counties.[5] Two analytic questions are posed:

1. How has the distribution of manufacturing employment changed since the 1960s?
2. How does the share of manufacturing growth in exurban counties compare with the shares of urban, suburban, and rural counties?

Two sets of policy implications are also posed:

1. What are the implications for the competition for manufacturing jobs?
2. What are the appropriate federal, state, and local policy roles?

Analytic Assessment

We use economic base analysis to assess exurban industrialization in the context of alternative landscapes. On the basis of work by Arthur C. Nelson and by Nelson and Kenneth J. Dueker,[6] we classify all 3,000+ counties and parishes in the contiguous 48 states as follows (see Figure 16.1):

- *Large Urban* if the county was a central county of a metropolitan statistical area (MSA) of more than 500,000 in 1985.
- *Suburban* if the county was not a central county but a metropolitan county in 1960.
- *Small Urban* if the county was a central county of an MSA of less than 500,000 in 1985.
- *Inner Exurban* if the county was a metropolitan county but not a central or suburban county in 1990.
- *Outer Exurban* if the county was classified, by using the Nelson and Dueker scheme, as exurban but was nonmetropolitan in 1990.
- *Rural* if the county in 1990 was nonmetropolitan and not exurban according to the Nelson and Dueker scheme.

We use the county as the level of analysis, principally because data are available in suitable detail only at this level. This will have the effect of underestimating exurban industrialization trends, because many western urban and suburban counties are as large as entire eastern states and, despite having their own exurbs, are, therefore, counted as either urban or suburban. Data for the analysis derive from *County Business Patterns* for 1964 and 1987.[7] The 1964 *County Business Patterns* provides the earliest possible year for analysis.[8] Some data for counties and certain industries are suppressed to ensure confidentiality; in such situations, we estimate employment for suppressed industries.[9]

Since 1972, industries have been classified by the 1972 edition of the *Standard Industrial Classification* (SIC) manual and its 1977 supplement.[10] For purposes of analysis, those definitions are reasonably comparable to the definitions used in 1964. Table 16.1 defines the SIC codes for manufacturing industries for which data were assembled and analyzed. We use location quotient (LQ) and shift-share analysis to evaluate the pattern of industrial employment distribution and change in the contiguous 48 states between 1964 and 1987. We chose these approaches for their ability to describe the relative intensity of industrial activity change in terms that are easy to translate and understand. We evaluate industries at the two-digit SIC level; analysis was attempted but abandoned at the four-digit level because of data-suppression problems and our inability to interpolate at this level of detail. We considered evaluating trends within regions but decided not to since our primary interest is in identifying national trends. We are aware that the New England, Mountain,

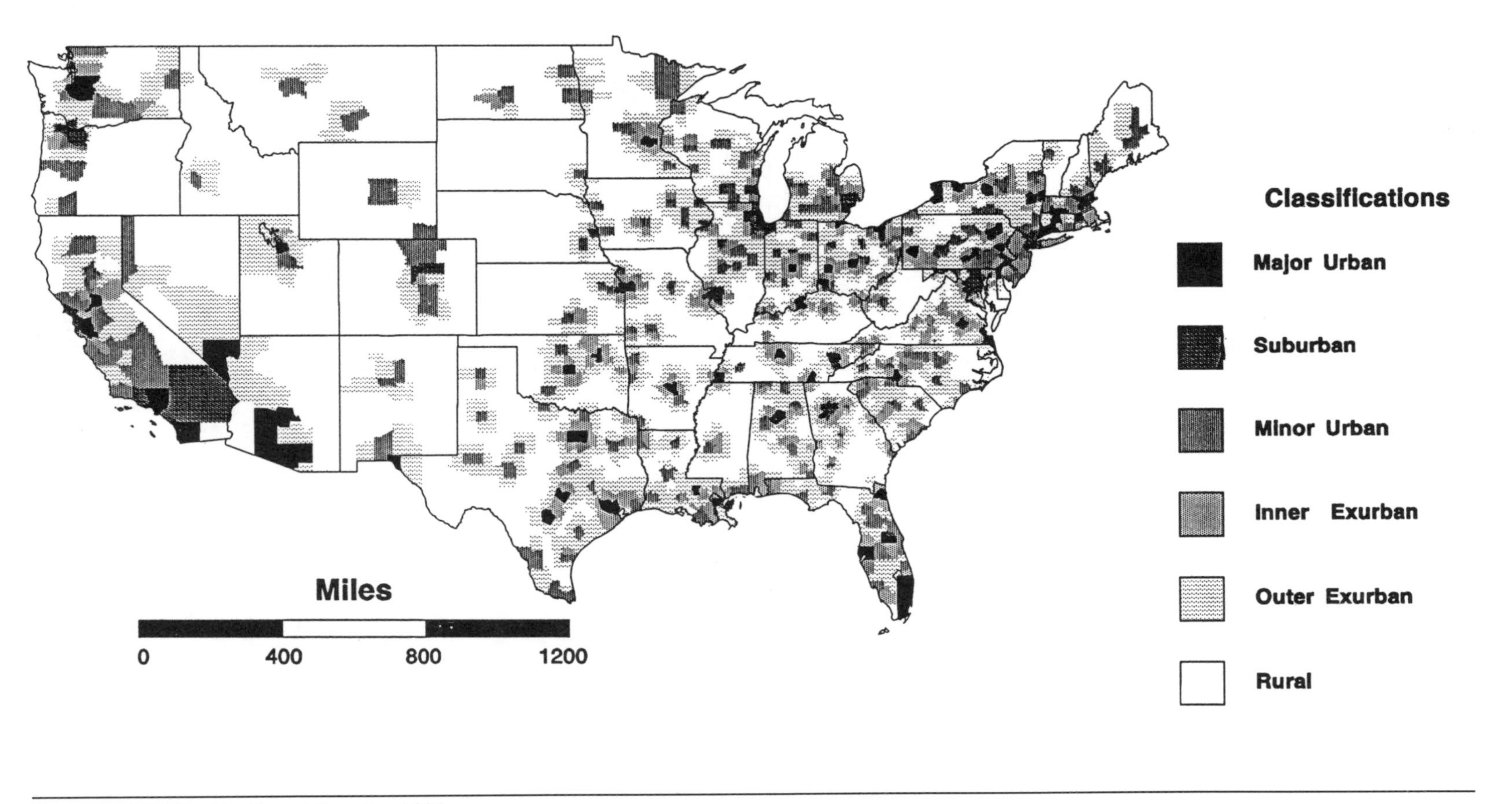

Figure 16.1. Classification of Counties, 1992
NOTE: Contiguous 48 United States; County Classification 1992.

TABLE 16.1 Two-Digit Standard Industrial Classification (SIC) Code Definitions for Manufacturing Industries

SIC Code	*Manufacturing Industry Definition*
20	Food and kindred products
21	Tobacco manufacturers
22	Textile mill products
23	Apparel and other textile products
24	Lumber and wood products
25	Furniture and fixtures
26	Paper and allied products
27	Printing and publishing
28	Chemicals and allied products
29	Petroleum and coal products
30	Rubber and miscellaneous plastics products
31	Leather and leather products
32	Stone, clay, and glass products
33	Primary metals industries
34	Fabricated metals products
35	Machinery, except electrical
36	Electric and electronic equipment
37	Transportation equipment
38	Instruments and related products
39	Miscellaneous manufacturing industries

and Pacific census regions have few exurban counties.

We now address our two analytical questions.

How Has the Pattern of Employment Distribution Changed since the 1960s?

Generally speaking, since the 1960s, exurban counties have gained concentration in many manufacturing industries, and inner-exurban counties have gained concentration in the service industry. These findings are derived from analysis of changes in LQs between 1964 and 1987. The LQ measures the degree to which employment in a given industry in a given geographic space exceeds what that employment would be if local employment were prorated among industries according to industry percentages of national employment.

An LQ of 1.00 indicates that county employment is proportionate to national employment; LQs greater than 1.00 indicate higher concentration of industry employment relative to the nation, whereas LQs lower than 1.00 indicate lower concentration. By comparing LQs at two points in time—1964 and 1987, in our study—one can detect general trends in the location of employment. For example, with an LQ of 0.96 in 1964 (see Table 16.2), suburban counties had roughly mirrored the nation in manufacturing employment concentration. By 1987, the LQ fell to 0.84, indicating that suburban counties lost concentration relative to other regions. Our adaptation of LQ analysis is reviewed in Appendix A.

Tables 16.2 and 16.3 report the 1964 and 1987 LQs for the two-digit manufacturing industries, divided into durable and nondurable goods sectors.

Total and Nondurable Goods Manufacturing Industries

An LQ above 1.0 signifies that the employment percentage of a given industry in a given county class is above the industry's national percentage. Table 16.2 shows the 1964 and 1987 LQs for all major industrial classes. In

TABLE 16.2 Location Quotients, Total and Nondurable Goods Manufacturing Industries 1964 and 1987

	Large Urban		*Suburban*		*Small Urban*		*Inner Exurban*		*Outer Exurban*		*Rural*		*Exurban*	
SIC[a] Code	*1964*	*1987*	*1964*	*1987*	*1964*	*1987*	*1964*	*1987*	*1964*	*1987*	*1964*	*1987*	*1964*	*1987*
20-39	0.91	0.83	0.96	0.84	1.05	1.05	1.24	1.21	1.18	1.53	0.81	1.15	1.20	1.39
20	0.99	0.75	0.64	0.56	1.22	1.22	1.01	0.97	1.16	1.79	1.10	2.01	1.11	1.42
21	0.98	0.95	0.40	0.12	1.88	1.91	0.79	1.05	0.92	1.13	0.48	0.28	0.87	1.10
22	0.30	0.26	0.47	0.30	1.29	0.99	2.75	2.07	2.63	3.85	0.47	1.01	2.68	3.05
23	0.67	0.84	1.63	0.58	0.58	0.59	1.20	1.07	1.73	2.27	1.45	2.23	1.53	1.73
26	0.74	0.64	0.77	0.75	1.35	1.33	1.23	1.19	1.38	1.74	1.31	1.62	1.32	1.49
27	1.24	1.13	1.46	0.99	0.83	0.90	0.48	0.90	0.46	0.87	0.23	0.75	0.47	0.88
28	0.87	0.72	1.15	1.14	1.06	1.09	1.52	1.68	0.97	1.22	0.63	0.85	1.17	1.42
29	0.86	0.78	1.03	0.72	1.36	1.51	1.63	1.53	0.68	1.00	0.60	0.83	1.03	1.24
31	0.40	0.58	1.26	0.63	1.16	0.96	1.56	1.39	1.88	2.23	1.47	2.50	1.76	1.85

a. SIC = Standard Industrial Classification.

TABLE 16.3 Location Quotients, Durable Goods Manufacturing Industries 1964 and 1987

	Large Urban		*Suburban*		*Small Urban*		*Inner Exurban*		*Outer Exurban*		*Rural*		*Exurban*	
SIC[a] Code	*1964*	*1987*	*1964*	*1987*	*1964*	*1987*	*1964*	*1987*	*1964*	*1987*	*1964*	*1987*	*1964*	*1987*
24	0.32	0.31	0.25	0.36	0.92	0.91	1.11	1.16	2.65	3.15	5.15	4.18	2.08	2.26
25	0.75	0.63	0.59	0.62	1.19	1.10	1.30	1.52	1.95	2.11	0.70	1.46	1.71	1.85
30	1.01	0.61	0.88	0.98	0.92	1.15	1.41	1.51	1.12	1.84	0.59	0.95	1.23	1.69
32	0.77	0.64	0.72	0.73	1.16	1.15	1.65	1.57	1.61	1.92	0.58	1.17	1.63	1.76
33	0.89	0.63	0.95	1.03	1.22	1.25	1.80	1.62	0.81	1.54	0.53	0.72	1.17	1.58
34	1.18	0.92	0.85	0.92	0.94	0.97	1.12	1.33	0.88	1.28	0.25	0.74	0.97	1.31
35	1.11	0.89	0.77	0.91	1.14	1.12	0.88	1.09	0.99	1.31	0.51	0.94	0.95	1.21
36	1.00	0.93	1.25	1.11	0.92	1.06	1.15	1.13	0.89	1.07	0.54	0.62	0.98	1.10
37	1.16	1.14	0.92	0.78	1.24	1.08	0.86	0.93	0.59	0.86	0.35	0.59	0.69	0.89
38	1.26	1.08	0.97	1.16	0.76	0.97	1.19	1.00	0.72	0.76	0.19	0.46	0.89	0.87
39	0.99	1.11	1.42	0.86	0.64	0.79	1.24	1.06	1.07	1.17	0.38	0.78	1.13	1.12

a. SIC = Standard Industrial Classification.

1964, inner-exurban counties had an overall manufacturing LQ of 1.24, the highest LQ among all county classes for overall manufacturing activities. In 1987, however, outer-exurban counties had the highest overall manufacturing LQ-1.5 of all county classes. The overall exurban county class (including inner- and outer-exurban subclasses) increased its manufacturing LQ from 1.20 in 1964 to 1.39 in 1987. Total manufacturing employment in the exurban county class increased by more than 1,000,000.

Exurban counties have high LQs in several nondurable goods industries, and the LQs for many increased between 1964 and 1987. For example, the LQ for food and kindred-products industries increased in outer-exurban counties from 1.16 in 1964 to 1.79 in 1987. Total employment in this industry in exurban counties increased by more than 20%, or by nearly 75,000.

The textile industry has especially high LQs in exurban counties, and the LQs again increased between 1964 and 1987 overall. The LQ did fall in inner-exurban counties, from 2.75 in 1964 to 2.07 in 1987, but it rose in outer-exurban counties, from 2.63 in 1964 to 3.85 in 1987. For all exurban counties (including inner- and outer-exurban counties), the LQ rose from 2.68 in 1964 to 3.05 in 1987. However, total employment in the textile industry fell by more than 10%, or nearly 50,000. The reason the textile industry LQ could rise while textile employment was falling is that national textile employment was falling—along with the national textile share of total employment, or Ei/E.

Exurban employment in apparel also declined—by 10%, or about 30,000. Apparel is another industry with declining employment. The LQs for apparel increased for exurban counties as a whole, from 1.53 in 1964 to 1.73 in 1987, and for outer-exurban counties in particular, from 1.73 in 1964 to 2.27 in 1987. The rural counties showed large gains in both LQ—from 1.45 in 1964 to 2.23 in 1987—and employment. Rural apparel employment increased more than 25%, or nearly 25,000.

The paper and allied-products industry strengthened its LQ in outer-exurban counties, from 1.38 in 1964 to 1.74 in 1987, and in rural counties, from 1.31 in 1964 to 1.62 in 1987. Employment in the paper and allied-products industry increased by about 20%, or about 20,000, in the exurban county class (including inner- and outer-exurban counties) and by about 33%, or about 12,000, in the rural county class.

The chemicals and allied-products industry increased in LQ for all exurban counties, from 1.17 in 1964 to 1.42 in 1987. Exurban employment in chemicals grew by more than one-third, or nearly 65,000.

The exurban class (including inner- and outer-exurban counties) LQ for the leather and leather-products industry increased from 1.76 in 1964 to 1.85 in 1987. The industry's LQ for rural counties increased from 1.47 in 1964 to 2.50 in 1997. Industry employment fell, however, by nearly 70,000 in the exurban county class.

Durable Goods Manufacturing Industries

In all but a few durable goods manufacturing industries, exurban LQs have been increasing (see Table 16.3). There is one notable exception—instruments; it will be reviewed last.

Lumber and wood-products industries have been concentrated in rural counties, but the rural LQ fell, from 5.15 in 1964 to 4.18 in 1987. In outer-exurban counties, the lumber and wood products industries' LQ rose from 2.65 in 1964 to 3.15 in 1987. Employment in rural counties was essentially stable, increasing slightly more than 5%, or 8,000. Employment in outer-exurban counties rose by nearly a third, or more than 60,000.

The furniture and fixture industry increased its exurban LQ from 1.71 in 1964 to 1.85 in 1987. Rural counties showed a larger increase in LQ, from 0.70 in 1964 to 1.46 in 1987. Furniture and fixture employment in exurban counties rose nearly 50%, or more than 60,000; employment in rural counties nearly tripled, increasing by more than 20,000.

The LQ of the rubber and plastics industry in exurban counties increased from 1.23 in 1964 to 1.69 in 1987. Employment increased more than 150%, or more than 160,000.

The LQ of the stone, clay, and glass industry was fairly steady for the exurban counties, rising slightly from 1.63 in 1964 to 1.76 in 1987. However, the rural counties had a big jump in LQ, from 0.58 in 1964 to 1.17 in 1987. Stone, clay, and glass employment in exurban counties increased about 10%, or about 18,000, but it more than doubled in rural counties, increasing by about 16,000.

In the primary metals industries, exurban counties increased their LQ from 1.17 in 1964 to 1.58 in 1987. However, the industry LQ fell in inner-exurban counties from 1.80 in 1964 to 1.62 in 1987, increasing in outer-exurban counties from 0.81 in 1964 to 1.54 in 1987. Employment remained constant in outer-exurban counties but fell by nearly a third, or more than 50,000, in inner-exurban counties.

The fabricated metals industry increased its LQ in exurban counties from 0.97 in 1964 to 1.31 in 1987. The industry's LQ climbed even more sharply in rural counties, from 0.25 in 1964 to 0.74 in 1997. Exurban employment increased by nearly two-thirds, or about 165,000; rural employment nearly quadrupled, increasing by more than 35,000.

The machinery industry had similar trends. The exurban LQ rose from 0.95 in 1964 to 1.21 in 1987; the rural LQ rose even more, from 0.51 in 1964 to 0.94 in 1987. Exurban employment increased by about 67%, or more than 180,000. Rural employment increased about 150%, or nearly 50,000.

Exurban and rural counties increased their LQs in most durable goods industries, but exurban communities merely held steady in the instrument and related-products industry. Its LQ was 0.89 in 1964 and 0.87 in 1987, despite nearly doubled employment (an increase of about 50,000). The instrument and related-products industry increased its LQ in rural counties, from 0.19 in 1964 to 0.46 in 1987. Instrument and related-products employment in rural counties increased nearly 500%, but only by around 10,000. Amy Glasmeier has written about the high-tech potential of rural America, noting (among other factors) that rural areas near urban high-tech centers offer both attractive lifestyles to labor and low land and labor costs to high-tech firms.[11] Under the SIC code, the instrument and related-products industry is considered high-tech. Although Glasmeier's analysis included "exurban" counties, we concur with her overall premise, but note that the change in employment in rural counties is numerically small.

The rise in LQs of durable goods manufacturing industries, in exurban and rural counties, is consistent with 1970s literature, which suggests that, to some extent, exurban and rural locations are preferred over urban and suburban locations. Yet access to markets remains important. Reasons for this tension between isolation and market accessibility include increasingly strict environmental and land-use regulation in urbanized areas, adequate supplies of suitable labor in exurban and rural counties, adequate highway networks and freight opportunities, and lower land prices.

How Does the Share of Manufacturing Growth in Exurban Counties Compare with the Share in Urban, Suburban, and Rural Counties?

Shift-share analysis can be used to further describe changes in distributional patterns of the economy. We find that exurban counties accounted for the lion's share of the shift in manufacturing employment change from 1964 to 1987. Central and suburban counties were big losers in the distribution of manufacturing employment change.

The shift-share method is commonly used to compare a given region's employment distribution at one time with its distribution at another. It is also used to compare one region to another in terms of (a) percentage share of national employment and (b) changes in share.

Shift-share analysis takes a region's absolute growth and divides that growth into three elements: the national, industry-mix, and local share factors. The three add up to total growth as an absolute amount, not a percentage. Each element is a supposed "source" of growth, but the term should not be taken literally; analysts know that the three sources are not literal in a cause-and-effect sense but rather are conceptual categories describing influences that might (or might not) be at work. Shift-share is thus a descriptive, not an evaluative, technique. (Critics of shift-share argue that it misleads people into thinking growth, or lack of growth, is attributable to the three shift-share factors. Its critics maintain that shift-share analysis is superficial at best and, at worst, diverts attention from true or more fundamental causes of growth.) Although we explain how all three factors are calculated, and report on all three, our discussion will concentrate on the local share factor—which we call "county-class" share, since it highlights trends in manufacturing activity by county class.

Appendix B describes how we adapted shift-share analysis to this study. Table 16.4 displays the shift-share factors by county class and major industry groups for 1964 to 1987, including nondurable goods manufacturing industries. Table 16.5 displays results for durable goods industries.

Total and Nondurable Goods Industries

Inner-exurban, outer-exurban, rural, and small-urban counties gained in manufacturing overall; large-urban and suburban counties lost (see Table 16.4). In the nondurable goods industries, the employment share of large-urban and suburban counties declined, except in one industry. Large urban counties increased their share of apparel and other textile products. Those counties offer important product development and marketing advantages in the industry, especially for high-fashion products.

Again, with the exception of one industry, the nondurable goods manufacturing share increased in small-urban, inner- and outer-exurban, and rural counties. The exception is the textile industry. Small-urban and inner-exurban counties declined in textile employment. Since the 1970s, textile industries have moved away from population centers to exurban and rural locations, or overseas.

After considering the few exceptions above, we find that inner- and outer-exurban counties had the largest nondurable goods finding consistent with our theme of exurban industrialization.

Durable Goods Manufacturing Industries

In general, large-urban and suburban county classes lost or held constant their durable goods manufacturing employment share (see Table 16.5). Small-urban counties gained or held constant in durable goods manufacturing industries, except for one industry—transportation equipment. This industry requires considerable amounts of land for its expansive research and development facilities and often requires testing transportation equipment in remote locations. Exurban and rural counties increased their transportation equipment shares. On the whole, exurban counties showed the bigger increase in durable goods share. The general increase for exurban counties is consistent with our thesis.

Summary and Policy Implications

As we suspected, manufacturing activity has moved away from large-urban and suburban counties, mostly into exurban counties; the distribution of this change varies, however. Inner exurban counties realized large increases in LQs principally in the durable goods manufacturing industries. Outer-exurban and rural counties saw large LQ increases in nondurable goods manufacturing industries. On the other hand, the LQ findings differ somewhat from the shift-share findings. All exurban counties realized large gains in share of employment, in both durable and nondurable goods manufacturing industries, whereas all other county classes lost share or held about constant.

Two sets of policy implications emerge. One concerns the competition for manufacturing jobs, and the other concerns the role of federal, state, and local policy in managing this competition.

Implications for the Competition for Manufacturing Jobs

Competition for the relatively few new plants that locate in any area, including exurban

TABLE 16.4 Shift Share by County Class, Total and Nondurable Goods Manufacturing Industries 1964 and 1987

County Class	*SIC[a] Code*	*National Share*	*Industry Share*	*County-Class Share*	*SIC[a] Code*	*National Share*	*Industry Share*	*County-Class Share*
Large urban	All	5,373,287	(4,257,994)	(788,197)	26	157,868	(138,755)	(23,750)
Suburban		2,174,830	(1,831,906)	(490,857)		63,447	(55,766)	(6,519)
Small urban		2,614,971	(2,285,414)	185,757		121,059	(106,403)	3,891
Inner exurban		1,351,474	(1,240,253)	363,443		48,253	(42,411)	8,818
Outer exurban		2,205,734	(1,999,359)	421,041		92,730	(81,503)	8,632
Rural		534,179	(478,310)	308,812		31,313	(27,522)	8,929
All exurban		3,557,208	(3,239,612)	784,484		140,983	(123,914)	17,450
Large urban	20	548,187	(568,848)	(132,765)	27	404,144	(81,317)	(58,225)
Suburban		137,226	(142,398)	(25,326)		184,225	(37,068)	(125,691)
Small urban		287,053	(297,871)	13,870		114,122	(22,962)	27,443
Inner exurban		103,908	(107,824)	16,151		29,053	(5,846)	66,860
Outer exurban		204,576	(212,287)	68,239		47,852	(9,628)	54,050
Rural		68,986	(71,586)	59,831		8,436	(1,697)	35,563
All exurban		308,484	(320,111)	84,390		76,905	(15,474)	120,910
Large urban	21	25,888	(36,304)	(258)	28	233,397	(192,875)	(46,859)
Suburban		4,079	(5,720)	(2,024)		118,678	(98,073)	(10,144)
Small urban		20,923	(29,341)	848		119,631	(98,861)	11,224
Inner exurban		3,850	(5,400)	1,536		75,089	(62,052)	29,785
Outer exurban		7,648	(10,725)	296		81,958	(67,729)	7,825
Rural		1,422	(1,994)	(398)		18,966	(15,673)	8,170
All exurban		11,498	(16,125)	1,832		157,047	(129,781)	37,610
Large urban	22	97,491	(117,453)	(11,155)	29	46,574	(53,975)	(3,559)
Suburban		58,261	(70,191)	(19,977)		21,564	(24,990)	(6,911)
Small urban		175,115	(210,973)	(30,074)		30,978	(35,900)	4,666
Inner exurban		164,295	(197,937)	(15,335)		16,319	(18,912)	1,758
Outer exurban		268,943	(324,013)	59,064		11,673	(13,527)	2,739
Rural		17,010	(20,493)	17,476		3,627	(4,204)	1,306
All exurban		433,238	(521,950)	43,729		27,992	(32,439)	4,497
Large urban	23	295,728	(347,492)	104,405	31	48,051	(79,306)	10,011
Suburban		302,348	(355,271)	(183,191)		59,077	(97,504)	(13,781)
Small urban		116,978	(137,453)	6,557		59,150	(97,624)	(3,565)
Inner exurban		107,131	(125,883)	5,200		35,010	(57,783)	964
Outer exurban		262,758	(308,752)	28,850		72,373	(119,448)	166
Rural		78,459	(92,192)	38,178		19,963	(32,948)	6,207
All exurban		369,889	(434,635)	34,050		107,383	(177,231)	1,130

a. SIC = Standard Industrial Classification.

areas, at any given time will result in winners and losers. How can exurban communities successfully compete against other exurban communities? How can nonexurban communities compete against exurban communities?

In this context, several researcher have suggested that once a firm chooses a general region in which to locate, local conditions become important.[12] The most important factors are site availability and building area, rental rates, and structural value. In their study of speculative industrial buildings in exurban Georgia, Arthur C. Nelson and Phillip Cosson found that availability of land with infrastructure and buildings with space available at regionally competitive rents attracted manufacturing firms.[13] To these

TABLE 16.5 Shift Share by County Class, Durable Goods Manufacturing Industries 1964 and 1987

County Class	*SIC[a] Code*	*National Share*	*Industry Share*	*County-Class Share*	*SIC[a] Code*	*National Share*	*Industry Share*	*County-Class Share*
Large urban	24	63,496	(42,081)	(400)	35	619,072	(416,815)	(166,714)
Suburban		19,203	(12,726)	9,915		165,603	(111,499)	27,257
Small urban		78,068	(51,738)	3,840		269,094	(181,178)	7,660
Inner exurban		41,295	(27,367)	14,410		90,689	(61,060)	61,397
Outer exurban		169,305	(112,205)	3,383		174,935	(117,782)	31,948
Rural		116,477	(77,193)	(31,150)		32,068	(21,591)	38,453
All exurban		210,600	(139,572)	17,793		265,624	(178,842)	93,345
Large urban	25	102,954	(58,992)	(23,407)	36	567,626	(371,775)	(45,250)
Suburban		31,445	(18,018)	(828)		276,199	(180,901)	(65,617)
Small urban		68,837	(39,443)	(3,419)		220,588	(144,478)	65,700
Inner exurban		32,915	(18,860)	19,672		120,858	(79,158)	29,832
Outer exurban		84,603	(48,477)	(9,690)		160,077	(104,845)	7,320
Rural		10,807	(6,192)	17,671		34,350	(22,498)	8,016
All exurban		117,518	(67,337)	9,982		280,935	(184,003)	37,152
Large urban	30	158,422	(5,283)	(125,042)	37	708,522	(648,443)	(2,314)
Suburban		53,668	(1,790)	4,232		216,135	(197,808)	(49,681)
Small urban		60,829	(2,029)	38,863		318,314	(291,323)	(32,563)
Inner exurban		40,977	(1,366)	22,881		97,142	(88,905)	32,168
Outer exurban		55,728	(1,858)	46,327		113,495	(103,871)	33,418
Rural		10,469	(349)	12,737		23,958	(21,927)	18,973
All exurban		96,705	(3,224)	69,208		210,637	(192,776)	65,586
Large urban	32	153,569	(147,018)	(26,930)	38	149,161	7,243	(40,836)
Suburban		55,034	(52,687)	(2,805)		44,279	2,150	12,334
Small urban		97,519	(93,359)	3,858		37,945	1,843	27,121
Inner exurban		60,809	(58,215)	8,959		26,131	1,269	(234)
Outer exurban		101,584	(97,250)	1,891		27,143	1,318	(5,950)
Rural		12,858	(12,309)	15,027		2,558	124	7,566
All exurban		162,393	(155,465)	10,850		53,274	2,587	(6,184)
Large urban	33	380,860	(539,854)	(71,645)	39	131,499	(117,585)	21,549
Suburban		156,660	(222,059)	2,500		72,858	(65,149)	(37,371)
Small urban		220,475	(312,515)	10,324		35,950	(32,146)	11,959
Inner exurban		142,766	(202,365)	6,701		30,505	(27,277)	486
Outer exurban		109,497	(155,208)	45,785		45,126	(40,351)	(3,725)
Rural		25,223	(35,753)	6,335		5,688	(5,086)	7,102
All exurban		252,263	(357,573)	52,486		75,631	(67,628)	(3,239)
Large urban	34	480,778	(301,066)	(145,053)				
Suburban		134,841	(84,438)	2,771				
Small urban		162,343	(101,660)	17,554				
Inner exurban		84,479	(52,901)	51,434				
Outer exurban		113,730	(71,218)	40,473				
Rural		11,541	(7,227)	32,820				
All exurban		198,209	(124,119)	91,907				

a. SIC = Standard Industrial Classification.

factors, Barry Rubin and Margaret Wilder added community leadership and a highly developed business-social structure.[14] In this respect, industries will be attracted to exurban communities that create business-social climates in which such firms are immediately welcomed.

The quality of the labor force is probably also important and influences current trends in manufacturing location. The highest value-added industries, such as electronics, remain attracted to urban and suburban areas with large supplies of skilled, educated labor, but so are most industries locating in exurban areas. It is likely that those communities boasting reasonably high-quality educational and vocational training will attract more manufacturing firms than those lacking such assets. By contrast, industries attracted to rural areas are, on the whole, lowest in value-added and least demanding of skills or higher levels of education.

An unknown factor could be working against exurban locations, however. The movement of some manufacturing firms from central cities to exurbia may reflect the cost of redeveloping obsolescent sites more than the innate attractiveness of exurban areas. As cities and their suburbs clean up abandoned sites, create industrial parks protected from potentially conflicting land uses, and extend property tax abatement to manufacturing firms while continuing investments in public education, some may become more attractive to firms that would otherwise locate in exurban areas. Moreover, rising transportation costs, in the form of higher fuel taxes and costly safety and pollution control measures, could make some exurban areas less attractive.

More problematic is the sustained improvement in labor productivity, which requires greater skill and adaptability in tasks than may be available in exurban areas, relative to urban and suburban areas. Two recent decisions of major German automobile manufacturers to locate in South Carolina and Alabama have contributed to this caveat; in both cases, such firms located not in exurban areas, but in small-urban counties near larger MSAs possessing important educational and vocational training facilities.

What Are the Appropriate Federal, State, and Local Policy Roles?

Certain directions for federal, state, and local economic development policy are suggested. First, federal and state economic development assistance should be prohibited from intervening in the locational decision making of foreign-based firms, which tend to locate within the United States anyway. Such decisions mean an overall efficiency gain for the U.S. economy, but because of tax concessions and tax-based financial incentives, the total efficiency gain on the economy may be reduced. The magnitude of such incentives, made by a particular state using both state and federal resources, could more than offset economic gains. In the cases of the German automobile plants cited, tax incentives threaten realization of all national efficiency gains.

Second (and related to the first), federal and state economic development policies should not result in one class of counties using such resources to lure firms that, based on our analysis, are inappropriate. Again, the rationale is to prevent the reduction of efficiency gains that would normally accrue to the national economy, when locational decisions are not distorted.

Third, state and local governments need to overhaul fiscal relations to prevent the same set of communities from using incentives to distort otherwise efficient locational decisions. In California, for example, the municipality that lands a producer of goods subject to a sales tax allows the "winner" to retain all local option share of the taxes.[15] In Georgia, by contrast, the state redistributes a share of sales taxes on a per capita basis among local governments; within counties, local option sales taxes are redistributed per capita among all eligible jurisdictions, such as municipalities and the county. The Twin Cities (Minneapolis-St. Paul) Metropolitan Council and the local governments in northeastern New Jersey go one better by engaging in property tax base sharing among all counties and municipalities. Such efforts reduce the possibility that local government incentives will distort locational decisions.

Fourth, *good* incentives should be used, where appropriate to the county class. Generally speaking, a good incentive is one that removes an obstacle to net improvement in efficiency gains otherwise accruing to the national economy. Suppose, for example, that a firm's most economically efficient location is not available because of ownership patterns and poor infrastructure. Economic development policies could be used to acquire that site, but only to the extent that economic costs of using a second-best site are higher than the incentive costs. Job training programs geared toward particular firms may also be warranted, if such programs are needed to train the labor force, no matter where the firm locates. Certain tax incentives may prove appropriate, especially when used to correct for distortions induced by existing state or local fiscal policies.

As the next century arrives, exurban governments may find themselves involved in more costly bidding wars than seen in recent years. Policymakers at all levels need to understand which industries are attracted to exurbia. Federal and state policy may be needed to help certain locales improve their attractiveness to certain industries—and, at the same time, to prevent the use of scarce federal or state resources to compete.

APPENDIX A
Location Quotient Analysis

The location quotient (LQ) is the industry's percentage of total county (or other subnational) employment divided by the industry's percentage of total national employment. When a locality's LQ is above 1.0 for a particular industry, county employment in that industry is relatively high, or above the national average. LQs can be used to discern changes in relative employment by industries within regions. The LQ formula used in our analysis is

$$LQ_i = \frac{\left(\frac{e_i}{e}\right)}{\left(\frac{E_i}{E}\right)}$$

where

e_i = the employment in industry *i* in a given class of counties;
e = total employment in a given class of counties;
E_i = national employment in industry *i*; and
E = total national employment.

APPENDIX B
Shift-Share Analysis

Shift-share analysis attributes change in employment to three factors, or elements of a region's growth in a specific industry: (a) the national growth rate for industry in general; (b) the amount by which the national growth rate, for the specific industry being measured, exceeds or falls short of that for industry in general; and (c) the amount by which the locality's growth rate exceeds or falls short of the national growth rate for that industry.

An example will illustrate how these elements interlock. Suppose that industry *i*, in a given region, grew in employment from 100 to 180 between 1964 and 1987. Absolute growth for the industry was

$$180 - 100 = 80.$$

Next, suppose that the national growth rate for the period was 50%, or 0.50. If industry *i* in the region had grown at the national rate for overall industry, industry *i* would have grown by

$$0.50 \times 100 = 50.$$

This amount, 50, is called the *national share,* or the share of total growth (total = 80) attributable to any tendency for industry *i* in the region to grow (or decline) at the national growth rate in general. Alternatively, the national share effect is the amount by which industry *i* would have grown in the region if it had grown at the national growth rate. The formula for the national share is

$$\text{National share} = Ei\left(\left[\frac{US^*}{US}\right] - 1\right),$$

where

Ei = regional employment, E, in the ith industry at the beginning of the period, which is 1964 in this case;
US* = total national employment at the end of the period, which is 1987 in this case; and
US = total national employment at the beginning of the period.

The next factor is the *industry mix share.* Suppose that the national growth rate for industry *i* (as opposed to overall industry) was 70%, or 0.70. Industry *i* grew nationally at a faster rate than industry in general; the difference in growth rates was

$$0.70 - 0.50 = 0.20.$$

If industry *i* in the region had grown at the industry's national growth rate, 0.70, its growth would have *exceeded* the national share amount (50) by

$$(0.70 - 0.50) \times 100 = 0.20 \times 100 = 20.$$

This amount, 20, is the industry-mix share or effect. It is the amount by which industry *i*'s growth in the region would have exceeded the national share (50) if its growth rate regionally had been the same as its growth rate for the nation. The formula for the industry-mix share is

$$\text{Industry-mix share} = Ei\left[\left(\frac{USi^*}{USi}\right) - \left(\frac{US^*}{US}\right)\right],$$

where

USi = national employment in the ith industry at the beginning of the period, and
USi* = national employment in the ith industry at the end of the period.

The third factor—the third part of industry *i*'s growth in the region—would ordinarily be called the regional share. Since our "regions" are classes of counties (e.g., exurban counties), we will call this factor the *country-class share*—that portion of industry *i*'s growth attributed to its being faster or slower in the region than for the nation. We have already assumed that industry *i* grew from 100 to 180 in the region during the period: a growth rate of 80%, or 0.80. This regional growth rate exceeded industry *i*'s national growth rate by

$$0.80 - 0.70 = 0.10.$$

Because industry *i* grew faster regionally than nationally, its growth exceeds the sum of (a) the national share and (b) the industry-mix share by

$$0.10 \times 100 = 10.$$

This amount, 10, is the county-class share—the amount by which industry *i*'s actual growth exceeded the amount of growth it would have had in the region if it had grown at the industry's national growth rate. The formula for the county-class share is

$$\text{County-class share} = Ei\left(\left[\frac{Ei^*}{Ei}\right] - \left[\frac{USi^*}{USi}\right]\right),$$

where

Ei^* = county-class employment in the ith industry at the end of the period.

Notes

1. Calvin L. Beale, "The Recent Shift of United Stated Population to Nonmmetropolitan Areas, 1970-75," *International Regional Science Review* 3 (1977): 113-22; David L. Brown and John M. Wardwell, eds., *New Directions in Urban-Rural Migration* (New York: Academic Press, 1980); Don A. Dillman and Daryl J. Hobbs, *Rural Society in the U.S.: Issues for the 1980s* (Boulder, CO: Westview, 1982); G. V. Fuguitt and P. R. Voss, *Growth and Change in Rural America* (Washington, DC: The Urban and Land Institute, 1981); R. E. Lonsdale and H. I. Seyler, *Nonmetropolitan Industrialization* (Washington, DC: V. H. Winston & Sons, 1979); Gene F. Summers, Sharon D. Evans, Frank Clemente, E. M. Beck, and Jon Minkoff, *Industrial Invasion of Nonmetropolitan America* (New York: Praeger, 1976); Gene F. Summers, "Industrialization," in *Rural Society in the U.S.*, ed. Don A. Dillman and Daryl J. Hobbs (Boulder, CO: Westview, 1982), pp. 164-74.

2. Richard A. Engles and Robert A. Forstall, "Is America Becoming More Metropolitan?" *American Demographics* 3 (December 1981): 18-22.

3. Arthur C. Nelson, "Regional of Exurban Industrialization," *Economic Development Quarterly* 4 (1990): 320-33; Arthur C. Nelson, "Characterizing Exurbia," *Journal of Planning Literature* 6 (1991): 350-68; Arthur C. Nelson and Kenneth J. Dueker, "Exurban Living through Advanced Technology," *Journal of Urban Planning and Development* 115 (1989): 101-13; Arthur C. Nelson, William J. Drummond, and David S. Sawicki, *Exurban Industrialization* (Washington, DC: Economic Development Administration, U.S. Department of Commerce, 1992); Arthur C. Nelson and Kenneth J. Dueker, "The Exurbanization of America and its Planning Policy Implications," *Journal of Planning Education and Research* 10 (1990): 91-100.

4. Nelson and Dueker, "Exurban Living through Advanced Technology."

5. See Nelson, "Regional Patterns of Exurban Industrialization"; Nelson and Dueker, "Exurban Living through Advanced Technology."

6. Nelson, "Regional Patterns of Exurban Industrialization"; and Nelson and Dueker, "The Exurbanization of America."

7. U.S. Department of Commerce, *County Business Patterns* (Washington, DC, 1964 and 1967).

8. Before 1964, data were not stored on tapes in a form readily accessible and suitably documented.

9. Since *County Business Patterns* reports number of employees by firm-size category, we constructed a distribution of firms by firm employment size. Within each firm-employment-size category, we estimated the average number of employees. The solution leads to some estimation error, but it is reasonable, given limitations of data for purposes of ensuring confidentiality.

10. U.S. Department of Commerce, *Standard Industrial Classification Manual* (Washington, DC, 1972 and 1979).

11. Amy Glasmeier, *America's High-Tech Potential* (New Brunswick, NJ: Rutgers University, Center for Urban Policy Research, 1991).

12. See Paul R. Blackley, "Urban-Rural Variations in the Structure of Manufacturing Production," *Urban Studies 23* (1986): 471-83; John P. Blair and Robert Premus, "Major Factors in Industrial Location," *Economic Development Quarterly* 1 (1987): 72-85; Roger W. Schmenner, *The Manufacturing Location Decision: Evidence from Cincinnati and New England* (Washington, DC: U.S. Department of Commerce, Economic Development Administration, 1978).

13. Arthur C. Nelson and Phillip Cosson, "Evaluating the Effectiveness of a Speculative Industrial Budding Program," *Economic Development Review* 7 (1990): 51-4.

14. Barry M. Rubin and Margaret Wilder, "Urban Enterprise Zones: Employment Impacts and Fiscal Incentives," *Journal of the American Planning Association* 55 (1989): 418-32; Barry M. Rubin and Margaret Wilder, "Targeted Redevelopment through Urban Enterprise Zones," *Journal of Urban Affairs 10* (1988): 1-18.

15. Through legal action, "losers," may strike a deal with "winners" to share sales tax revenues. The process is cumbersome, however.

PART V

Politics, Planning, and Evaluation

Chapter 17

The Politics of Local Economic Development

Harold Wolman with David Spitzley

❀❀❀

During the past 10 years, there has been a burgeoning of literature on the politics of local economic development, much of which has been published in *Economic Development Quarterly* (*EDQ*). The 10th anniversary of *EDQ*, therefore, seems a propitious time to review this literature and to set forth what we know—and do not know—about this topic. Accordingly, this article reviews the literature on the politics of local economic development to (1) set forth what we substantively know, (2) critique the literature, (3) discover what we do not know that is worth knowing (and what we think we know that we do not), and (4) set forth, based on these findings, an agenda for future research.

My concern here is limited to local or urban economic development. By *local,* I mean substate, although most of the literature is concerned with the politics of economic development within metropolitan areas and much of it with the central cities within these areas. Defining the term *economic development* is more difficult. With rare exceptions, the literature does not bother to define the term and thus, not surprisingly, encompasses a wide variety of activities and policies that sometimes appear to have little relation to each other.

Theorists distinguish between economic growth and economic development, a distinction, however, largely missing from the politics of local economic development literature. Kindleberger and Herrick (1977), in their work on economic development, separate the terms as follows:

> Economic growth means more output, while economic development implies both more output and changes in the technical and institutional arrangements by which it is produced and distributed. . . . As with humans, to stress "growth" involves focusing on height or weight (or GNP), while to emphasize development draws attention to changes in functional capacities—in physical coordination, for example, or learning capacity (or ability of the economy to adapt). (p. 3; see also Beauregard 1993, 271)

Reprinted from *Economic Development Quarterly,* Vol. 10, No. 2, May 1996.

They also make clear that while the concept of economic development implies improvement of material well-being, it also may refer to improvements in the distribution of income and "greater participation of broadly based groups in making decisions about the directions, economic and otherwise, in which they should move to improve their welfare" (Kindleberger and Herrick 1977, 1).

Despite these distinctions and nuances, Kindleberger and Herrick (1977) operationally define economic development as increases in national income. (They write: "Although it is clear from the discussion above that we consider economic development a multivariate concept, this book nevertheless will follow convention and generally use as a proxy measurements based on national income" [p. 7].) And, indeed, this is the common convention: What most economists mean by economic development is an increase in area employment, income, or both. My own preference is to think of economic development as an increase in the economic well-being of area residents, usually manifested by positive changes in the level and distribution of area employment and per capita income.

The literature on the politics of local economic development, however, goes far beyond this. The term *economic development* is often used to refer to what is essentially land development or physical development, frequently project based, even if these efforts are not directed at or do not increase income or employment. This is a source of confusion, not just in the literature but in policy as well. As Swanstrom (1985) notes in his scathing critique of "growth politics":

> For all their talk about jobs, programs such as urban renewal, tax abatement, and UDAGs [Urban Development Action Grants] are not serious efforts at economic development. They are subsidies for real estate deals—showy brick and mortar developments, complete with ribbon cutting ceremonies. Even if the subsidies actually were effective in causing new real estate developments (a dubious promise, as we have seen), their economic effects would still be limited. (p. 236)

So we begin with evidence of confusion: Most of the literature on the politics of local economic development is not actually concerned, at least conceptually, with development but with growth; some of that literature is focused more on project-based land and physical development rather than on increases in income or employment, the more conventionally accepted operational definition of economic development. Since it is frequently unclear whether the term *economic development* is being used in its narrower sense to refer to efforts to increase the employment and income of area residents or in its broader sense to include land and physical development efforts, and, conversely, whether the term *development* is meant to be confined to economic development in terms of employment and income or is meant more broadly to include land development, I have cast a relatively broad net over the literature. (I have, however, ignored a rather large body of literature whose primary concern is residential growth and development.)

As the above discussion implies, the literature on the politics of local economic development is quite diverse. There are many case studies of local economic development decision making; surveys of the attitudes of economic development practitioners and elected officials; qualitative in-depth interviews of practitioners and elected officials; quantitative studies using cross-sectional data and frequently employing multivariate techniques; pieces that rely on exposition, argument, and analysis; and those that are concerned with broad-range theory development. Most of the literature is drawn from the disciplines of political science, sociology, and urban planning, with some infrequent contributions by economists.

I will review this literature by summarizing its major findings and by assessing and critiquing it. I begin by examining the broad forces that, it is claimed, impel local governments to engage in economic development activity and those that have caused such activity to increase in recent years. Next, I examine the outcomes local governments appear to be seeking when they engage in economic development activity. I then turn to the question of what accounts for variation in the extent and kind of economic development activity that local governments pursue.

The next section moves from a general concern with local government to a concern with actors at the local level. I explore literature relating to how local officials think about local economic development, why they undertake local economic development activity, and the political calculus of local economic development decision making. I conclude this section by asking whether local economic development can be considered *rational.*

In the next section, I ask how much choice local officials actually have in economic development policy making. Can their activity actually make a difference in affecting the area economy, and to what extent is local economic policy itself constrained by business interests, broadly construed?

The last section is concerned with the actual politics of local economic development. I examine literature on the interests involved—particularly the role and degree of dominance of business—and the local decision-making process itself, including the extent of conflict in the process and the openness of the process to citizen input and participation. I end by summarizing my major conclusions and presenting a set of directions for future research.

I begin the review by asking what drives local governments to engage in economic development activity. In so doing, I draw on a literature that is primarily theoretical and reliant on exposition and argument.

The Forces Behind Local Economic Development Activity

The literature addresses two related concerns: What are the factors that impel local governments to engage in economic development activity, and why has local economic development activity increased so substantially in recent years?

Factors Impelling Local Governments to Engage in Economic Development Activity

Much of the theorizing about why local governments engage in local economic development activity is related to its structural features—in particular, the economic and fiscal problems posed by the mobility of capital across fixed geographic boundaries within a highly fragmented system of local governments. Peterson (1981), in his book *City Limits,* propounds the most complete and influential explanation. Peterson argues that given the mobility of capital across local government boundaries, cities have a *unitary* (i.e., single and overriding) interest in the well-being of their economy and in attracting economic activity:

> Cities constantly seek to upgrade their economic standing. . . . [they] seek to improve their market position, their attractiveness as a locale for economic activity. . . . it is only a modest oversimplification to equate the interests of cities with the interests of their export activities. (p. 22-3)

> Cities, like private firms, compete with one another so as to maximize their economic position. (p. 29)

> Developmental policies are those local programs which enhance the economic position of a community in its competition with others. (p. 41)

Peterson (1981) argues that local government leaders take this unitary interest of local government into account (1) because economic prosperity is required for protecting the fiscal base of local government and thus for permitting elected officials to deliver a reasonable quality of public services at reasonable tax levels, (2) because economic development activity is politically popular, and (3) because most local officials have a sense of community responsibility and want to do what is good for the community.

Swanstrom (1985) makes essentially the same argument:

> The *mobility of wealth* and the relative interdependence of local governments in the American economic and political system bring into being a *governmental marketplace*: local governments compete to attract mobile wealth in order to keep their tax base up and service demands down. (p. 31; see also Bowman 1988)

The importance of the fiscal imperative for local economic development is also stressed by Bowman and Pagano (1992) and Pagano and Bowman (1995). Like Peterson (1981), they conceive of a city as an organization with its own set of systems maintenance interests; unlike Peterson, they suggest that the propensity for local governments to engage in local economic development activity is contingent—that is, some pursue economic development, whereas others do not. In their 1995 book, Pagano and Bowman write:

> Our contention is that city leaders pursue economic development in response to changing (often deteriorating) fiscal conditions. (p. 3)
>
> A city engages in problemistic search for the appropriate response to the city's need to survive, to thrive and to reestablish a balance. In particular, the equilibrium that cities seek to maintain is between their revenue-generating capacity and their service delivery needs. . . . Threats to a city's revenue stream disrupt the tax-services balance and most assuredly trigger the search for a development policy to redress the imbalance. (p. 26)

The above discussion focuses on the structure of local governments, the mobility of capital, the competition among cities for mobile capital, and the consequences for the city's fiscal situation if capital should exit the city. This explanation, while coherent and convincing in the American context, depends on a local fiscal system that is structurally dependent on generating tax revenue from locally based economic activity and from city residents. In a system characterized by fiscal equalization grants from higher levels of government (e.g., the United Kingdom, Germany, and many other Western nations) and/or with larger local government units and less fragmented systems of local government, mobile capital and residents do not pose the same fiscal threat; thus this same process is unlikely to be the impetus for local economic development activity. Yet, such activity is reputedly as widespread in European local governments as in local governments in the United States.

There are other explanations offered for the inevitability of economic development activity by local governments. Molotch (1976) and Logan and Molotch (1987) point to the existence of an urban *growth machine,* which pursues development through intensified land use that increases land values and thus land rents to members of the growth machine. For a variety of reasons, the growth machine constitutes an elite able to persuade local government to pursue policies consistent with growth machine ends. Molotch (1988) elaborates his reasoning:

> In capitalist societies . . . urban areas become the arena in which property entrepreneurs use government and other civic institutions to maximize returns on their investments. The best way to make money from such places is to increase the intensity of economic activity occurring within one's turf. . . . While property entrepreneurs within a locality compete with one another to push development in the direction of their own property instead of someone else's, all such actors stand to gain in common if activity levels increase in the locality as a whole. Areawide intensification, ordinarily in the form of increments to the basic economy which, in turn, generate labor in-migration and other economic growth (e.g., wholesale and retail trade), benefits the investments of all local property entrepreneurs. Local real estate investors thus make up a reliable core for the growth machine elite. Their common urban program is to attract more jobs, people, and thus rents. (p. 26)

Thus the impetus to pursue development policies (note that the focus here is on land development, with economic development, if any, occurring in its wake) comes not from internal forces (the need to maintain fiscal equilibrium, given capital mobility and competition among cities) but from external pressure put on local government by a powerful set of community elites with a shared interest in obtaining the increased rents from intensified land development. Local elected officials are themselves a component of the growth machine—partly because most accept the dominant ideology of growth, partly because some may personally profit from increases in land rents, and, more broadly, because of the need or desire for campaign contributions from elements of the growth machine, particularly property developers and speculators (Logan and Molotch 1987, 66-7).

Others see local economic development activity resulting from certain leadership characteristics of urban government, thus emphasizing agency more than structure. Stone (1989) argues that cities can be characterized by different types of *urban regimes* and that development-oriented regimes are frequent because development activity meets regime needs for *social production.* He contends that urban regimes are best understood as governing coalitions that come together for social production purposes—that is, for accomplishing some public purpose or policy objective. Local elected officials engage in development-related activity because business usually has the preponderance of resources in a community necessary to meet social production ends (and thus is a likely regime member) and because business interests are primarily concerned with development, particularly downtown development (again, land development rather than economic development).

Pagano and Bowman (1995) contend that leadership vision and city image explain why even fiscally healthy cities frequently pursue economic development activities. (Their emphasis on economic development activity as an imperative for fiscally stressed cities has already been referred to above.) They argue that leaders see their city as fitting within a system of cities in an urban hierarchy to which they compare themselves. If the leadership elite is concerned either with improving their city's position within the group of cities relevant to it or with moving up to a new set of cities within the hierarchy, they are likely to undertake development activity to bring about the desired result. The authors write:

> Cities are concerned about their images. City leaders, as a group, share a collective vision of what the city could, or should, become. In pursuit of this vision, cities mobilize public capital. They do this in an effort to move the city toward a desired end state. . . . Images can take the form of economic development goals for a city. City leaders have a vision for their city, and economic development is a large part of that vision. Development projects, be they industrial or commercial, are often promoted because they will provide jobs and generate revenues that benefit the city. But, just as important, these projects are tangible symbols of survival, transformation or ascendancy. (p. 48)

Why Local Government Economic Development Activity Has Increased

To this point, I have discussed forces and factors that explain why local governments are driven to engage in economic development activity. These forces are not time specific; if operative, they should presumably have resulted in local economic development activity at any time, certainly in the past fifty years and perhaps longer. Yet, it appears that while economic development has long been a function of local government, the great explosion in such activity has occurred only recently, within the last two decades (see Kantor and David 1988, 230). Why has this been the case, and does this cast doubt on some of the explanations suggested above?

Four sets of complementary reasons are set forth to explain this relatively recent perceived increase in local economic development activity. First, the mobility of capital has increased substantially during the past decades and is now international in scope, leading to increased competition among cities to maintain their economic and fiscal bases (Friedland 1983; Kantor and David 1988; Clarke and Gaile 1989). Second, it is argued that slow national economic growth during the past two decades has resulted in similarly slow growth or even decline in many urban economies, with consequent pressure to take action to provide jobs for residents and fiscal resources for local governments (Blair, Fictenbaum, and Swaney 1984; Jones and Bachelor 1993). Third, international economic restructuring has resulted in particularly hard economic times for cities dependent on traditional manufacturing employment, leading elected officials to engage in economic development activities for both employment and fiscal reasons (Fainstein et al. 1983; Judd and Parkinson 1990). Finally, cutbacks in aid from the national government—a result of slow economic growth, rising deficits, and ideology—have reduced national assistance for local economic development and have thrown local governments back on their own resources if they are to undertake economic development activity (Clarke and Gaile 1989).

The increasing mobility of capital is stressed by Clarke and Gaile (1989):

> As the rate of capital and labor mobility accelerates, and the global competition for investment tightens, local communities become more vulnerable to external decisions that can dramatically influence their economic well-being. The complexity and rapidity of these economic changes threaten the stability of local revenue sources. . . . This revenue imperative—the effort to increase the stability and lower the vulnerability of the local revenue base—prompts many local officials to seek new types of economic activity to provide more local jobs. (p. 575)

The impact of the international economic restructuring process on local government economic development activity is elaborated on by Judd and Parkinson (1990):

> To understand the pressures on individual cities to revitalize their local economies, it is essential to appreciate the scale and pace of the restructuring of the international economy. . . . From 1973 to 1981 . . . industrial employment went into a nosedive. Statistically, it was compensated for by gains in service jobs, but those jobs were not necessarily, or even generally, located in the same regions or metropolitan areas that suffered sharp declines in manufacturing. . . . Faced with the erosion of their most important economic sectors, old port and industrial cities and the regions that had been built upon a base of industrial production or resource extraction responded, initially, in a predictable way: Their elites sought favorable policies from the central state and . . . [w]hen it became clear to urban leaders that they would not be bailed out by national policies, then their efforts . . . became redirected to finding ways to regenerate their local economies. (pp. 15-7)

The process Judd and Parkinson (1990) describe is one that affected all Western nations, and the responses they chronicle are largely aimed at regenerating local economies to increase employment rather than at responding to fiscal concerns. Local elected officials respond to the increasing unemployment and underemployment of their residents (voters) by trying to increase local employment opportunities. Blair, Fictenbaum, and Swaney (1984, 64) refer to the emergence in the United States of a "market for jobs" that "differs from the traditional 'job market' where individuals and firms compete for labor services. In the market for jobs, cities, states and other governmental entities seek to purchase jobs for their current or future residents." The authors argue that local workers are the primary intended beneficiaries of this market for jobs. This emphasis on employment-related objectives for local economic development activities contrasts with prior explanations focusing on fiscal concerns. As previously noted, fiscal objectives are likely to have much lower explanatory power in systems in which local fiscal well-being is not tied as exclusively to the performance of the local economy as in the United States.

The Outcomes Sought by Local Economic Development Activity

The question of what outcomes local officials are indeed pursuing—or think they are pursuing—when they engage in economic development activity is an interesting and important one. As noted earlier, economic development traditionally is viewed in terms of increasing employment, income, or both rather than increasing local revenues. It is important to recognize that these goals are not necessarily mutually consistent. Increases in local revenues may occur without increases in local employment (through more intensive land use or employment of nonlocal residents), and increases in employment of local residents may occur without increases in local revenue (if local residents are employed outside the city boundaries and the city has no income tax). Local officials frequently do not see or pay attention to this potential divergence. Yet, the political response to local economic development policies—as well as evaluation of the activity—may rest critically on which of these outcomes is expected and the extent to which such expectations are fulfilled.

There is substantial disagreement in the literature on what outcomes or objectives American local governments are actually pursuing in their economic development programs. Pagano and Bowman (1995), Schneider (1989), and Jones and Bachelor (1993) all argue that local economic development activity is essentially fiscally driven. Pagano and Bowman, for example, state flatly: "The decision to mobilize a development tool is anchored in a tax-service disequilibrium and fundamentally unrelated to employment and income issues. . . . City behavior is designed to address the tax-service disequilibrium, not the number of persons employed" (pp. 25-6). They buttress their contention by examining forty development decisions in ten cities, concluding that "of the 40 projects examined, interview data and project files indicate that 27 were intended to generate revenue" (p. 96).

Contrary to this point of view, Blair, Fictenbaum, and Swaney (1984, 64) argue, as noted, that the surge in local economic development activity reflects the emergence of a market for jobs in which local governments compete with one another "to purchase jobs for their current or future residents." Riposa and Andranovich (1988) review 20 case studies of economic development decision making by various authors and conclude that the justification for most local government programs was at least partly with reference to job creation.

Surveys of local elected officials and practitioners suggest the possibility (if not the probability) that many do not distinguish between the two objectives. In a survey of local elected officials, Furdell (1994) reports that when presented with a list of 10 possible goals for local economic development activity, 65.6% named "increasing jobs located in the city" as one of the top three goals; 34.7% identified it as the top priority. "Increasing the tax base and tax revenues" received the next highest support; 42.7% called it one of their top three goals, and 13.1% called it their top priority. However, only 24.3% selected "increasing employment for city residents anywhere where jobs are" as one of the top three priority goals; only 5.9% chose it as the top goal. Since this last goal reflects the pure employment objective in its concern for employing city residents, suspicion is strong that those who chose the first goal (increasing jobs located in the city) either had a fiscal objective in mind—since jobs located in the city are much more likely to yield fiscal benefits to the city government than are jobs for city residents located outside of the city—or made no distinction between the employment and fiscal objectives.

Bowman's (1987, 10) survey of mayors finds similar results but does not distinguish between increasing jobs in the city and increasing jobs for city residents. She finds that 69.5% of her respondents indicated that "increasing employment opportunities" was among their top three economic development goals, with 25.8% calling it their top priority goal, as compared with 56.1% who chose "improving the city's tax base" as among their top three goals, with 23.0% naming it their top priority. In a survey of local economic development practitioners, Levy and Stephenson (1991) find that 36% of those who worked in an economic development agency within a general purpose unit of local government named increasing employment as top priority, whereas 28% named increasing the property tax.

What Accounts for Variation in Local Economic Development Activity

There are testable propositions embedded in the theoretical literature described above, particularly relating to the posited importance of fiscal stress, slow economic growth, and economic restructuring. As this literature suggests, despite the imperatives driving local governments to engage in economic development activity, there may be variation in the extent to which they do so. I now turn to an examination of the empirical literature that deals with why local governments engage in different levels and kinds of economic development activity. In the course of this discussion, the importance of the factors theorized above will be examined. However, before this literature can be assessed, we must ask how local economic development activity (the dependent variable) is characterized and categorized in the literature and the extent to which local governments engage in the various forms of this activity.

Kinds of Local Economic Development Activity

There are a variety of schemas employed in the literature to classify the various forms of local economic development activity. Most of these involve a priori classification based on criteria salient to the author. However, there are also a few efforts to derive empirically based classifications.

Of the a priori schemas, the most common involves classification by purpose or strategic intention. Thus Matulef (1987), for example, sorts local economic development tools or activities into five categories according to purpose: economic revitalization, support services for economic revitalization, project coordination, financial assistance, and capacity building. Feiock (1987) identifies four categories: promotion, service coordination, business nonfinancial assistance, and incentives.

Many authors classify policies, tools, and/or activities along a continuum related to innovation versus tradition. Several authors make use of a distinction, first offered by Eisinger (1988), between supply-side policies and demand-side policies. Eisinger argued that

> subnational economic development policy has undergone a recent shift from an almost exclusive reliance on supply side location incentives to stimulate investment to an approach that increasingly emphasizes demand factors in the market as a guide to the design or invention of policy. (p. 10)

By supply factors, Eisinger (1988) refers to traditional incentives to attract economic activity into an area, while by demand factors, he has in mind efforts to discover, expand, develop, or create new export markets for local goods and services; strategies to promote new business creation and small-business expansion; and government assistance to local business in new product development and market expansion, through subsidizing research and development and through strategic investment. Eisinger identifies the supply-side approach as traditional and the demand-side approach as entrepreneurial.

Following Eisinger (1988), Reese (1992), for example, divides policies into supply-side (e.g., tax incentives, debt financing schemes, infrastructure investment, regulatory policy, tax increment financing arrangements, enterprise zones, and land and site development) and demand-side policies (e.g., business incubators, venture capital financing, research and development support, small-business support, development of export markets, and job training programs). Based on a survey of local economic development officials in Michigan, she concludes that the state's cities continue to rely heavily on supply-side techniques rather than the more entrepreneurial demand-side techniques.

The distinction between supply-side and demand-side techniques, frequently referred to in the literature, is somewhat mystifying. There would appear to be a relatively limited number of tools available to local government to bring about an increase in demand for goods and services produced in the area. These tools include import substitution strategies such as buy-local requirements for local government or efforts to persuade local businesses to purchase from local suppliers, export promotion activity, and assistance to local firms in developing new markets and marketing strategies for their products. But many of the tools listed on the demand-side by Eisinger (1988) and Reese (1992), for example, seem unambiguously to be on the supply-side of the product market, such as job training programs, business incubators, research and development subsidies, and strategic government investment for product development or new technology—as well as efforts to encourage new local business formation and local entrepreneurship, itself usually considered a supply-side function. All of these affect either the cost or the nature of the goods and services local businesses will supply rather than the external demand for them.

Clarke and Gaile (1989, 1992) employ, more convincingly, a distinction between conventional and entrepreneurial approaches. Conventional strategies center on public intervention to attract economic activity; entrepreneurial strategies involve greater governmental flexibility, innovation and risk, efforts to stimulate new enterprise, use of government authority to shape market structure and opportunity, joint public-private ventures, and public strategic investments. Based on a survey of economic development officials in large (over 100,000 population) American cities, Clarke and Gaile (1989, 189) conclude that those policy strategies adopted after 1980 "reflect a shift away from conventional economic development orientations toward market-based or entrepreneurial approaches."

Goetz (1990) divides economic development policies according to intended outcomes. Type I, or traditional, policies are aimed at increasing private-sector investment in the local economy, either by enhancing the public sector's capacity to facilitate private investment (e.g., creating business advisory councils, provision of technical assistance, or consolidating permit approval processes) or by attracting local investors to the area (e.g., tax abatements, or land or wage subsidies). Type II policy

> *mandates* public benefits by requiring private sector developers to provide a wide range of direct economic and social goods. . . . Type II policies refer to programs in which local jurisdictions require private developers to provide a service or public benefit in exchange for development rights. (pp. 170, 172)

Type II policies are thus similar to what are widely termed *progressive* policies (see Miranda, Rosail, and Yeh 1992). As the emphasis on development rights suggests, many of these policies relate to land development more than to economic development as we have defined it. Goetz surveys 395 city, county, and regional institutions to determine frequency of use of eight Type I and seven Type II strategies and finds that 72% used at least one Type II policy.

Another set of criteria used to categorize local economic development activity employs the visibility and distribution of costs and benefits (Rubin and Rubin 1987; Clingermayer and Feiock 1990; Sharp and Elkins 1991; Feiock and Clingermayer 1992; Cable, Feiock, and Kim 1993). Sharp and Elkins (1991, 132), for example, divide economic activities into those with visible tax costs (e.g., tax abatements and use of local taxes for economic development), those with relatively invisible tax costs (e.g., tax-exempt bonds, loan guarantees, and federal grants), and those with highly visible benefits (e.g., infrastructure improvements).

In addition to these a priori classifications, both Reese (1993a) and Fleischmann, Green, and Kwong (1992) employ factor analysis to generate empirically derived categories of economic development activities. Using data on which of 55 different tools were used by each of 967 cities collected from a survey of cities by the International City Management Association (ICMA), Reese identifies 13 conceptually distinct clusters of tools, encompassing 42 of the 55 tools. She then groups these, by inspection, into four broad categories:

Marketing Factors

- Traditional/Domestic (e.g., brochures, visits to prospective firms)

- Demand Side/Foreign (e.g., soliciting foreign business, developing export markets)

Financial Factors

- Tax Incentives (e.g., tax abatements)
- Loan Incentives
- Entrepreneurial (e.g., shared equity in projects)

Land- and Property-Management Factors

- Traditional Land Incentives (e.g., land acquisition, clearing, and sale of land)
- Land Support (e.g., water and sewer)
- Entrepreneurial Land Incentives No. 1 (e.g., industrial property management)
- Entrepreneurial Land Incentives No. 2 (e.g., transfer of development rights)

Governance/Infrastructure Factors

- Preservation (e.g., incentives for historic preservation)
- Transportation (e.g., improved streets, improved parking)
- Services (e.g., improved public safety, improved street cleaning)
- Red-Tape Reduction (e.g., improved building inspection)

Fleischmann, Green, and Kwong (1992) employ a similar approach that yields nine distinct categories: loan incentives, financial incentives, activities to attract and/or retain business, revitalization activities, regulatory reform, developmental land management, historic preservation, aesthetic improvement, and management of city facilities.

Explaining Local Government Use of Economic Development Policies

Why the variation among local governments in the extent to which they engage in economic development activity and in the kinds of activity in which they engage? A variety of hypotheses are offered and tested in the literature to explain such differences in economic development aggressiveness. These include propositions presented in the above discussion about local economic development imperatives, such as *fiscal need* (local governments experiencing greater fiscal stress are likely to engage in more active economic development activities); *economic growth* (slow economic growth is associated with high economic development activity); and *deindustrialization* (high losses in manufacturing employment are related to more economic development activity). A variety of other factors are suggested as well: *citizen need* (high unemployment, low per capita income, and high poverty rates lead to greater local economic development activity); *regional competition* (greater economic development activity on the part of other local governments in the region leads to higher levels of local economic development activity for a specific local government); *government structure* (centralization of executive power and unreformed governments both lead to greater local economic development activity because, it is contended, economic development activity is facilitated by a strong lead actor); and *population size* (larger cities are likely to engage in more economic development activity, ceteris paribus, because they have greater professional capability).

There is now substantial empirical literature devoted to examining and testing these hypotheses. Unfortunately, the literature is quite problematic, and it is necessary to discuss why prior to presenting results. The first problem is the dependent variable itself. Few of the studies make any effort to present a conceptual definition, but clearly most, in one way or another, are concerned with the extent of economic development activity by local government. This is usually operationalized in one of two ways. In the first method, the dependent variable is dichotomous: the adoption or nonadoption (or use or nonuse) of a specific local economic development tool, technique, or strategy. In most cases, several such tools are examined separately. Thus Feiock and Clingermayer (1992) attempt to explain the use of four separate tools characterized by different levels of visibility and benefit distribution: tax abatements, advertising, service coordination, and business-assistance centers. (For other studies that have operationalized the dependent variable in a dichotomous way, see Cable, Feiock, and Kim 1993; Sharp and Elkins 1991; Miranda, Rosail, and Yeh 1992 [see below]; Clingermayer and Feiock 1990; Feiock and Clingermayer 1986.)

The second method of operationalization involves a simple count of the different kinds of economic development tools used. Typically, this involves a survey instrument, sent to local authorities, asking the respondent to put a check next to each of a list of economic development tools, techniques, policies, or strategies used over some period. Thus Fleischmann, Green, and Kwong (1992) use a 1984 ICMA survey (several other studies use a later 1989 ICMA survey) that listed 64 separate economic development techniques and asked respondents to check each one used. This produced an interval-level variable, with a score for each city ranging between 0 and 64. (For other studies that use this method of operationalization, see Reese and Malmer 1994; Donovan 1993; Feiock 1992; Sharp 1991; Green and Fleischmann 1991; Goetz 1990; Rubin and Rubin 1987; Rubin 1986.)

There is no other way to say it: Both of these measures are very poor operationalizations of the concept of extent of economic development activity. The dichotomous variable simply gives an indication of whether a technique has been employed but says nothing about the extent to which it has been employed. The count of techniques used suffers in a similar fashion: It perhaps provides a measure of diversity of tools or complexity of approach—or, more broadly, the administrative and policy development capability of local governments with respect to economic development—but it is an inadequate measure of the extent of economic development activity. A local government employing only two tools but spending $10 million on each of them would presumably engage in more economic development activity than does one using ten tools but spending only $100,000 on each. Yet, the count-of-tools measure would provide just the opposite conclusion.

A reasonable measure of the extent to which a local government engages in economic development activity should be related to resource deployment toward that end—in terms of either dollars spent on economic development purposes or employees devoted to that function. The only research reviewed that uses such a resource-related measure is by Reese (1991), who, in a study of the use of tax-abatement policy by local governments in Michigan, employs dollars of taxes abated by local government as the dependent variable.

The reason that resource deployment measures are not used is straightforward and obvious: They are unavailable through the usual national data sources for local government expenditure and manpower (i.e., the U.S. Bureau of the Census's annual *City Government Finances* and *City Government Employment*). The only way to obtain them is by going to individual local governments and compiling them. This would be an extremely difficult, perhaps impossible, task, since a substantial portion of local government expenditure on economic development does not flow through the general-fund budget but through various enterprise funds and off-budget agencies.

When preferred operational measures are unavailable, it makes sense to move to second-best ones. But the measures used here are not second best but third or fourth best, and we must ask whether their relationship to the concept they purport to measure is so remote as to render them meaningless. In the case of the simple count of economic development tools, it is an empirical question whether the number of tools used is highly correlated with resource deployment measures. The answer is not known, nor do any of the studies provide (or even attempt to provide) persuasive arguments why this might be so. A priori, it seems plausible that communities that use a large number of separate tools may well be deploying more resources than those that use a small number; this suggests that, at most, the variable be measured ordinally (e.g., very few tools, some tools, large number of tools) rather than continuously, and even then with great caution.

Both Bowman (1988) and Miranda, Rosail, and Yeh (1992) make an effort to get around this problem by relying on respondent perceptions. Bowman asked respondents to rate how aggressive their city was in pursuit of economic development (answers could range from *Very aggressive, Fairly aggressive,* to *Not aggressive*). Miranda, Rosail, and Yeh asked respondents to characterize each of a variety of tools by frequency of use (answers could range from 1 = *Never* to 5 = *Very frequent*). Tools were then grouped into two categories according to whether they were Type I or Type II tools (see previous discussion), and a score was assigned to cities for each of the two categories by simply aggregating the individual scores of the tools in each category. Although there are obvious problems in relying on perceptions, the Miranda approach seems a preferable second-best method compared with those discussed earlier.

The next concern is the models examined for explaining variation in the extent of economic development activity across local governments. These models vary enormously, and many are severely lacking in theoretical grounding. Some examples should provide the proper flavor. Feiock and Clingermayer (1992) employ a model consisting of social and economic conditions of cities, government institutions, growth coalitions, and competition. Sharp's (1991) model includes economic distress, citizen accessibility, and reformedness. Donovan (1993) uses controversy, population size, fiscal stress, residentiality, city factors, and policy structure. Rubin and Rubin (1987) include citizen need, urban capacity, fiscal need, and process of growth. Rubin (1986) sets forth a complex model, including ecological variables related to the city's economic and fiscal condition and attitudes of the public decision makers, with the formation of a formal economic development organization as an intervening variable.

In an analysis of sixteen articles that attempt to explain variation in the extent of local economic development activity, I find that the most frequently included conceptual variables in explanatory models were local government institutional structure (nine times), fiscal stress (nine times), need or deprivation (five times), economic distress (five times, although this might overlap with the need variable depending on operationalization), openness or citizen access (four times), and city size (three times). Note that this is a count of explanatory variables at the conceptual level. In fact, some of these were used operationally to a greater extent, since an operational variable selected to measure economic distress might be the same variable operationalized in another study to measure need.

With so many different model specifications, it is extremely difficult to assess the impact of any of the variables across studies. This is partly because the context in which the variable is examined (i.e., the control variables for that variable) differs from study to study. But it is also because of the very different (and sometimes inexplicable) ways in which these variables are operationalized. Fiscal stress or fiscal condition conceptual variables, for example, are variously operationalized as property tax per capita (Donovan 1993; Sharp and Elkins 1991), per capita tax revenues (Cable, Feiock, and Kim 1993, for tax burden), property tax rate (Rubin and Rubin 1987; Rubin 1986), the ratio of local revenues to local expenditures (Goetz 1990), city bond rating (Goetz 1990; Feiock 1992), and percentage of total revenue derived from property tax (Sharp and Elkins 1991).

Most of these bear no relation to fiscal stress or condition. Tax collections per capita conflates wealth (higher property values lead to higher revenues at any given tax rate), preference for public services as opposed to private goods, and tax rate. By itself, tax collections per capita have nothing to do with either fiscal stress or tax burden. The proportion of local revenue derived from the property tax may be a measure of revenue diversification but is unrelated to fiscal need. The ratio of local expenditures to local revenues at first glance appears to be closely related to budgetary fiscal stress but, in fact, is not, since local governments may draw down on general-fund surpluses in any one year to pay for expenditures. In such a case, expenditures might exceed revenues to a varying extent, but this could well be evidence of a healthy fiscal condition rather than stress. Tax rate, so long as it is used consistently to apply to the same base (as was the case in the studies by Rubin and Rubin [1987] and Rubin [1986], since, in both cases, the municipalities examined were within the same state), is a reasonable measure of tax burden but not necessarily of fiscal stress. The best measure of fiscal stress, at least in terms of tax burden, is the one used by Sharp (1991), who employs per capita own-source revenue divided by the sum of median housing value (weighted by the percentage of city revenue derived from the property tax) and median family income (weighted by the percentage of city revenue derived from sources other than the property tax). This is a reasonable measure of community tax burden since it relates a community's actual revenue raised to its wealth. (Curiously, however, it is used as an operational variable for the concept of economic distress rather than tax burden.)

The concept of economic distress, condition, or need is sensibly measured by the unemployment rate (Feiock 1992; Reese 1991; Sharp 1991) but less understandably measured by population change (Goetz 1990; Miranda, Rosail, and Yeh 1992) and percentage of the population that is Black (Miranda, Rosail, and Yeh 1992). Need has less obvious meaning conceptually and is a grab-bag in terms of operational measures. Green and Fleischmann (1991) and Rubin and Rubin (1987) use the poverty rate, which clearly seems related to the concept. However, population (Clingermayer and Feiock, 1990; Bowman 1988) is, puzzlingly, used as an indicator of need, as is population change (Bowman 1988). Green and Fleischmann also use as indicators of need the percentage of jobs in the manufacturing sector (which is a measure of industrialization and might be a good measure of economic restructuring if it were cast into a change variable over time, but which is not related to need in any obvious and direct way) and the percentage of the population that is non-White. These last two variables may be related to need in that they are correlated negatively with income; if that is the case, it is much more sensible to directly use per capita income or median household income, as do Clingermayer and Feiock (1990) and Rubin and Rubin.

The operationalization of several less frequently used conceptual variables is also problematic. Competitiveness—the extent to which a local government is in active competition for economic activity with other local governments—is hypothesized to be positively related to local economic development activity by both Feiock and Clingermayer (1992) and Green and Fleischmann (1991). Both operationalize competitiveness as the average number of strategies used by other local governments in the same state (Feiock and Clingermayer 1992) or in the same census region (Green and Fleischmann 1991). This count-of-tools measure clearly suffers from the same difficulties described in the discussion of dependent variable problems. Donovan (1993) uses the extent of controversy over economic development issues as a conceptual variable in a two-stage model to explain variation in local economic development activity. However, extent of controversy is operationalized as the response of a single respondent (usually in the office of community development/redevelopment) in each community to the question "In general, how controversial would you say economic development issues are in your city?" (The response categories were *Not at all controversial, Sometimes controversial, Often controversial,* and *Always controversial.*) This provides the perception of the degree of controversy, but the perception of only one person—a slim reed on which to base a rather sophisticated concept.

Most of the studies that use adoption of a policy as a dependent variable employ logistic regression, while the count-of-tools studies generally use ordinary least squares analysis, although Rubin (1986) also employs a path model, and Donovan (1993) uses two-staged least squares. However, several of the earlier pieces rely exclusively on bivariate analysis, which, given the lack of controls, makes interpretation difficult.

Given these problems, what can we say with some degree of confidence about the findings from this body of research? Reese (1991) finds that the value of tax abatements (the only dependent variable in this entire set of studies that really makes sense as a resource-related measure of local economic development activity) in the state of Michigan is positively related to median income in a multivariate context; the author concludes that wealthier communities, having greater resources to do so, are more likely to provide greater amounts of tax abatements. On closer inspection, however, this finding is less impressive. For any property tax abatement granted for a specific firm, the value is likely to be higher in a wealthy community than in a poorer one, simply because property value is likely to be higher in that community. In short, wealth is explaining wealth rather than extent of economic development activity.

In the remainder of the studies, the dependent variable is either local government adoption (or use) of a tool or the total number of tools used. As I have argued, neither of these is a convincing measure of the extent of economic development activity. With that in mind, I turn first to the question of whether fiscal stress or unemployment is more important in explaining local economic development activity. Sharp's 1991 study is the only one that employs a useful operational definition of a fiscal variable (in reality, a measure of tax burden relating locally raised revenues to community wealth; see previous discussion). She finds a positive statistically significant relationship between fiscal stress and the number of economic development strategies employed by local governments, but she tests the relationship only in a bivariate context. Rubin (1986), in his study of economic development in small Illinois cities, finds that the tax rate (a measure of tax burden rather than fiscal stress) is indirectly related in the path model he tests to the number of economic development actions undertaken: the higher the tax rate, the greater the urgency felt by local economic development officials; the higher the urgency, the greater the number of economic development actions undertaken.

Sharp (1991) also finds a positive and statistically significant bivariate relationship between the unemployment rate and the number of economic development strategies used. Feiock (1992) finds the same relationship in a multivariate context. Other variables descriptive of community need also seem significantly related to the number of tools employed. Fleischmann, Green, and Kwong (1992) and Rubin (1986) find that the percentage of individuals in a local government with income below the poverty line was positively related to number of tools used, and Donovan (1993) and Rubin find that income was negatively related to the same dependent variable. Feiock and Clingermayer (1992) find per capita income negatively related to the adoption of tax abatement as an economic development tool. Cable, Feiock, and Kim (1993), on the other hand, find per capita income positively related to the adoption of tax abatement, albeit in a very simple four-variable model.

Fleischmann, Green, and Kwong (1992) and Feiock (1992) use multivariate analysis and find population size positively related to number of tools used. This is not surprising, if we assume, as I have argued, that the dependent variable is really measuring administrative capacity or sophistication, since larger cities are likely to possess greater amounts of these attributes.

Fleischmann, Green, and Kwong (1992) and Green and Fleischmann (1991) find that the degree of competitiveness, as measured by the average number of tools used by other local governments in the region, was positively related to the number of tools used by local government, although Feiock and Clingermayer (1992) do not find competition by other governments in the state to be significant in explaining local government adoption of tax abatement.

Rubin's 1986 study is of particular interest because of its use of both ecological and attitudinal variables. He finds that high tax rates and low per capita income lead to a perception on behalf of mayors and local development officials of urgency in terms of problems facing the community and that urgency leads to more economic development actions being undertaken.

A variety of studies examine governmental and political characteristics as explanatory variables. Reese and Malmer (1994) usefully point out that whether local governments provide tax abatements is, first of all, dependent on whether state legislation permits them to do so—an observation both simple and powerful but one ignored in nearly all of the other studies. They further find that local governments in states that do not permit abatements do not appear to compensate by using a greater number of other tools. Feiock and Clingermayer (1986) find that local governments with greater centralization of executive power (mayoral-council rather than council-manager form of government, presence of executive veto, partisan elections, and ward representation) were more likely to adopt tax abatement policy than were other local governments. They use chi-square analysis to compare the percentage of cities with and without each of the four separate structural features adopting tax abatement policy, controlling separately for income and city size. (See also Clingermayer and Feiock [1990] and Sharp [1991], both of whom have similar findings.) Feiock and Clingermayer explain these results by noting that mayors have both the electoral incentive and the institutional ability to focus their concern and efforts on economic development. However, using a more elaborate multivariate model, Fleischmann, Green, and Kwong (1992) find no relationship between presence of a mayoral-council form of government and the number of financial incentive tools used; nor do Donovan (1993), Green and Fleischmann (1991), or Feiock (1992) find any relationship between governmental structural characteristics and the number of economic development tools used. Since the latter findings all employ more extensive and elaborate multivariate models, doubt is inevitably cast on the findings of the earlier and less well specified models.

Fleischmann, Green, and Kwong (1992) do find a significant relationship between the size of a city's bureaucracy and the number of financial incentive tools employed, lending support to the contention set forth earlier that the count-of-tools approach is, in fact, a measure of local government capacity. Finally, Reese (1991) finds a significant negative relationship between a mayor's margin of victory and the value of total abatements granted. This suggests, not implausibly, that the political processes that incline mayors to engage in local economic development activity (e.g., public pressure, the desire to appeal to the electorate by doing something, the fear of being blamed if no action is taken) are more active when elections are closely contested.

How Local Officials Think About Economic Development

How do the actors involved in the economic development decision-making process view the problems faced, their ability to deal with them, the nature of the environment they operate in, and the political incentives and disincentives they confront? Most elected officials find economic development to be an important policy area. Bowman (1987) reports that in a 1987 survey of 326 mayors, 86% named economic development as one of their top three priorities. Furdell (1994) reports that in a survey of local elected officials for the National League of Cities, 85.8% responded that economic development was a "very important" local government responsibility (by contrast, only 47.5% said that "reducing poverty" was a very important local government responsibility). Furthermore, more than two-thirds indicated that local government action could have either a substantial or a significant impact on bringing about economic development (while less than 30% felt this was the case for reducing poverty).

Local economic development professionals were not as optimistic. Bowman (1987) asked a sample of economic development professionals more detailed questions about whether local government activity could play a major or minor role in bringing about various possible economic development outcomes. While 70.2% responded that local government activity could have a major role in improving the city's tax base, only 46.3% thought it could play a major role in affecting the number of jobs; 21.1%, in bringing about a self-sufficient economy; 10.4%, in affecting the level of personal income; and 8.7%, in improving distributional equity.

In his in-depth interviews with local economic practitioners, Rubin (1988) finds even greater pessimism over the efficacy of local economic development:

> They saw only a weak relationship between their efforts and resulting changes, between action and consequence. In social psychological jargon, they felt they lacked fate control. This problem was compounded by doubts about whether or not municipalities had the wherewithal to affect major economic changes, the value of the information on which decisions might be made and, more generally, the efficacy of the technology and tools used in economic development. (p. 237)

Rubin's (1988) interviews suggest that local economic development actors operate in an environment characterized by uncertainty, ambiguity, and turbulence (see also Reese 1993b, 1995). Rubin observes that "they see that their work environment is complex and undefined and involves an uncertain technology; they feel their jobs are not understood and that they must report to people who lack knowledge in their field" (p. 237).

The theme of uncertainty and ambiguity is a recurrent one in the literature, and although Rubin's (1988) interviews were conducted with practitioners, others stress its relevance to the environment in general, with implications for elected officials as well as practitioners (see Bachelor 1994; Jones and Bachelor 1993; Spindler and Forrester 1993; Wolkoff 1992; Wolman 1988). Spindler and Forrester (1993, 39) observe that environmental uncertainty leads to *political risk,* which they define as "the potential for positive or negative political (or bureaucratic) consequences resulting from a policy position or decision." Environmental uncertainty associated with economic development policy lies in the complexity of the processes affecting the local economy; the difficulty of predicting the actual results of economic development activities and the benefits that will accrue from them; and the concern that inaction in the face of local economic difficulties will lead to political attack if the economy does not recover, a local firm leaves, or a firm being publicly recruited locates elsewhere. But positive action does not necessarily alleviate uncertainty since it can attract criticism if the results are not successful or if the politician is seen as "knuckling under" to business. Finally, environmental uncertainty is exacerbated by the fundamental problem of information asymmetry (see Bachelor 1994; Jones and Bachelor 1993; Wolkoff 1992; Blair, Fictenbaum, and Swaney 1984; Friedland 1983): In any effort to attract economic activity to an area, the firm knows what it will take to do so, but local political actors know neither what it will take to attract (or keep) the activity nor (frequently) what is being offered by competing local governments.

Given these uncertainties, how do local economic development actors respond? Spindler and Forrester (1993) and Reese (1993b) argue that the classic response in the face of uncertainty is to adopt routines or decision rules, while Bachelor (1994) contends that local political actors will rely on *solution sets*—established policies or ways of doing things. Rubin's (1988) interviews lead him to question the extent to which economic development activity is instrumental as opposed to symbolic. Drawing on a 1987 mail survey he conducted of a random sample of economic development practitioners in cities with a population over 25,000, he states that more than half see their work either as driven by symbolic concerns or as formalistic (p. 236). He concludes that practitioners, to cope with an uncertain environment, vacillate between claiming credit for anything that has occurred regardless of whether their actions brought it about (i.e., "shoot anything that flies; claim anything that falls") and professionalism—that is, taking credit for properly undertaken actions.

Elected officials are also attracted by *credit-claiming* (as will be discussed), and many view being engaged in economic development activity as important to their electoral success. Furdell (1994) reports that when asked whether being associated with strategies to bring about economic development or to reduce poverty were more important to their electoral success, 48.2% of elected officials said bringing about economic development was more important, compared with only 2.9% who said reducing poverty was more important; the remainder called both equally important.

Given the electoral relevance of local economic development, it is important to inquire into the decision-making calculus of local elected officials.

Why Do Local Political Actors Undertake Local Economic Development Activity?

Three related questions will be discussed in this section: (1) What is the political calculus associated with local economic development decision making? (2) Why do local officials engage in certain forms of local economic development activity in the face of research results that show them unlikely to work? (3) Is local economic development decision making rational?

The Political Calculus of Local Economic Development Decision Making

Why do local political actors engage in economic development activity, even in the face of the environmental uncertainty described in the previous section? As has been noted, economic development is an important policy area for elected politicians. Politicians tend to view economic growth and development as a collective good that contributes to electoral success (Cable, Feiock, and Kim 1993). As Peterson (1981, 29) notes, "By pursuing policies which contribute to the economic prosperity of the local community, the local politician selects policies that redound to his own political advantage." Elkin (1987) expands on this and explains the attraction of development policies that have a physical dimension (even if they have no economic development payoff in a narrower sense):

> The first thing that public officials in the city, as officials elsewhere, must do is get elected. If they cannot achieve this they have little else to worry about. As do officials at other levels of government, city officials believe that their electoral prospects are markedly improved if they can secure a reputation for promoting innovative policies and if, in general, they are associated with publicly visible activities of almost any sort. . . . One of the easiest ways for public officials to gain the necessary reputation for innovation and to achieve visibility is by association with major land-use projects. . . . A major downtown mall or convention center can be advertised as taking the city into the new metropolitan age. Such projects are also visible in a way that few other things that happen in cities are, and such building is taken to be a sign that much else of note is going on in the city—even if it is not. (p. 37)

As this passage suggests, the electoral dimension of economic development decision making may have high symbolic content; Wolman (1988, 25) echoes Rubin (1988) in observing that local economic development activity may be "important as much for its symbolic content as for its effect" (see also Swanstrom 1985; Wolkoff 1992). Closely tied to the symbolic factor is the pursuit of credit-claiming and blame-avoidance opportunities (see Wolkoff 1983, 1991; Swanstrom 1985; Feiock and Clingermayer 1986; Wolman 1988; Noto 1991; Bachelor 1994). Feiock and Clingermayer (1986) observe that

> whether a development actually provides tangible benefits is, perhaps, relatively unimportant. What is important is that the use of these policies provides politicians with something for which they can claim credit. This notion of credit-claiming, popularized by David Mayhew (1974), has often been used in describing the behavior of members of Congress. Mayhew defines credit-claiming as "acting so as to generate a belief in a relevant political actor (or actors) that one is personally responsible for causing the government, or some unit thereof, to do something that the actor (or actors) considers desirable." . . . In city government, a council member or mayor may, for example, claim credit through the use of particularized benefits by luring a new plant to the city or by sponsoring new incentives for business investment. This kind of explanation has the advantage of accounting both for the popularity of development policies in both hard-pressed and well-off cities and for their use even when their benefits are unclear. (p. 212-3)

Many local economic development techniques lend themselves particularly well to credit-claiming. Local officials frequently justify tax abatements, for example, by arguing (however inaccurately) that any development that occurs would not have occurred in the absence of abatement and thus can be claimed as a policy success (Wolkoff 1983, 79).

Blame avoidance may also motivate economic development activity; local actors may wish to avoid the blame of standing by and doing nothing while the economy continues to deteriorate or while a firm decides to relocate to another community. But there are potential political costs as well as benefits in political involvement in economic development. Bachelor (1994, 604) notes that "these may be incurred whether or not a firm relocates, because officials may be criticized for being too generous with a particular business as well as for not having made more of an effort to prevent relocation of another" (see also Noto 1991). In addition, if economic development activity is not perceived as successful, Bachelor notes, it may lead to public questioning of the competence and ability of local officials. (p. 604)

Despite these possible costs, economic development activity is attractive to local officials for many reasons. In cities facing economic problems, fiscal problems, or both, there is likely to be strong public pressure on local officials to "do something." Even in nonstressed cities, the pressure to respond may be strong if other competitive jurisdictions are engaged in active efforts to attract economic activity. Such activity, particularly tax abatements and incentives, may be one of the few responses available to local governments, however remote the probability of success. Swanstrom (1985) sets forth the situation facing Cleveland in the 1970s:

> Politicians in Cleveland must convince their constituents that they are doing something about the severe problem of jobs and investment. Tax abatement had one overwhelming attraction: it was something the city government could do, even in its weakened state. Tax abatement purported to transform small amounts of public funds into huge amounts of private investment that would resurrect a dying city. In fact, supporters claimed that tax abatement created, as if by magic, something from nothing, for the purported cost to the public was zero (since the projects would not have occurred without the abatement, the city gained the amount of the new tax revenues rather than lost the amount of the abatement). (p. 148)

In addition, the costs of many of the standard local economic development activities are relatively invisible, in many cases not even showing up on the general-fund budget of a local government but in an enterprise fund or in the budget of a semiautonomous entity charged with financing economic development activity. Wolkoff (1991) suggests a variety of other reasons why economic development activities are attractive to local officials:

> The political time horizon of elected officials will typically be shorter than the time period over which project costs and benefits flow. To the extent that costs are borne by later generations of decision makers, the use of short time horizons will lead officials to pursue development beyond the point that even narrow jurisdictional interests would suggest. In any event, the costs of many direct subsidies are distributed over time and across many taxpayers. To the taxpayer, these costs take the form of lower services or somewhat higher taxes. In either case, these costs will be somewhat invisible. . . . Even when costs are evident, few economic development offices are able to keep meaningful track of these costs. Typically, development officials are charged with the responsibility of increasing development—budgetary consequences are accounted for elsewhere in the bureaucracy. (p. 517)

Why Do They Do It, Even Though We Tell Them It Does Not Work?

There are, as described above, strong reasons for local officials to engage in local economic development activity. Yet, the benefits of much of this activity are uncertain, and the activity itself, in the case of tax incentives, abatements, and subsidies, is widely portrayed in the research literature as being ineffective. Why do local officials nonetheless continue to provide such economic development incentives, and is it rational to do so?

I begin by noting again the frequent public pressure to do something, the pervasive environment of uncertainty that surrounds decision making, and the asymmetry of information that decision makers face. Wolman (1988) argues that the key factor is the politicians' calculation of risk. He discusses several possibilities. First, local officials may be ignorant of the research findings that fiscal incentives to attract economic activity do not work and instead straightforwardly pursue them in the expectation that they will prove efficacious. Second, the political pressure operating on the official to do something is so great that it cannot be ignored, particularly when combined with the opportunity for credit-claiming. The local official succumbs to this pressure (and the temptation), even though he or she knows the policy is not likely to work. Wolman, however, suggests a third possibility, premised on the politician actually having a reasonable understanding of the findings of the research literature:

> An informed understanding [of the literature] would be that fiscal incentives are quite unlikely to play a major role in firm location decisions at the interregional level, but may have some impact on the local level. The research findings are hardly definitive. Uncertainty abounds. Decision-making under uncertainty requires politicians to assess policy choices by calculating the benefits or costs of the outcome of policy action against the probability that they will actually occur. A politician might well reason that, although it is quite likely fiscal incentives will not affect location decisions, the political benefit that would accrue in the unlikely case the incentives were offered and they did work (or the political costs if they were not offered and they were to work for other cities) would be substantial and would indeed outweigh the much smaller political benefit of being correct and not offering incentives. This reasoning will obviously be abetted if the *fiscal* costs of being wrong (i.e., offering subsidies or incentives when they in fact have no effect) are either borne by others or are relatively small and unnoticeable. (p. 25)

Wolman then constructs a decision tree to show how, from a politician's point of view, it might be rational to offer fiscal incentives even though it were recognized that there were only a relatively small (25%) chance that they would work.

Wolkoff (1992) takes the argument a step further, not only showing that it might be rational for a politician to provide fiscal incentives but also examining the conditions under which it would actually be so. Using a game theory approach, he first divides firms into two types: those that have alternative profitable investment opportunities and would relocate to another jurisdiction unless subsidized (Type 1 firms) and those that lack relocation options and are thus tied to their present location (Type 2 firms). However, the situation is one of information asymmetry: While a firm knows whether it is Type 1 or Type 2, local policymakers do not. The game has three decisions. A firm must decide whether to request a subsidy, and its decision will be dependent on the cost of making an application weighed against the likelihood of success. The local government must then decide whether to grant the subsidy. Finally, the firm must decide, once the subsidy decision has been made, whether to remain or move.

The subsidy decision by the local government has economic, fiscal, and political costs and benefits. These include the cost of the subsidy, the investment the firm makes in the community if it stays, the loss in tax base (as well as wages and employment) if it leaves, and the political costs and benefits discussed earlier. As Wolkoff (1992) observes,

> This decision is complex because the community is unable to ascertain what type of firm is making the request and incorrect decisions are costly. Subsidizing Type 2 firms is costly because the community unnecessarily spends valuable tax dollars. Failing to subsidize Type 1 firms also has costs because of the potential weakening of the economic base. . . . The magnitude of the political costs the city incurs depends on the value of the firm to the community. But all subsidy refusals are costly because they send a negative message about the business climate. For Type 2 firms the political costs of refusing the subsidy consist solely of this negative message. The political costs of refusing to subsidize a Type 1 firm are much higher . . . because the departure of a local company leaves a politically expensive legacy of unemployed voters and depleted tax bases.

> Further, it places the governing ability of elected officials into question. (pp. 346, 350)

Each player (firm and local government) makes its decision based partly on the expected action of the other player. Since the local government is unsure whether a firm is Type 1 or Type 2, its rational strategy is to maximize the expected value of the various outcomes, based on the estimated value of the costs and benefits of each outcome and its estimate of the probability that the firm is Type 1 or 2. The game allows firms to make subsidy requests of varying amounts. Since local policymakers cannot differentiate between the two types of firms, they are better off arranging a situation in which firms signal their type by the value of the subsidy request they make. To encourage that outcome, it might seem that the best solution for a local government would be to grant subsidies to both Type 1 and Type 2 firms, provided that Type 2 firms could be persuaded to give themselves away by making only small requests. But why should they do so, when they know a larger request will be granted? Wolkoff (1992) shows that, paradoxically, the community might encourage Type 2 firms to show their hand by granting all small requests but not every high-subsidy request. As a consequence, Type 2 firms, preferring a certain windfall small subsidy to an uncertain large one, would request the former. Wolkoff concludes:

> By rejecting some subsidy requests, the community increases the likelihood of rejecting some firms having legitimate claims to large subsidies. Thus the locality would be refusing to subsidize firms that it knows will have relocation prospects, despite the policy objective providing subsidies to only those firms that would otherwise relocate. This seeming irrationality is perfectly consistent with the maximizing model developed here. The apparent irrationality at the micro level is resolved when one understands these decisions as being part of a more general subsidy strategy for all firms. In this case, the way to avoid offering high subsidies to undeserving firms is by rejecting the request of some Type 1 firms. (p. 352)

Wolkoff also notes it may be rational to provide subsidies to all firms that apply if all firms request the same level of subsidy and the request is relatively low.

Is Local Economic Decision Making Rational?

Both Wolman (1988) and Wolkoff (1992) show how seemingly irrational decisions may indeed be rational. However, the fact that there is a rational solution does not necessarily imply that local decision makers behave rationally. Bachelor (1994) argues that the constraints imposed on decision makers by uncertainty and inadequate information make it difficult for them to engage in rational behavior, even when political as well as economic costs and benefits are considered. Instead, they are likely to rely on solution sets (see Jones and Bachelor 1993), which are "characteristic patterns of solutions to problems and opportunities" (Bachelor, 1994, 613). Reese (1993b, 502) agrees, contending that "economic development policies are not likely to result from a rational, systematic weighing of likely costs and benefits. Rather, economic development policy is, to a large extent, driven by professional decision rules emanating from bureaucratic actors."

The apparent disagreement with Wolman (1988) and Wolkoff (1992) stems from two sources. First, neither Wolman nor Wolkoff argue that local officials actually engaged in the process of rational policy making and, even less so, in formal rational policy analysis. Their contention was that their decision outcomes were not necessarily irrational—and indeed, could be seen as rational, given the various political and economic costs and benefits they faced within a context of information asymmetry and uncertainty. In addition, their concern is specifically with nonroutine decisions on whether to provide financial incentives to specific firms, whereas Bachelor (1994) and Reese (1993b) appear to be concerned with a broader range of decisions that frequently occurs within the bureaucracy.

How Much Choice Do Local Governments Have?

The discussion thus far has been concerned with why local officials engage in economic development policy and the extent to which their engagement can be seen as rational. As noted, there are compelling reasons for their involvement. But how much are their activities constrained, and how much choice do they actually have? This question, widely discussed in the literature, has been the subject of considerable debate and controversy.

There are two separate strands of the discussion. First is the extent to which local policy can have an important impact on local economic development, defined in terms of bringing about growth in the area's employment and per capita income—that is, how constrained are local governments by the external economic forces over which they have little control? The second strand asks a somewhat different question: To what extent is local government policy determined by economic forces rather than political choice—that is, must local government policy be favorable to business or to growth machine coalitions out of a concern for retaining mobile capital within city boundaries? These are clearly two different questions. The second has to do with the effective discretion local governments have in choosing economic-development-related policy; the first is concerned with whether whatever policy is chosen will actually have an impact on local economic development. Unfortunately, these strands are frequently confounded so that the participants often appear to be talking past rather than to each other.

Can Local Economic Development Policy Bring About Local Economic Development?

We have seen that although local elected officials are quite optimistic about the ability of local economic development policy to have a beneficial impact on the local economy, local economic development practitioners are considerably less so. This review is not the place to examine the economic impact of local economic development policy. However, it is frequently argued that the ability of a local economic development activity to bring about changes in local employment and income, at least in the short run, is constrained by two factors. First is the inability to affect the external demand for the goods and services produced in the area (although in the longer run, local government action might be able to improve the area's performance through improving the productivity of its factors of production—e.g., providing a higher skilled labor force through better education and training policies). Second is the fact that local governments—even those of large cities—do not comprise a functional local economy but are part of a larger metropolitan-wide economy.

As a consequence, some analysts argue that local government economic development policy is unlikely to make much difference. Sassen (1990), one of the leading writers on economic restructuring, states forcefully that "major economic forces are largely beyond a city's control" (p. 239) and that "economic forces carry much greater weight than local politics in determining the shape of local development" (p. 235). However, she does contend that "under certain conditions, local governments or local initiatives can resist the tendencies of economic restructuring" (p. 238). Others give considerably more scope to the efficacy of local government policy. Parkinson and Judd (1988, 2) contend that "most cities are not the helpless pawns of international finance, industry and commerce. They are in a position to mediate and direct their own destinies."

How Much Choice Do Local Governments Really Have in Economic Development Policy Making?

Kantor and David (1988, 5) agree with the overwhelming importance of external economic forces but then make the link to the second strand of the argument by contending that city policy choices are constrained by these forces:

> The tremendous mobility of capital forms the major barrier to local economic development. So long as localities cannot firmly tie many business enterprises to the community's infrastructure and labor force, intergovernmental competition pressures city governments to provide business incentives. (p. 229)

In short, capital mobility within the context of a jurisdictionally fragmented metropolitan economy forces local governments to provide economic policies favorable to business. This parallels Peterson's (1981) argument (discussed earlier) that economic development is the unitary interest of the city and that cities are structurally compelled to pursue developmental rather than redistributional policy choices.

A variety of writers have reacted vigorously to this argument, which, in essence, says that economic forces rather than political choice determine the nature of local economic development policy. (This debate in some respects parallels an earlier and broader debate about the impact of economic and other environmental factors relative to political factors in state and local public policy making.) Stone (1987), one of the strongest critics of this point of view, argues that

> local government officials make genuine choices, albeit within structural boundaries. Local decision makers do not simply follow the imperatives that emanate from the national political economy; they must also interpret those imperatives, apply them to local conditions, and act on them within the constraints of the political arrangements they build and maintain. (p. 4)

Swanstrom (1985) also argues the same case, observing that "growth politics should not be seen as some form of economic determinism" (p. 5) and that

> city governments do not simply respond, in knee jerk fashion, to the pressures of growth; growth pressures must be filtered through *local political systems* with particular rules, power arrangements, and perceptions of the governmental marketplace. The needs of the political system may clash with the needs for economic growth. (p. 33)

Sanders and Stone (1987) expand on this argument:

> He [Peterson] has conceptualized developmental politics in an inappropriate way. As governing officials make decisions, they must heed political as well as economic imperatives. Moreover, the political imperatives of coalition building and conflict management may be more immediate than the economic imperative of enhancing the city's economic positions. . . . The realization that growth and redevelopment produce losers as well as winners makes the political imperative paramount. Coalition building and conflict management become urgent, and they do so under conditions in which particular allocational consequences are salient. (p. 538)

Peterson's (1987) response is, in effect, to agree that politics exist but that in the end, using an economistic approach, local government policy choices can be best understood by conceiving of local government as having a unitary interest in developmental policy that is driven by the city's economic needs:

> Unitary models, such as are presented in *City Limits,* can always be supplemented by more detailed case studies of bargaining, conflict, and compromise. Whether the additional details add materially to the explanation is another matter. . . . It is not good enough to show that conflict occurs, the easy-enough task that Sanders and Stone set for themselves. One also needs to show that one can add significantly to the explanation of outcomes by adding bargaining and conflict variables to the explanatory model. (p. 540-1)

Wong (1988) and Kantor and Savitch (1993) provide intermediate positions that seem eminently reasonable. Wong argues that economic constraints are real but that political factors can operate either to reinforce or to work against these constraints in various policy arenas (developmental, redistributive, and allocational). Thus "urban policy making can be the result of political choice as well as economic consideration." (p. 1)

This literature suggests that economic forces clearly condition the kind of economic development policies local government pursues but that within the framework set by these external forces, political choice exists. Whether the policy adopted is likely to have any real impact on the local economy in terms of conventionally defined economic development (increases in the employment and per capita income of local residents), as opposed to physical development and/or fiscal strengthening, is a different question and the subject of a literature that is not reviewed here.

The Local Politics of Economic Development

The Interests Involved

Politics at all levels involves the interplay and clash of various interests, usually reflected, if they are to be effective, through organized groups and institutions. What interests are involved in local economic development (or, more generally, development, since much of the literature does not distinguish between economic development and land development), and what sort of coalitional patterns can be discerned?

Kantor and David (1988) begin by noting that city officials, for reasons already discussed, are likely to favor local economic development. They observe that business interests are varied, but the most salient of these is the downtown business community, "comprised typically of banks, retail stores, newspapers, major corporations, real estate groups, and others who own large fixed investments in the central business district" (pp. 241-2). Organized labor is also usually supportive of economic development. As Kantor and David note,

> Labor's stake in economic development generally comes down to the bargaining power that organized and unorganized workers can expect to gain or lose as a result of city development strategies; programs that generate more jobs for the local workforce bring with it [sic] greater bargaining power in the labor market, in turn opening up opportunities for higher wages, better job security, and more amenable working conditions. Consequently, the interests of workers in urban development can often parallel that of business, though for different reasons. Large downtown reconstruction projects have often generated considerable support from unions even when they have displaced low-income residents or threatened the jobs of other workers. (p. 243)

Molotch (1976, 1988) and Logan and Molotch (1987) argue that economic development interests coalesce into a growth machine bonded by a common interest in obtaining rent from area growth. Molotch (1988, 26) observes that "The people who make up growth machines are not just the owners of land and buildings, but include others who have their fortunes tied to a specific area's growth." These include local financial institutions, local newspapers, and those whose livelihood serves the needs of growth machine entrepreneurs (real estate lawyers, accountants, property management firms, advertising agencies, etc.) as well as prominent local institutions such as universities and foundations whose fortunes are tied to place prosperity. Molotch notes that major corporations often play only a subsidiary role because, while operating locally, they can generate their investment return anywhere. (Indeed, Jones and Bachelor [1993, 207] argue that in automobile-dominated cities such as Detroit and Flint, politicians must "lure" the major auto corporations into participating in city projects.)

Levy (1990) details an even broader *development coalition*:

> In most communities the essentials of a coalition for development are present and waiting for the economic developer. Property owners, in general, are likely to be for development because it holds out prospects for reduced tax burdens. Those who own property whose market value will rise as development increases demand have a second reason to favor development. This may be equally true of the owners of vacant land with commercial potential, of

> owners of stores and commercial buildings, and of apartment owners. Homeowners may perceive that economic growth increases the demand for housing, and that therefore they will ultimately profit from it. . . . The banking community is generally for development since it means more loan business. . . . organized labor is almost inevitably for development. (p. 14)

Neighborhood interests, on the other hand, if organized, are more likely to form the core of opposition to development schemes, primarily because, in Molotch's (1976) terms, their interests are related to the *use value* rather than the *exchange value* of property. As Kantor and David (1988) observe,

> Efforts to promote the locality's economic base can often threaten the social and economic survival of these [neighborhood] enclaves. The most frequent residential victims of urban development tend to be lower-income neighborhoods whose geographical location in many large cities (adjacent to commercial and industrial areas), low property values, less skilled workers, and weak political resources make them highly vulnerable to business expansion schemes, highway improvements, and other forms of economic "modernization." (p. 244)

Neighborhood group opposition may be strengthened if it is organized into groups of similar racial or ethnic characteristics. Middle- and upper-income homeowners may also be opposed to economic development schemes if these bring undesirable consequences of growth (e.g., congestion, pollution, reduced amenities) to their neighborhoods, and groups concerned with environmental quality may also be opposed.

While supportive of the general thrust of the above generalizations, Schumaker, Bolland, and Feiock (1986) argue there is actually more variation than these generalizations suggest:

> [D]etailed case studies of urban redevelopment in specific cities . . . reinforce this notion of significant cross-community variations in the patterns of community conflict on growth policies. Even within communities, specific economic development issues can spawn conflicts within the business community or within the middle class, between core and peripheral areas or neighborhoods of the city, between groups seeking capital accumulation and those wishing social consumption, and even among lower-income neighborhoods. (Fainstein et al. 1983, 255, quoted in Schumaker, Bolland, and Feiock 1986, 27)

In their study of 12 economic development issues during a five-year period in Lawrence, Kansas, Schumaker, Bolland, and Feiock (1986, 37) found substantial variation in the makeup of pro- and antigrowth coalitions from issue to issue. However, they identify general patterns across the 12 issues: Progrowth coalitions consisted of elected officials, public administrators, local business interests, private elites, upper-status activists, high-property-value neighborhoods, public-sector employees, and Republicans; antigrowth coalitions consisted of neighborhood organizations, ad hoc protest groups, and poorer neighborhoods.

Schumaker, Bolland, and Feiock (1986, 35) also find that those most frequently involved in the 12 issues were the mayor (11 issues), the city manager (11), the local newspaper (10), the chamber of commerce (9), organized developers, realtors, and/or landlords (8), elected council members (7.9), business elites (6—the average involvement score of the five most active business elites), neighborhood organizations (6), city government administrators (5.8—average of the five top administrators), neighborhood leaders (5.4—average involvement of the five most active neighborhood leaders), ad hoc protest groups (5), and building trade unions (5).

The actors who initiated the most issues were organized developers, realtors, and/or landlords (6 issues) and the chamber of commerce (4), with the major political officials—the mayor and city manager—initiating only 1 issue each. The most active participants in opposition to the 12 issues were neighborhood organizations (in opposition to 6), neighborhood leaders (an average of 5.2 for the five most active leaders), and ad hoc protest groups (5). The mayor supported 7 of the issues and opposed 4. Schumaker, Bolland, and Feiock's (1986) study of Lawrence, Kansas, is one of the few efforts to empirically examine participation across a range of issues; the authors, of course, cannot generalize their findings to other cities, nor do they make any effort to do so. They do suggest, however, that the participation and coalitional patterns, as well as the degree of influence, are likely to vary both within and across cities.

The Role of Business

Virtually all of the literature emphasizes the central role of business in local economic development politics and policy making. In his study of Atlanta, Stone (1989, 219) observes, for example, that "The Atlanta case illustrates the way in which business enterprises are able to shape a city's policy agenda. . . . business has a somewhat unique ability to make things happen and play a central role in the governing coalition." Stone argues that such governing coalitions—or urban regimes—are "informal arrangements by which public bodies and private interests function together in order to be able to make and carry out governing decisions" (p. 6). Urban regimes, while not inevitably directed toward undertaking development policies in the interest of business, nonetheless appear to do so quite frequently. In addition to the development-oriented *corporate* or *activist* regimes, Stone (1987) also identifies *caretaker* and *progressive* regimes; Orr and Stoker (1994) point to *human development* regimes as well as *downtown development* regimes.

What accounts for the strength of the interest of business in the development process, and how overwhelming is that influence? Earlier writers stressed the power of business to, in effect, dictate to and control local elected officials (see Hunter 1953; Mills 1956). This approach has now been largely abandoned in favor of shared ideology, structural, and systemic explanations. Much recent writing on this subject derives from Lindblom's (1977) discussion of the *privileged position* of business as inherent in the ideology of capitalism. Thus Barnekov and Rich (1989) argue that it is the pervasive ideology of *privatism* that leads public officials to pursue economic activities that benefit business.

Kantor and David (1988), Elkin (1987), Jones and Bachelor (1984), and Stone (1980, 1989) stress the structural relationship of business and local officials that creates what Stone (1980) calls a *systemic* bias toward business. Elkin, for example, writes:

> To see the politics of growth as an exercise of power by businessmen manipulating public officials is to miss the sense in which it is a product of mutuality of interest. And that mutuality grows out of the structural features that define the city's political economy; given the manner in which officials get elected, the prerogatives of private controllers of assets, the limits on a city's ability to affect and exercise property rights, and the need for cities to raise money in private credit markets, city officials will naturally gravitate toward an alliance with businessmen. (p. 42)

In particular, local governments are structurally dependent on business because of the mobility of capital, the relatively small size of urban jurisdictions, and the perceived dependence on business investment, not only for providing jobs and income to city residents (i.e., ensuring good economic performance) but also for generating adequate local public revenues for providing public services without overburdening residents (voters) with high tax burdens. As Kantor and David (1988, 250) state, "Urban economic dependency assures business a position of privilege, and this limits development choices by the prospective threat of disinvestment." Or as Jones and Bachelor (1993, 243) succinctly note, "Businessmen provide the majority of livelihoods for the community's citizens, and businesses have the power to deny that livelihood by relocating."

Both Kantor (1987) and Fainstein et al. (1983) also contend that the cooperation of business is required to bring about acceptable local economic performance and adequate local public revenues, each of which is desired to improve chances of electoral success. As Fainstein et al. observe, "The system of finance compels every local state [i.e., local government] at least to maintain its revenue base by attracting investment which contributes to the market value of real property" (p. 251). Thus, according to Kantor, "officials at all levels are constrained to promote their economies and induce business investment" (p. 495). And Jones and Bachelor (1984, 1993) note that they must seek such investment in a situation of information asymmetry that places them at a severe disadvantage. The fact that local officials do not know what level of public action is required to maintain or attract a specific firm (knowledge that, of course, is available to the firm) provides a decision-making process that produces a *corporate surplus.*

Stone (1989) argues that it is the need of elected officials to gain the cooperation of business in order to get things done that explains the dominance of business. Setting forth what he calls a social production model, he contends that if elected officials who wish to pursue entrepreneurial policies are to be successful, they have no choice but to seek the cooperation of the business sector, for business is the only actor that has sufficient resources to accomplish public purposes. As a consequence, entrepreneurial elected officials are driven to create urban regimes that depend heavily on business.

Jones and Bachelor (1993) observe that the need for politicians to gain the cooperation and participation of business has turned the original business control of local government development decisions on its head:

> Businessmen are, on the one hand, less interested in exerting control over local government. On the other hand, politicians are more interested in influencing businessmen. This new dependency in the city means that governmental officials will attempt to influence the decisions of businessmen, rather than the other way around. (p. 251)

How Dominant Is Business? The Relationship Between Local Government and Business

The above discussion leads to the obvious question: How dominant is business? Do business interests have such great advantages that they virtually always win? Some of the literature, particularly that in the growth machine tradition, suggests that this is indeed the case (see Molotch 1988; see also Barnekov and Rich 1989; Fainstein et al. 1983). However, other authors (Kantor and Savitch 1993; Jones and Bachelor 1986, 1993; Swanstrom 1985) suggest that business influence, while substantial, does not necessarily always dominate. These authors emphasize that local government and business are involved in an exchange (bargaining) relationship and that local government has some resources it can use to its advantage. Swanstrom (1985) explains the bargaining resources available from the perspective of a central city:

> Central cities have a strong market position with regard to certain forms of economic activity; they need not approach the bargaining table with hat in hand. The reason for this is that not all economic activity is hypermobile. While there are deep centrifugal forces in the economy dispersing manufacturing and routine back office service sector jobs, at the same time there are deep centripetal forces pushing certain forms of service activity into central business districts. (pp. 236-7)

More generally, Kantor and Savitch (1993) argue that local governments have bargaining advantages under certain market conditions (e.g., when sites within the local government are very appealing to local investors, when businesses in the community have fixed capital in place with substantial sunk investment costs, when there are substantial economies of agglomeration); under conditions of strong popular control (e.g., competitive parties, programmatic parties, multiple channels for citizen participation); and wherein mechanisms for public intervention and support from other levels of government are high (e.g., when the local government is not so dependent on locally raised revenues). They conclude:

> When bargaining relationships put local government at a persistent disadvantage, the public sector tends to absorb greater costs and risks of private enterprise. . . . When bargaining relationships put local governments at a persistent advantage . . . public actors tend to impose costs for the privilege of doing business in the locality or to place other demands on the private investor. They may levy differential taxes on businesses located in high-density commercial districts, charge linkage fees on downtown development (which can be invested elsewhere), demand amenity contributions that are applied toward the enhancement of city services, or impose inclusionary zoning, requiring developers to set aside a number of low-or-moderate rental units in market-rate housing. (pp. 234-5)

Under what conditions—or what kinds of local governments—will the latter types of policies, frequently referred to as progressive policies (see Clavel 1986; Clavel and Kleniewski 1990), most likely be produced? Jones and Bachelor (1993) conclude that

> certain governmental arrangements add measurably to the probability of successful negotiation with businessmen: these include the sheer size of the jurisdiction, a mayoral form of government, and a professional bureaucracy. In general, any arrangement that aids a mayor in mobilizing power among the various constituencies that make claims on government also aids in negotiating with external organizations. (p. 251)

Goetz (1990) finds that progressive policies—which he terms Type II policies—are more likely to be adopted by larger communities, those with strong community-based political activity, and jurisdictions with lower growth rates. Miranda, Rosail, and Yeh (1992) find that progressive development policies are unlikely to be adopted by cities experiencing economic distress and that adoption rates are related to "cultural" factors: They are more likely to be adopted in communities with a high proportion of renters, a low proportion of married households, and a low proportion of children (in short, "yuppie" areas, or communities dominated by higher-education institutions and their students).

An interesting perspective on the actual beneficiaries of local economic development programs is provided by Bowman (1987) in the results of a National League of Cities' survey of economic development professionals. Respondents were asked to indicate whether each of a set of participants were major beneficiaries, minor beneficiaries, or not a beneficiary of city-sponsored economic development activity. The proportion of respondents naming actors as major beneficiaries was 74.4% for the central business district, 66.9% for local developers, 66.3% for the local labor force, 59.6% for existing business firms, 46.7% for new business firms, 43.9% for the unemployed workforce, 43.6% for low-income neighborhoods, and 35.3% for local commercial banks.

The literature reviewed in political science, sociology, and planning yielded virtually no empirical studies of the actual distribution of costs and benefits of local development activities. Bartik (1991), an economist, in his study *Who Benefits from State and Local Economic Development Policies?,* comes to a surprising conclusion in light of much of the above literature. Focusing not on the impact of specific policies but on that of economic growth—presumably the desired product of these policies—he concludes:

> The empirical estimates indicate that faster local growth has stronger effects on the annual real earnings of Blacks (20% greater effect than the average) and on less-educated individuals (15% greater effect for someone with three less years of schooling). . . . The greater effects on Black and less-educated individuals are large enough that local economic development policies probably have progressive ef-

fects on the distribution of income. However, state and local economic development policies can hurt lower-income groups if the cost per job created is too high, or if they are financed in a highly regressive manner. (p. 206)

Bartik's (1991) conclusion reflects a different perspective in which winners and losers are not viewed as immediately affected institutional and organized interests (e.g., business, neighborhood groups) but along a different dimension related to the ultimate impact on individuals.

The Local Decision-Making Process

How are economic development decisions actually made within the local public sector, and what are the relative roles played by elected public officials and economic development professionals? The question is difficult, given the variety and complexity that characterize the organization of the economic development function at the local level.

First, as Levy (1990, 21) notes, the economic development function can be lodged in four different institutional forms or some combination of these forms. It can exist within the general purpose local government, either as (or within) a line or a staff agency; it can be in a semipublic agency that, while created by legislative act, is not part of the structure of the general purpose government and has some degree of autonomy, including the right to enter into contracts and, in many cases, the power to condemn land or to issue tax exempt bonds; it can be in some form of public-private partnership, frequently a not-for-profit partnership in which the public sector is represented through membership on the board; or it can be, particularly in smaller communities, a purely private group such as the chamber of commerce.

Within the general purpose local government, there are five different organizational possibilities (see Fleischmann and Green 1991): There can be a separate Department of Economic Development; the function can be lodged in a special unit within the chief executive's office; it can be placed within a broad Department of Community Development, which has responsibility for related functions such as planning and housing; it can be within a traditional line department such as public works; or its functions can be scattered across a variety of departments.

Clarke and Gaile (1992) conducted a survey of all American cities with population above 100,000 and conclude:

In most cities mayors take the lead in promoting economic development. Many cities also indicate that deputy mayors have special responsibilities in this area. Economic development line agencies reporting to the mayor and mayor/city manager or city council have lead responsibility for policy formation in most cities; over 50% place lead responsibility in economic development line agencies that are separate from community development or planning agencies. Thus, the design of economic development policy rests with elected officials or those appointed by them. In contrast, economic development line agencies often share implementation responsibilities with special authorities or quasi-public organizations. Cities often turn to organizations such as citywide development corporations or special authorities such as redevelopment agencies and port authorities to carry out economic development projects. Many of these organizations have special financing authorities or resources not available to line agencies. (p. 193)

The importance of the mayor is a theme frequently emphasized (see Schumaker, Bolland, and Feiock 1986; Feiock and Clingermayer 1986; Jones and Bachelor 1986, 1993). A mayor is seen as a focal point for leading an economic development effort and for negotiating deals with development interests, a potential *political entrepreneur* able to step forward and create effective coalitions (see Schneider and Teske 1993), and a provider of leadership, a critical factor (see Pagano and Bowman 1995; Judd and Parkinson 1990; Mollenkopf 1983; Fosler and Berger 1982). Schumaker, Bolland, and Feiock (1986) argue:

Policy entrepreneurs are important to the success of growth policies. But such entrepreneurs may not always emerge. When mayors are not

> directly elected and when they have few formal powers . . . they may not have the resource base to put together pro-growth coalitions. (p. 43)

Feiock and Clingermayer (1986) contend that a mayor within a system of centralized executive power (i.e., a directly elected mayor with veto power, a small weak council, partisan elections) is likely to be the critical actor in the local economic development process:

> A strong executive is particularly advantageous to business interests, since these interests need only deal with a single, central authority, rather than with a number of elected officials, and can do so with the knowledge that authority can deliver on any promises that he/she makes. This insight is particularly relevant to the making of development policy, since most such policies require some negotiation between representatives of the city and state and federal officials, or business and union leaders, or some combination of all of these actors. (p. 215)

However, the dominant role of mayors suggested by these authors is effectively challenged by Reese (1995), who makes an apt distinction between "big events" and routine decision making:

> Descriptions of economic development policy making in major cities from Detroit to Atlanta highlight the role played by elected officials in determining economic development policy. These cases, however, have several elements in common. They occur in large communities and involve a "big event": a major new auto plant, a river front market or aquarium, or a significant attraction such as a World's Fair. Such situations are often characterized by "peak bargaining" where mayors are central in negotiations with the leaders of a small number of other major interests or "command posts" and expensive capital projects and issues of current high salience are involved (Jones and Bachelor 1993). However, most economic development decisions are not big events. . . . Economic development policy activity is a combination of a few "big events" and many routinized decisions and day to day actions. And such routines are the province of professionals and bureaucrats. (pp. 6-7)

To cope with the environment of uncertainty in which they work (see earlier discussion) and the consequent risk of making the wrong decision (see Spindler and Forrester 1993, 39), economic development professionals rely heavily on these routines (Reese 1995, 1993b; Spindler and Forrester 1993; Rubin 1990). Spindler and Forrester (1993, 41) write that "Routines to avoid risk include collaboration between economic and political elites, a closed agenda and decision making process, and concealment of policy costs and beneficiaries to avert public scrutiny." In policy terms, they point to three specific routines: Adopt new policies as they emerge in competing local governments; provide universal incentives rather than those based on need; and adopt symbolic economic development programs. Rubin (1990) also notes how environmental turbulence drives practitioners to engage in symbolic and formalistic behavior.

The Degree of Conflict

One of the more vigorous debates in the literature on local economic development politics concerns the extent to which local decision making and the political processes surrounding it are characterized by overt conflict. Peterson (1981), once again, is at the center of the debate; consistent with his theory of the unitary interest of local government, he sees the politics of development as highly consensual, supported by virtually all elements of the community:

> Plans to attract new industry to a community, to extend its transportation system, or to renew distressed areas within the city are characteristic types of developmental policies. Such policies are often promulgated through highly centralized decision-making processes involving prestigious businessmen and professionals. Conflict within the city tends to be minimal, decision-making processes tend to be closed until the project is about to be consummated,

> [and] local support is broad and continuous. (p. 132)

In a later (and widely ignored) passage, Peterson (1981) makes clear that this generalization is not universal:

> The consensual quality of development policies does not hold in each and every case. Apart from factions and groups which may put their separate interests ahead of that of the community, under some circumstances community leaders may fundamentally disagree about the overarching interest of the community. Especially in smaller communities and in suburban areas, where economic and status interests may bifurcate, community leaders may split into "growth" and "no-growth" factions. While some express concern for the economic base of the community, others argue that the community should hold out against urbanization or attempt to survive as a residential enclave apart from the centers of commercial and industrial activity. If there is no agreement on the city's overarching interests, consensus yields to bitter, antagonistic ideological conflict. (p. 149)

Elkin (1987; see also Stone 1989; Kantor and David 1988) argues that conflict is absent not because there is a broad consensus but because oppositional views are screened out of the decision process. Elkin observes that

> efforts to promote investment are likely to prompt sufficient dissent to turn the thoughts of at least some citizens to political action. . . . Here is where the disposition of officials to induce investment and the importance of institutional arrangements become manifest, for they both work to forestall protest in the first place, for example, by screening development decisions from public view. (pp. 43-4)

Other writers (see, for example, Sharp and Bath 1993; Donovan 1993; Sanders and Stone 1987) vigorously attack the view that economic development decisions do not involve conflict, using Peterson (1981) as a convenient straw man (and, in the process, ignoring his codicil about the existence of conflict). These authors cite case studies indicating that conflict around economic development issues is not unusual and is frequently severe. Citing Swanstrom (1985), Jones and Bachelor (1986), and Fainstein et al. (1983), Sharp and Bath (1993, 214) write, "With respect to Peterson's thesis, there is substantial case study evidence of citizen mobilization around economic development issues, especially in the form of controversies over particular development policies and projects."

Sharp and Bath (1993), Sharp (1990), and Wong (1988) all contend that the degree of controversy is contingent and more likely to occur in some circumstances than others. Sharp argues that there are three distinguishable styles of developmental politics: a style dominated by the business elite, largely lacking in public controversy and consistent with Peterson's (1981) description of developmental politics; a populist style in which the elite is challenged, fitting better the pluralist model of decision making; and a mixed model somewhere between the two (p. 261). Sharp suggests that this *populist* form of conflictual developmental politics is more likely to arise when the potential unequal outcomes of development policy are more visible, when neighborhood organizations are already active, and in institutional forms that do not submerge conflict (i.e., in cities with mayor-council systems rather than council-manager systems).

Empirical work also suggests that conflict and controversy are not absent in economic development policy making (in addition to the case studies cited earlier, see Schumaker, Bolland, and Feiock 1986; Goetz 1990; Clarke and Gaile 1992). Donovan (1993) asked local economic development officials in a seven-county area around Los Angeles to respond to the question "In general, how controversial would you say economic development issues are in your city?" Officials in 36.3% of the 132 responding cities indicated that economic development issues were always or often controversial; 45.5% responded that they were sometimes controversial; only 18.2% said they were not at all controversial. Donovan also tries to account for the existence of controversy using the respondent's perception of the degree of controversy over economic development decisions in the locality as the dependent variable for community controversy. Controversy was positively related to the percentage of professionals residing in the city and to the percentage of renters. However, it was less likely to occur in cities with mayor-council systems (contrary to Sharp's [1990] hypothesis) and was unrelated to the respondent's perception that neighborhood groups were important (again not supporting Sharp's hypothesis).

Openness to Public Participation

A question closely related to the degree of controversy and conflict is the openness of the process and the extent of participation by citizens and citizen groups. Kantor and David (1988, 244) echo Peterson (1981) by arguing that developmental policy making "is usually not overtly conflictual because the process of decision making often is not very open and participatory. . . . There is a powerful tendency for urban governmental authorities to insulate the process of economic policy making from popular involvement" (see also, for example, Spindler and Forrester 1993; Barnekov and Rich 1989).

Kantor and David (1988, 246-9) argue that local governments engage in a variety of strategies for limiting public participation, including placing economic development decision making in independent or quasi-independent authorities outside the realm of general-purpose decision-making bodies to which citizens and citizens groups have more structured access, and *social control* strategies designed to isolate decision making from mass political pressures.

However, a variety of case studies suggest that the public is not always shut out of the decision-making process. Based on their empirical analysis of data from the 1984 ICMA survey of local governments on economic development activity, Sharp and Elkins (1991, 137) observe that "our analysis provides a number of partial challenges to contemporary assumptions that the realm of development policy is insulated from popular pressure."

From the survey responses, Sharp and Elkins (1991) are able to create an ordinal scale measuring the degree of citizen involvement in economic development policy making. (However, the scale appears to be more a measurement of whether mechanisms for citizen involvement existed than a measurement of the extent to which they were used.) Local governments that engaged in economic development planning—a minority of respondents—were asked whether the city used each of the following mechanisms: (1) appointed advisory committees representing the entire community, (2) open meetings or public hearings, (3) citizen surveys, and (4) elected neighborhood commissions. This measure is then used as an independent variable to examine whether citizen involvement affected a local government's propensity to adopt economic development policies with varying degrees of visibility of costs and benefits. However, the analysis controls only for fiscal stress and does so using two different measures of stress—per capita property tax revenue and property tax revenues as a proportion of total local revenue—neither of which, as has been argued, is a reasonable operationalization of the concept of fiscal stress.

Sharp and Elkins (1991) find that under conditions of high fiscal stress—that is, high property tax revenues per capita (which is probably a measure of wealth rather than fiscal stress)—there is a significant and positive relationship between the degree of citizen involvement and the adoption of economic development policies that minimize apparent tax costs (e.g., tax-exempt bonds and loan guarantees) and that maximize apparent benefits (e.g., infrastructure improvement). They find a significant but negative relationship between citizen involvement and tax abatement, an activity which, they argue, has highly visible tax costs. The authors conclude that "when local officials eschew tax abatement and adopt alternative policies that minimize tax costs and maximize benefits, they are responding to the public, or at least are constrained by the anticipated public reaction to tax abatement." (p. 137)

Cable, Feiock, and Kim (1993) employ the 1989 ICMA survey to perform a very similar analysis, which, unfortunately, suffers from many of the same problems: use of the same index for citizen involvement and operationalization of property tax stress as property tax revenue per capita. They also employ a third control variable, fiscal capacity, operationalized as per capita income. They find that "The probability of costly economic development policy adoption diminishes in cities with greater openness. . . . Availability of mechanisms for popular decision making decreases the probability of providing loan subsidies, direct loans and cash contributions for economic development" (p. 94). However, in contrast to Sharp and Elkin's (1991) findings, openness does not decrease the probability that tax abatement policies will be adopted. The authors then examine each of the participation mechanisms individually and find that adoption of tax abatement policy is significantly and positively related to the use of open public meetings by the local government but significantly and negatively related to the existence of an advisory board appointed to represent the entire community. They conclude:

> Tax abatement is characterized by a highly concentrated distribution of benefits and a diffuse distribution of costs, creating an incentive structure that encourages participation by highly organized and motivated groups of beneficiaries. Proponents of tax abatement are overrepresented and opponents underrepresented in public meetings lending the appearance to policy makers of broad based community support for adoption. Absent a crusading local media or disgruntled antitax entrepreneur, public debate of tax abatement is likely limited and structured by its proponents. . . . In contrast, the results also suggest that advisory committees appointed to represent the entire community may faithfully accomplish what public meetings cannot. (pp. 96-7)

Despite the shortcomings of the research, it seems clear that the availability (and presumed use) of public participation mechanisms is related to the probability of adoption of various kinds of economic development policies. What is not clear is whether the availability (and use) of these mechanisms is the causal mechanism for adoption or whether it is itself an intervening variable—a reflection of the kind of community that is likely to oppose visible and costly economic development policies.

Summary and Conclusion

What do we know, in light of the literature reviewed and its limitations, about the politics of local economic development? The literature itself, as is evident from the above review, presents some difficulties. First, there is not a consistent definition—or, indeed, understanding—of what is meant by the term *economic development,* so that different phenomena are frequently being examined under the same rubric. In particular, the literature sometimes includes studies of physical-development and land-use projects as economic development, while a narrower definition would include only those efforts to increase area employment and income. The confusion is understandable since, as we have seen, policymakers themselves disagree over the objectives they wish to accomplish through what they call economic development policy.

Much of the interesting scholarly work related to the politics of local economic development consists of theory and theoretical exposition concerned with questions such as the forces that impel local governments to engage in economic development, why local economic development activity has increased over time, the extent of conflict in local economic development politics, or how dominant is business. Unfortunately, in many cases careful empirical research on these questions is either lacking or unconvincing. There are a large number of single-city case studies that either serve as the source for theoretical propositions or are illustrative of them, but there are few multicity empirical studies from which confident generalizations can be made (although in a few cases, findings are consistent enough across well-done case studies that generalizations are possible).

The quantitative empirical research suffers from a series of problems that have been detailed above. In particular, the inability to convincingly measure the extent of economic development activity by local governments, as well as inadequate operationalization of other variables, makes questionable the findings of much of the research. And despite the interesting theoretical and speculative literature on why local government economic development activity has increased, there is no research that empirically tests these propositions. Such research would require either time-series analysis or cross-sectional models for two different time periods. (Indeed, there are virtually no empirical results that demonstrate that local economic development activity has increased over time.) Finally, the quantitative research has not examined, or examined convincingly, some of the most important political variables that theory suggests might contribute to local economic development outcomes, such as the presence or absence of urban regimes of various types.

Given these caveats, what *do* we know about the politics of local economic development? Local policymakers appear split over whether the outcomes they are pursuing relate to employment creation or fiscal enhancement, although there is some reason to suspect that the latter rationale may be stronger (since less than a quarter of local officials called increasing employment for city residents anywhere where the jobs are—as opposed to within the city boundaries in which tax yields can be increased—a top priority for local economic development activity).

In terms of what accounts for the amount of economic development activity local governments engage in, if economic development activity is measured by a simple count of the tools employed or by whether a particular tool has been used, there is evidence that it is positively related to fiscal stress, the unemployment rate, and the poverty rate and negatively related to income. The number of economic development tools a local government employs also appears to be positively related to the degree of competition from neighboring governments, as measured by the number of tools employed by other governments in the region.

Interviews with local policymakers (both elected officials and bureaucrats) indicate that local economic development activity occurs in an environment of substantial uncertainty and turbulence. Local elected officials appear considerably more optimistic about the efficacy of the economic development actions they undertake than are local economic development professionals. The uncertainty and risk involved in economic development decision making lead professionals to resort to standard decision rules and formalistic or symbolic behavior.

Local politicians have strong incentives to involve themselves in economic development policy making. They perceive their electoral prospects, as well as their ability to finance publicly desired levels of services at reasonable tax rates, to be tied to local economic performance. Furthermore, they believe that economic performance is held hostage to the mobility of capital across local government borders and that competitor local governments are actively seeking to encourage such mobility. Particularly in cities experiencing economic distress, fiscal distress, or both, the public expects local officials to do something; local economic development activity provides a visible means of responding to that expectation. The uncertain ties between action and result also provide opportunity for credit-claiming—that is, taking credit for any desirable event and claiming that it resulted from local economic development activity.

Given the political and economic incentives and the costs and benefits facing local policymakers as well as the environmental uncertainty and information asymmetry, decisions to engage in economic development activity, such as the granting of tax abatements, can be shown to be rational even though such activity may be fiscally costly and have a relatively low probability of working. However, a rational outcome, from the point of view of the decision maker, does not necessarily imply a rational process.

The scope for bringing about economic development (increases in local employment and income) through local policy is obviously limited by forces operating in the national and international economies over which local government has little or no control. While local government can do relatively little to affect, in the short run, demand for the products the area produces (in the longer run, public decisions can be made that affect the kinds of goods and services that are produced locally), it has a greater ability to affect supply-side factors, such as the skill levels and productivity of the labor force, that can make the area a more attractive locale for economic activity.

Much of the literature on the politics of local economic development is concerned with the role of business. There is general agreement that local governments have strong incentives and pressures operating on them to pursue economic development policies favorable to business interests. These incentives and pressures include structural factors (e.g., the mobility of capital across a fragmented system of local government, competition among various governmental units for mobile capital, the fiscal dependence of local governments in most states on property tax revenue), the pervasive ideology of privatism, the dependence on business resources for entrepreneurial urban regimes to accomplish development objectives, and pressure from the growth machine to engage in development activity that yields rents to its members.

However, there is disagreement about the extent to which local governments have effective discretion to pursue economic development policies not directed toward the immediate interests of business, the extent to which business is a dominant player in local economic development politics, and whether business (or business interests) always wins. The answer emerging from the literature is, not surprisingly, that it depends. Although business interests are likely to be strong nearly everywhere, local governments may have some discretion to adopt policies not immediately favorable to business (particularly in communities with high percentages of professionals, renters, and young people and in which there are active neighborhood groups), business is not always a dominant or major player, and it does not always win. It must be said that while the fact of business participation and even dominance is attested to in a variety of case studies, there has been virtually no systematic research on how often business wins or even what is meant by business winning. Bartik's (1991) conclusion that the poor and minorities benefit from economic growth suggests that there are different dimensions of this question and that it may be the case that business winning (presumably over neighborhood interests or simply in terms of providing some windfall to business) is not incompatible with poor and minority residents also winning, if disproportionate increases in income are evidence of that.

Within local government, the mayor's role is likely to be particularly important on big-event issues, whereas routine issues are the province of the bureaucracy. Strong mayoral leadership within a strong-mayor structure provides at least the potential for political entrepreneurship and the development and implementation of effective economic development policy.

The degree of conflict, controversy, and public participation characterizing local economic development politics is the subject of considerable dispute, although the tendency to caricature some of the arguments lends perhaps more vehemence to the debate than appears necessary. While it is usually in the interests of development-oriented regimes to reduce conflict, controversy, and public participation, they are not always successful in doing so. The conditions under which they successfully (and unsuccessfully) avoid conflict and controversy present a fertile field for empirical research.

Indeed, this review suggests a series of next steps for research on the politics of local economic development. First, it is critically necessary to develop improved operational measures of the extent of local economic development activity, collect the appropriate data, and reexamine the question of what factors are associated with varying levels of activity. It also seems desirable to address empirically the extent to which local economic development activity has increased over time and the factors associated with that increase.

It is time to move beyond some of the more narrow questions that have dominated research and have now been largely resolved or played out. I particularly have in mind questions focused on the extent of choice available to local government in economic development decision making and whether the economic development process is characterized by conflict or consensus. The research has amply established that local governments can and do exercise choice, albeit bounded by some constraints, and that local economic development policy making sometimes involves controversy and conflict. It is time to focus attention on the circumstances under which more or less conflict occurs rather than whether it ever occurs; on the participants and their interests in the conflict; and on how the conflict is resolved and in whose favor. Addressing these questions through multicase and multicity empirical research rather than an exclusive focus on single-issue and single-city case studies would be particularly useful.

Likewise, more empirical research is called for on the behavior of actors (including local officials) in the economic development policymaking process. An extension of the work of Schumaker, Bolland, and Feiock (1986) (examining the extent of participation of different actors as well as their coalitional behavior) from one city—Lawrence, Kansas—to a larger number would greatly increase understanding of the actual political behavior of actors in local economic development.

Similarly, empirical research on the attitudes of local decision makers (contrasting locally elected and appointed officials) would be useful. What are the mental worlds of these local officials concerned with economic development? How do they conceive of the local economy, the forces that affect it, their role in affecting it, and the process of economic development? What objectives do they think they are pursuing in economic development policy? What explains variations in these objectives? To what extent do differences in objectives lead to differences in economic development policies or tools employed? What are the political calculations of local officials as they relate to local economic development, and what are their motivations? Much of the existing literature on these topics has been primarily speculative. To extend knowledge in these areas will require qualitative in-depth interviews as well as surveys of a broad range of decision makers in a multicity context to provide attitudinal variables for multivariate analysis.

The role of business has been at the center of much of the literature on the politics of local economic development. However, most of this literature seems directed at business participation in one particular (usually quite large) development project. Less empirical work (as opposed to speculative theorizing) has been done on the role of business in day-to-day and more routine economic development decision making at the local level. An empirical multicity analysis, if possible using quantitative data, on the political role of business, the degree to which it dominates the process, and the extent to which it wins is badly needed to move reasoned and compelling argument forward to generalizable knowledge.

It should be added that while there is substantial discussion in the literature on "winners" and "losers," there is a marked paucity of systematic empirical research on the distributional outcomes of local economic development policy. Such research needs to be pursued along a variety of dimensions—not only in terms of business or downtown interests versus the neighborhoods but also in terms of the differential effect of local economic development activity on the level and distribution of employment and income by class, ethnicity and racial groups, and gender.

Finally, some effort must be directed toward integrating the two traditions of research on the politics of local economic development—the case study approach and the quantitative empirical research. The best case studies are frequently theoretically rich but, given the inherent limitations of single-city case studies, difficult to generalize from. Too much of the quantitative empirical research, on the other hand, is insufficiently informed by the insights and theoretical development of the best case studies. Sophisticated (although sometimes ambiguous) concepts such as urban regimes have arisen from theory-oriented speculation and theoretically driven case studies. But these concepts need to be carefully clarified, and efforts need to be made to develop operational definitions that will enable a more rigorous and systematic testing, through both multicity case studies using a common framework and careful, quantitative, but theory-driven empirical research.

References

Bachelor, L. 1994. Regime maintenance, solution sets, and urban economic development. *Urban Affairs Quarterly* 29:596-616.

Barnekov, T., and D. Rich. 1989. Privatism and the limits of local economic development policy. *Urban Affairs Quarterly* 25:212-38.

Bartik, T. J. 1991. *Who benefits from state and local economic development policies?* Kalamazoo, MI: Upjohn Institute.

Beauregard, R. 1993. Constituting economic development: A theoretical perspective. In *Theories of local economic development,* edited by R. D. Bingham and R. Mier. Newbury Park, CA: Sage.

Blair, J. P., R. H. Fictenbaum, and J. A. Swaney. 1984. The market for jobs: Locational decisions and the competition for economic development. *Urban Affairs Quarterly* 20:64-77.

Bowman, A. O'M. 1987. *The visible hand: Major issues in city economic policy.* Washington, DC: National League of Cities.

———. 1988. Competition for economic development among southeastern cities. *Urban Affairs Quarterly* 23:511-27.

Bowman, A. O'M., and M. Pagano. 1992. An analysis of the public capital mobilization process. *Urban Affairs Quarterly* 27:356-74.

Cable, B., R. Feiock, and J. Kim. 1993. The consequences of institutionalized access for economic development policy making in the U.S. *Economic Development Quarterly* 7:91-7.

Clarke, S. E., and G. L. Gaile. 1989. Moving towards entrepreneurial state and local development strategies: Opportunities and barriers. *Policy Studies Journal* 17:574-98.

———. 1992. The next wave: Postfederal local economic development strategies. *Economic Development Quarterly* 6:187-98.

Clavel, P. 1986. *The progressive city.* New Brunswick, NJ: Rutgers University Press.

Clavel, P., and N. Kleniewski. 1990. Space for progressive local policy. In *Beyond the city limits,* edited by J. Logan and T. Swanstrom. Philadelphia: Temple University Press.

Clingermayer, J. C., and R. C. Feiock. 1990. The adoption of economic development policies by large cities: A test of economic, interest group, and institutional explanations. *Policy Studies Journal* 18:539-52.

Donovan, T. 1993. Community controversy and the adoption of economic development policies. *Social Science Quarterly* 74:386-402.

Eisinger, P. K. 1988. *The rise of the entrepreneurial state.* Madison: University of Wisconsin Press.

Elkin, S. 1987. *City and regime in the American republic.* Chicago: University of Chicago Press.

Fainstein, S., N. Fainstein, R. Hill, D. Judd, and M. P. Smith. 1983. *Restructuring the city: The political economy of urban redevelopment.* New York: Longman.

Feiock, R. C. 1987. Urban economic development: Local government strategies and their effects. *Research in Public Policy Analysis and Management* 4:215-40.

———. 1992. The political economy of local economic development policy adoption. Paper presented at the annual meeting of the Urban Affairs Association, April 29-May 2, Cleveland, OH.

Feiock, R. C., and J. Clingermayer. 1986. Municipal representation, executive power, and economic development policy activity. *Policy Studies Journal* 15:211-29.

———. 1992. Development policy choice: Four explanations for city implementation of economic development policies. *American Review of Public Administration* 22:49-63.

Fleischmann, A., and G. P. Green. 1991. Organizing local agencies to promote economic development. *American Review of Public Administration* 21:1-15.

Fleischmann, A., G. P. Green, and T. M. Kwong. 1992. What's a city to do? Explaining differences in local economic development policies. *Western Political Quarterly* 45:677-99.

Fosler, R. S., and R. Berger. 1982. Public-private partnership: An overview. In *Public-private partnership in American cities,* edited by R. S. Fosler and R. Berger. Lexington, MA: Lexington Books.

Friedland, R. 1983. *Power and crisis in the city.* New York: Schoeker Books.

Furdell, P. 1994. *Poverty and economic development: Views of city hall.* Washington, DC: National League of Cities.

Goetz, E. 1990. Type II policy and mandated benefits in economic development. *Urban Affairs Quarterly* 26:170-90.

Green, G. P., and A. Fleischmann. 1991. Promoting economic development: A comparison of central cities, suburbs and non-metropolitan communities. *Urban Affairs Quarterly* 27:145-54.

Hunter, F. 1953. *Community power structure.* Chapel Hill: University of North Carolina Press.

Jones, B. D., and L. W. Bachelor. 1984. Local policy discretion and the corporate surplus. In *Urban economic development,* edited by R. D. Bingham and J. P. Blair. Beverly Hills, CA: Sage.

———. 1986. *The sustaining hand.* Lawrence: University Press of Kansas.

———. 1993. *The sustaining hand.* 2d ed. Lawrence: University Press of Kansas.

Judd, D., and M. Parkinson. 1990. *Leadership and urban regeneration.* Newbury Park, CA: Sage.

Kantor, P. 1987. The dependent city: The changing political economy of urban economic development in the United States. *Urban Affairs Quarterly* 22:493-520.

Kantor, P., and S. David. 1988. *The dependent city.* Boston: Scott, Foresman/Little, Brown.

Kantor, P., and H. Savitch. 1993. Can politicians bargain with business? *Urban Affairs Quarterly* 29:230-55.

Kindleberger, C., and B. Herrick. 1977. *Economic development.* International student edition. London: McGraw-Hill Kogakusha.

Levy, J. M. 1990. *Economic development programs for cities, counties, and towns.* New York: Praeger.

Levy, J. M., and M. O. Stephenson. 1991. Selling the community: How economic developers view their role. Paper presented at the annual meeting of the American Political Science Association, September, Washington, DC.

Lindblom, C. 1977. *Politics and markets.* New York: Basic Books.

Logan, V. R., and H. L. Molotch. 1987. *Urban fortunes: The political economy of place.* Berkeley: University of California Press.

Matulef, M. L. 1987. Strategies for economic revitalization. Paper presented at the annual meeting of the American Society for Public Administration, March 28-April 1, Boston.

Mills, C. W. 1956. *The power elite.* Oxford, UK: Oxford University Press.

Miranda, R., D. Rosail, and S. Yeh. 1992. Growth machines, progressive cities and regime restructuring: Explaining economic development strategies. Paper presented at the annual meeting of the American Political Science Association, September, Chicago.

Mollenkopf, J. 1983. *The contested city.* Princeton, NJ: Princeton University Press.

Molotch, H. 1976. The city as a growth machine. *American Journal of Sociology* 2:302-30.

———. 1988. Strategies and constraints of growth elites. In *Business elites and public policy,* edited by S. Cummings. Albany: State University of New York Press.

Noto, N. 1991. Trying to understand the economic development official's dilemma. In *Competition among state and local governments,* edited by D. Kenyon and J. Kincaid. Washington, DC: Urban Institute Press.

Orr, M., and G. Stoker. 1994. Urban regimes and leadership in Detroit. *Urban Affairs Quarterly* 30:48-73.

Pagano, M. A., and A. O'M. Bowman. 1995. *Cityscapes and capital.* Baltimore: Johns Hopkins University Press.

Parkinson, M., and D. Judd. 1988. Urban revitalization in America and the U.K.—The politics of uneven development. In *Regenerating the cities: The U.K. crisis and the U.S. experience,* edited by M. Parkinson, B. Foley, and D. Judd. Manchester, UK: Manchester University Press.

Peterson, P. E. 1981. *City limits.* Chicago: University of Chicago Press.

———. 1987. Analyzing developmental politics: A response to Sanders and Stone. *Urban Affairs Quarterly* 22:540-7.

Reese, L. A. 1991. Municipal fiscal health and tax abatement policy. *Economic Development Quarterly* 5:23-32.

———. 1992. Local economic development in Michigan: A reliance on the supply-side. *Economic Development Quarterly* 6:383-93.

———. 1993a. Categories of local economic development techniques: An empirical analysis. *Policy Studies Journal* 21:492-506.

———. 1993b. Decision rules in local economic development. *Urban Affairs Quarterly* 28:501-13.

———. 1995. Local economic development activities in the U.S. and Canada. Unpublished manuscript.

Reese, L. A., and A. B. Malmer. 1994. The effects of state enabling legislation on local economic development policies. *Urban Affairs Quarterly* 30:114-35.

Riposa, G., and G. Andranovich. 1988. Whose interests are being served in economic development policy? Paper presented at Urban Affairs Association conference, March, St. Louis.

Rubin, H. J. 1986. Local economic development organizations and the activities of small cities in encouraging economic growth. *Policy Studies Journal* 14:363-88.

———. 1988. Shoot anything that flies; claim anything that falls: Conversations with economic development practitioners. *Economic Development Quarterly* 2:236-51.

———. 1990. Working in a turbulent environment: Perspectives of economic development practitioners. *Economic Development Quarterly* 4:113-27.

Rubin, I., and H. J. Rubin. 1987. Economic development incentives: The poor (cities) pay more. *Urban Affairs Quarterly* 23:37-62.

Sanders, H., and C. Stone. 1987. Developmental policies reconsidered. *Urban Affairs Quarterly* 22:521-51.

Sassen, S. 1990. Beyond the city limits: A community. In *Beyond the city limits,* edited by J. Logan and T. Swanstrom. Philadelphia: Temple University Press.

Schneider, M. 1989. *The competitive city.* Pittsburgh: University of Pittsburgh Press.

Schneider, M., and P. Teske. 1993. The progrowth entrepreneur in local government. *Urban Affairs Quarterly* 29:316-27.

Schumaker, P., J. Bolland, and R. Feiock. 1986. Urban economic development and community conflict. In *Research in urban policy,* edited by T. Clarke. Beverly Hills, CA: Sage.

Sharp, E. B. 1990. *Urban politics and administration.* New York: Longman.

———. 1991. Institutional manifestations of accessibility and urban economic development policy. *Western Political Quarterly* 44:129-47.

Sharp, E. B., and M. Bath. 1993. Citizenship and economic development. In *Theories of local economic development,* edited by R. D. Bingham and R. Mier. Newbury Park, CA: Sage.

Sharp, E. B., and D. R. Elkins. 1991. The politics of economic development policy. *Economic Development Quarterly* 5:126-39.

Spindler, C., and J. Forrester. 1993. Economic development policy: Explaining policy preferences among competing models. *Urban Affairs Quarterly* 29:28-53.

Stone, C. N. 1980. Systematic power in community decision making. *American Political Science Review* 74:978-90.

———. 1987. Summing up: Urban regimes, development policy, and political arrangements. In *The politics of urban development,* edited by C. N. Stone and H. T. Sanders. Lawrence: University Press of Kansas.

———. 1989. *Regime politics: Governing Atlanta.* Lawrence: University Press of Kansas.

Swanstrom, T. 1985. *The crisis of growth politics: Cleveland, Kucinich and the challenge of urban populism.* Philadelphia: Temple University Press.

Wolkoff, M. J. 1983. The nature of property tax abatement awards. *Journal of the American Planning Association* 49:77-84.

———. 1991. Restricting interjurisdictional competition: Is there a will and a way to cool the fires? *State Tax Notes,* December 9, unpaginated.

———. 1992. Is economic decision making rational? *Urban Affairs Quarterly* 27:340-55.

Wolman, H. 1988. Local economic development policy: What explains the divergence between policy analysis and political behavior? *Journal of Urban Affairs* 10:19-28.

Wong, K. 1988. Economic constraints and political choice in urban policy making. *American Journal of Political Science* 32:1-18.

Chapter 18

Shoot Anything That Flies; Claim Anything That Falls

Conversations with Economic Development Practitioners

Herbert J. Rubin

❀❀❀

This article contains excerpts from conversations where urban economic development practitioners discussed the problems of their jobs.[1] It is easy to be sympathetic with the difficulties faced by these quasi-public administrators. Much of their work entails the risk and insecurity of being in the private sector, yet receives the detailed scrutiny of those working in the fishbowl of the public sector. Seen in isolation these interviews provide a portrait of individuals valiantly attempting difficult and often thankless tasks. However, interpreted more broadly, these interviews suggest reasons cities are willing to make concessions to business that seem to have little economic development impact.[2]

The model suggested is not one of dominance by power elites or corruption or conspiracy. Rather the model involves the far more subtle process of system bias[3] in which a search for administrative certainty and task closure leads to the public sector's favoring business interests. System bias occurs when public administrators, in an attempt to make their jobs more manageable, make ordinary decisions that systematically favor one interest group over another. As Stone indicates, "Bias is more likely to become apparent as the eventual outcome of a large number of interrelated decisions—many of which have low visibility."[4] System biases emerge as a reasonable accommodation to the daily uncertainty practitioners, both economic developers as well as other public administrators, confront in their jobs.

According to the system bias model, business interests are favored because businesses, especially those seeking to expand or relocate, provide demands that are clearly defined and

AUTHOR'S NOTE: An earlier draft of this article was presented at the annual meeting of the American Society for Public Administration, March 1987. I would like to thank Irene Rubin for providing several critical readings.

Reprinted from *Economic Development Quarterly,* Vol. 2, No. 3, August 1988.

bureaucratically obtainable. Low-key, apparently reasonable demands provide a set of achievable goals for administrators working in an uncertain environment. As Stone claims, "The key is not business over nonbusiness, however, but bureaucratically and therefore predictable over nonbureaucratic and therefore unpredictable."[5]

Recently Buss and Redburn have made similar arguments on how the possession of information in a complex environment can yield power to business and industrial developers.

> Developers are often able to control the economic development decision process for several reasons. First, economic development is extremely complex, requiring detailed knowledge and experience in management, legal affairs, financing, construction and land economics in order to be successful. . . . Second, developers tend to be well integrated into both public and private sector networks that make decisions about development. Developers, or more likely their surrogates, are often members of public planning or zoning boards and commissions, local economic development corporations (EDCS), Private Industry Councils (PICs), or economic development task forces appointed by mayors, county commissioners or Chambers of Commerce.[6]

The interviews with the economic development practitioners are supportive of a model of system bias. First, they see that their work environment is complex and undefined and involves an uncertain technology; they feel that their jobs are not understood and that they must report to people who lack knowledge about their field. Furthermore, their work involves a difficult effort in bridging the gap between the public and private sectors.

In response, they seek out the appearance of some certainty in their task by adopting a philosophy of "shoot anything that flies; claim anything that falls." To do so, practitioners can define their work as responding to a set of business demands (or requests). Working to actualize these requests provides the practitioner with a checklist[7] of activities that replaces the broad, undefinable task of promoting community economic development with a set of concrete sequential steps. Certainty (at least, of what to do) is substituted for the uncertainty that permeates their work life.

If these speculations are correct, a compassion for the difficulties economic development practitioners face on their jobs turns into a concern for the consequences of their actions.

The Cause: Environmental Uncertainty

Public debate on the appropriate role of the public sector in bringing about economic change indicates that the task, at best, is an uncertain one.[8] In the interviews the practitioners presented a multifaceted view of how they perceive such uncertainty and the frustration they feel about not being sure of how to actually accomplish their jobs.

Most important, they saw only a weak relationship between their efforts and resulting changes, between action and consequence. In social psychological jargon, they felt they lacked fate control. This problem was compounded by doubts about whether or not municipalities had the wherewithal to affect major economic changes, the value of the information on which decisions might be made and, more generally, the efficacy of the technology and tools used in economic development.

The uncertainty about the causal mechanisms underlying economic development was paralleled in their perceptions about the role they played in their cities. First, several felt that their jobs were not understood by others in the community. To compound the problem, their work entailed bridging the gap between the public and private sectors, leading to questions about whom they were representing: Were they truly acting in the public interest?

Fate Control

Many of the practitioners were frustrated because they and their fellow citizens had little, if any, control over decisions that affected their local economies. This feeling was most clearly articulated by an economic development official from a Chamber of Commerce, who, when asked what bothered him most about his job, responded:

> I guess the uncontrollable factors that you work with. For example, the downturn in the economy. We couldn't do anything about it. We lost one of our major manufacturers. We lost a major employer . . . and it is just . . . frustrating because it is something you can't deal with. You just ride it through. You do the best you can in the circumstances but it is frustrating. . . . We are trying to get employers, and here is a company that pulls out and goes to Mississippi. . . . The problems you can wrestle with aren't so bad, but being victimized by those things over which you have no control is frustrating. (Chamber 1)

Or, as a city economic development official complained, "There is no answer to economic development. When a place is in the dumps, it's in the dumps. And, to some extent, economic development can help it, but it can't solve it" (Planner 1).

Similar sentiments about the uncontrollability of the tasks were echoed by the planning and economic director in a rapidly growing community. To him, success as well as failure occurs because of exogenous events over which he has little control.

> You can be just great because . . . the Illinois Department of Transportation makes some kind of decision. . . . What does that mean? Twice as many people will ride downtown. . . . Potentially gasoline stations, barber jobs, newspaper stands, the restaurants will have twice as many people stopping in, and I had nothing whatsoever to do with the decision. Now, all at once, the downtown can boom. So, a lot of things we do is luck from other people. (Planner 2)

He continued by describing that, often, change was attributable to locational factors over which no one had any control:

> But there are economic advantages that some communities have that sort of seems just because they are located where they are . . . they happen to be in a growth path or an interstate or an airport was located near where they were. (Planner 2)

In many ways economic development seemed a self-perpetuating process. As the city official in charge of economic development in a declining city lamented, "People want to go where it is a successful area. . . . And, then, the places where they really want it [ecnomic development]—nobody wants to go there because it is not successful" (City ED 2).

Is the Technology Available Effective for Economic Development?

The practitioners discussed the different incentive programs used by localities, yet questioned whether or not such programs really had an effect. Perhaps there was an acceptance that such programs must be attempted even if their consequences were not fully known. To do something is better than to remain inactive.

This ambivalence was seen in the responses of an economic development practitioner from one of the larger cities:

> Some people will argue . . . the kind of tools a municipal government has are relatively marginal. . . . Another school of thought will be, if you really believe business climate is the magic door to salvation, local government has a lot to do with it and how it performs and the image it gives off and the way in which it interacts with business. (Planner 3)

What cities can do is to be flexible enough with codes and regulations so as not to hamper business development. These concerns were reflected by the head of an economic development commission, who saw the tradition-bound practices of local government as a handicap to developmental activities.

> This city is so backwards. . . . We have adopted . . . codes . . . totally out of date. . . . A good example is [company name]. . . . They went into an existing building and wanted to install state-of-the-art-computer work, which meant not putting computer wires into . . . conduits.

> The city insisted that they use conduits. Well, the economic development commission intervened as well as the Chamber of Commerce, and we got the city to change the code for that installation only. (Development Group 3)

Even trying to find out whether or not their tactics are working can prove frustrating. "I think part of the frustration is, at times, companies are terribly reluctant to tell you why you won or why you lost. You know, so, you get a general idea, understanding of the process, that yes we did the right kind of thing. What we could have done differently it's difficult to say." (Development Group 2)

In a humorous contrast, another planner joked about the limits of local action in bringing about economic development and offered the following possibility for city economic development efforts:

> We could get about 20 guys and two or three cars and go out to the interstate . . . and we put up a road block, and every time a Fortune 500 semitrailer would come along we'd stop it and hold it hostage until the Fortune 500 business agreed to build a plant in our town, and that's how we got economic development. (Planner 2)

More often, though, the practitioner had to learn to cope with failure.

> But I also think you have to learn to accept rejection in this business, maybe more so than you do in an awful lot of other professions. . . . But you make so many calls and you're going to make *x* number of sales. And that percentage may be different for different communities. But you usually lose more than you win. (Development Group 2)

Even the apparently simple task of getting and providing information was fraught with uncertainty.

> We have these people who market these studies that will tell you what industry to target, but the problem is they are using data that is at least five years old and . . . all the factors are going to constantly change. (Development Group 2)

In addition, there was concern that having knowledge about problems might create difficulties since, in many cases, there is little that can be done to rectify the problem. One practitioner expressed concern about what a city would do if it had early information to predict which businesses were about to relocate.

> One of the projects . . . I've worked on personally quite a bit is an early warning system for plant closings. . . . That is a good policy thing but it raises too many questions what local government can actually do once they get a case. . . . There is a danger that information will create false hopes and expectations. (Planner 3)

Underlying much of this sense about an uncertain environment was their own experiences that indicated many economic development decisions were based upon epiphenomenal factors.

> You know, you get down to the bottom line as to why did the company go in a location. I would dare to bet more often than not it doesn't boil down to what I said or did. . . . It boils down to where the company executive can moor that 23-foot sailboat. . . . I think more often than not that decisions are made on those types of factors than they are on good hard economic data. (City ED 3)

Even when practitioners felt they were making progress, people in the localities might create difficulties.[9] For example, working with local business people, especially downtown retail merchants, can be frustrating. An economic development director (ED) for a small city complained:

> We have merchants associations that are concerned about the health of the downtown but they are concerned in the limited sense. . . . The goal of every politician is to get elected. The goal of every business person is to make it through the year. . . . There is a furious debate taking place right now because the merchants are quite frightened that even one parking space might be lost. You see, parking is the be all and end all of economic development in this town. (City ED 1)

Her frustrations were echoed by the head of GEDCO, a public-private economic development group in a small manufacturing town:

> [Dealing with] the downtown businesses is the most exasperating experience. Different in attitude. They just like the status quo. We redo the street scape. They won't even pull weeds in front of their store. They want the city to come out and pull the weeds. (Development Group 1)

Such frustrations also occurred when dealing with board members of economic development commissions who might be more concerned with what was good for their businesses than what might be good for community economic development.

> Well, I had five bankers on my board here. I said, "This guy wants revenue bonds— They say, "We will make a good deal for him; the five of us will get together and present a package. . . . [The proposed developer] told them to shove it, and when I picked up the phone and asked him what it meant, he said, "This [bid] is not competitive. . . . You can't have five bankers on a committee. Number one, you don't need that many. Number two, they are not competitive when they work as a group." (Development Group 4)

Similar feelings were expressed by a senior planner and the economic development person for a rapidly growing suburb. He felt that the composition of the economic development commission

> certainly causes a lot of funny situations. You know, I think in their minds, where does their personal self-interest and the interest of the city start or stop? And you know if you used the same set of criteria on the business development commission as you would on the zoning board or the village staff, none of them could work there. The chairman of their business development commission is the head of the big local bank. . . . I'm sure that the first bank they hear about is his. That's probably one reason why his company, the bank, is very happy to let him be chairman of our commission. And, I think especially on these economic development things, it is virtually the same everywhere. And, I think there is a conflict there, but I don't see any change or any solution to it. (Planner 1)

These problems were further complicated whenever the practitioners felt that local government held antibusiness attitudes. A city economic development person complained about problems his board and the city council (he blurred them in the interview) created for him. He was concerned with their antibusiness feeling and felt that the public officials believed

> that a business is not supposed to make money, and that is an attitude I don't understand. . . . The mentality then goes on into commission meetings: Why do you need this? Why should we help you? Shit, that is an embarrassing question to be asking in a business. . . . You know, profit is a dirty word, and why that is, I don't know, but it is. (City ED 3)

Another planner from a suburb indicated the role conflicts between an economic developer, who wanted to provide incentives for business growth, and his city manager, who was not going to compromise to lure in business:

> The firm says, we would like to have a reduction in property taxes, we would like to have a low-interest loan, we'd like for you to help us get employment assistance through the private industry council, and we'd like you to dedicate a street and give a certain zoning and rush through the review process. Right. Seven or eight items. And the village manager would say, well, it's interesting to hear your requests. Now, if you want to locate in our village, you are going to live up to our standards. Maybe we are willing to let you locate here. Not one [laughter]. And that is economic development from the village manager's point of view. Again, I'm overcharacterizing a little bit. (Planner 2)

In addition, there are the inherent problems of dealing with government, whether caused by the time horizons of politicians or the more general delays of bureaucratic decision making.

> I think there are some inherent conflicts there because politicians tend to be more short-run oriented, and a lot of things we do as professionals take longer. I think the average decision to relocate a firm, I think, they estimate is three years from conception to the time they decide to move. . . . So these things take a long time—politicians tend to be more short-run oriented because of reelection pressures—so there is some inherent pressure there. (City ED 2)

Such uncertainty in their work environment has a direct effect upon practitioners' self-images as hardworking professionals by exaggerating the difficulties of succeeding at their work.

Career Isolation

Probably because of the uncertainty of the technology and the lack of a connection between action and developmental outcomes, the practitioners felt that they were isolated in their work, that very few people know or understand what they are doing for the community.

> None of the people I work for have any experience in my kind of business, whereas if you work for a corporation . . . people know what you do in your job . . . but, by and large, you find [that] people who sit on these development boards haven't the foggiest notion of what makes this thing fly; same with government officials. (Development Group 1)

This same theme was echoed by a economic development person working at a Chamber of Commerce, who mentioned as a major problem

> being the only person in town doing what you are doing. The minister has got another minister to talk to if it is a dull congregation or if he is having [a] fund-raising problem . . . but when you are dealing with a particular situation looking for a solution, there really is, sometimes, nobody. Sometimes you are forced to make a decision without, maybe, having all the comfort of full-informed consultation that you would like to have. (Chamber 2)

Even their bosses on the economic development commission might have little idea of what the practitioners do to promote development.

> I am not overly excited about the composition of the commission, who devote one hour a month to a meeting, maybe some additional work on the side. . . . And, I think that a lot of people look at this type of activity, whatever that may be, as being very glamorous, you know, stars and lights, and bells and whistles, and things like that. That you simply put an ad in whatever the appropriate publication is, because they don't know, and as a result of that, the largest company in the U.S. is going to read it, and they are going to bring in 500 jobs next week. That, I think, is their impression. (City ED 2)

Linkage Person

To compound the uncertainty in their work, the economic development person fills an unusual position, with the responsibility to link the public and private sectors. As a link person, reticulator, or meshing person,[10] the economic developer has to work with private profit-seeking firms to accomplish a collective, public goal of bringing about community improvement. As the director of one economic development commission stated:

> Well, my position, I think, is unique in that I don't work for the city. I work for the commission: I was hired by the commission; I'm paid by the commission. The commission is funded by community development block grant funds solely . . . so I don't answer to the mayor. . . . Although the enabling legislation gave us broad powers, in essence, we can't do anything without city council approval, which is fine. (Development Group 3)

Uncertainty was compounded whenever there was a battle between the public and private sectors over control of economic development. The present head of GEDCO, a nongovernmental economic development organization, described the vagaries of his position over the last decade. The organization had just been refunded as the principle economic agency for the city replacing a city-run program. This reversed the process that occurred a decade ago when the city itself took over economic development. The director described how he was persuaded to leave GEDCO and join the city's effort.

> The city manager wanted to form a community development department, and he wanted me to head it, and I wasn't particularly enthused about working in a governmental environment. But he indicated he was going to reduce his contribution to GEDCO, which was half its budget, so I said, "You never explained the benefits of the job to me like that before, so [laughter] I think I'll take it." So the economic development efforts shifted over to the city, and GEDCO went on with a volunteer director. (Development Group 1)

Uncertainty over where an economic development program should be housed reduced the stability of the organizational environment. As an economic development effort fails, or as an economic situation in a community changes, there is a tendency to change organizational form to combat the problem. This response was criticized by a former director of an economic development corporation in an small, nonsuburban city who lamented:

> In some towns it creates nothing but confusion. Sometimes they have a Chamber of Commerce group function with an economic development program; people aren't satisfied. Then it spins off and forms a private corporation and they do it, and then the city gets pissed off, and they set up a community development organization, and the county [also] does it; so in an area with limited funds, you have four organizations wasting money. And they go through cycles like this, and it is utterly stupid. (Development Group 4)

The problem of linkage was compounded for the public sector employee who had to bridge the gap between the private sector and other city officials. Often this person wore multiple hats, simultaneously working in overlapping positions, some of which were seen as curtailing business development, others of which were intended to promote development. As a city planner lamented:

> I'm the planner, I'm the economic developer. I'm the transportation guy, capital programming guy, board and commissions guy, zoning, 15 or 20 different things. . . . You got a growth community, you're an economic developer. (Planner 1)

However, in such circumstances, conflicts between the different roles are inevitable.

> Well, they had a little bit of a problem defining exactly what E. D. is, how much of my role is business advocacy and how much of my role is business policeman. I do business licensing, which is not an advocacy arrangement. It is setting up hoops for business to jump through. (City ED 1)

These sentiments were echoed by the planners who were also responsible for economic development. As one said, describing his relationship with the business community, "You know, on the one hand, you are trying to induce them in and, on the second hand, you tell them to stop, you don't like their plan. On the one hand, you are regulatory and, on the other, you are booster. Yah. There is some conflict" (Planner 1).

Such conflicts might simply reflect the different responsibilities of the public administrators. However, far more serious issues of personal and professional ethics can occur. As one planner, who shortly after the interview left his city job to join the private sector, stated:

> There was a good reason why this stuff stayed in the Chamber. The Chamber doesn't have to pass the test of integrity or implied conflict of interest or what not. When I go out and negotiate a joint venture deal for the village . . . there aren't any rules. There are several million dollars out on the table. There is no public hearing. There is no due process. Pure old-fashioned power. I got, you need. Here's a time schedule. What's your bottom line, what's mine? How can we help each other? And, I don't say any-

> thing is bad on it. That's the wave of the future. That's how we build roads here.
>
> You know you are walking a thin line between private and public, and the economic developer really is asking to be looked at by a very different microscope. When I am director of planning I have no budget to wine and dine people. . . . An economic development guy gets a budget, and he goes on entertainment, and nobody is supposed to ask what he does with it. I have to think about that because one of the nice things about the planning profession [is] . . . there is a lot of good feeling and a lot of nice things about saying, "Look, I really work for a higher ideal. I'm clean and I have a white suit and [I] have a microchip and I slay mountains and I look out for the public interest." (Planner 1)

The Consequences: A Search for Certainty

Coping with uncertainty is far from easy for the economic development people. They are ambivalent about the actions they must take, vacillating between efforts at "credit claiming," that is, taking credit for apparent achievements, and a search for professionalism, that is, taking credit for properly undertaken actions.

At times, the practitioners portrayed their work and the expectations of their bosses as an effort to "shoot anything that flies, claim anything that falls," that is, to look for credit, no matter what their real role was in bringing about the change. However, they felt that it was wiser for them personally to credit successes to businesses and politicians rather than claim it themselves.

Still, most of the respondents are professionals at their tasks and know full well how much they can actually accomplish. They would rather be judged on their efforts and the professionalism of their efforts, process tasks that they can control, rather than on the uncontrollable outcomes. They express cynicism at the credit claiming while at the same time realizing its importance. And, in many ways, they express a somewhat idealistic view of the possibilities of what they might accomplish.[11]

To further complicate the picture, many of the practitioners indicated satisfaction with being part of the "inner group." They identify with business, in part, because business people seem to know better what they want and present a more coherent set of goals than local government, a set of goals that provide them with a focused set of tasks to perform.

Credit Claiming and Proceduralism

In an environment in which success is difficult to achieve, and the relationship between action and outcome is uncertain, there is a temptation to credit claim. This idea was most dramatically expressed by the head of a public-private economic development organization who stated that

> there is a phrase that many people won't admit to, but "shoot anything that flies; claim anything that falls" [laughter]. I think a lot of people busy themselves with these retention surveys, whatever, and going to trade shows, and ostensibly you are doing something even if it is the wrong routine. . . . God, he is out there marketing. A lot of people think that Wall Street is just waiting to hear about [this city] . . . but I think most boards are enthralled with the marketing aspects. You send out some material, and they are going to love you. (Development Group 1)

That is, to show they are doing something, there is a tendency to try to claim credit for actions that supposedly bring about economic development, irrespective of the perceived effectiveness of such techniques.

The economic development practitioners argued that their use of such ineffective techniques was a response to board pressure. A city economic development official complained when describing his advertising campaign, a technique most practitioners thought accomplished little.

> We do some advertising, I guess, more in defense than really an offensive program. What I mean by that is *Plant Site and Area Development*, *Business Facility*, all have their state,

their community reviews. At that time, we put something in. I'm criticized more by absence than I am by their presence. An executive of the economic development commission will pick up *Site Selection Handbook.* "Here is article on . . . Where is this city's ads?" . . . I personally don't believe in a lot of promotion and attraction efforts. (City ED 3)

They lamented that such concerns with action and promotion efforts turn economic development into pure boosterism.

And, unluckily, boosterism gets involved in economic development a lot more than maybe it should. You just want to run out and pat people on the hand and buy lunches and have a slide show. I think, to some extent, it discredits us all because it is a quick shot with little impact, but it is flashy, and it is so appealing to the mayor and these guys, you know. . . . Joe went out and had lunch with this corporative president on Tuesday and he is already doing things. (Planner 1)

More generally, by setting up incentive programs to lure in businesses to communities, practitioners could show they were taking action even if they were aware that such programs were often ineffective.[12]

Usually Somebody Else Gets the Credit

Though admitting much of what occurred was either formalistic or "shooting anything that flies," the practitioners were hesitant to claim personal credit. Perhaps the reluctance is simply a realistic self-assessment that economic development practitioners really are accomplishing little on their own.

Even if we are successful in our negotiations . . . we will not get credit for it. The company will get credit for it as having chosen [city] because that is the way the game is played. You must give credit to the company. . . . But, basically, I don't think, unless you have unraveled some very unusual problem and found a unique solution to it where otherwise you won't have gotten it. It is pretty hard to take credit for anything. (Chamber 1)

In part, this lack of credit is attributable to a realistic self-assessment of how much they contribute to development versus capital investors.

You can't be in a position where you want every development to have your name on it. I may be the facilitator, but I am not the one investing the money. That's my board of trustees. That is a private developer. It is not me. At no point do I lay two stones atop one another with my own hands. So you have to be able to step back and give credit where credit is due. Although you may have sweated many long hours, remember nothing gets done in town on your say so. (City ED 1)

However, the lack of public credit they can take is in part attributable to the need to maintain political support.

We've taken the position that we don't get elected. . . . We normally attempt to get the mayors or the political officials involved in taking credit for participating and taking credit. . . . But we also recognized that its important that people also know that this organization was involved. . . . The press release is written on our stationery with quotes from the mayor. (Development Group 2)

However, such credit to the politicians might be assigned with a degree of cynicism:

[The mayor] likes it when he can announce something. Like when we bought our first building when we had our first demolition contract. He loves to get his picture taken with bulldozers. He likes to make announcements, and, therefore, he likes me to communicate everything through him. Fine. I firmly believe in that. I don't want to get in the newspaper. It's not my fault the writers put my name in all the time. I don't call them. That's what makes a lot of the city council professionally jealous. (Development Group 3)

Still there is a compensating factor. In many ways the economic development person's success is the success of the city.

> If you have a high profile economic development program in place, you are high profile with it. . . . Even if your name isn't attached to it. Well, Oakbrook. I don't know who the E. D. person is in Oakbrook, but Oakbrook is in the paper every day. . . . And their value is assured to their next employer. It's either get your name out there or get your town's name out there. (City ED 1)

Idealism and Professionalism

At least, rhetorically, some of the practitioners expressed a rather strong sense of idealism. They indicated that their work had the potential of achieving a common, collective good. An economic development professional with specialized certification in his field reflected on the personal rewards the job provided.

> I think the rewards, the psychic rewards, the ego rewards, are very, very strong. . . . I think it is as much quiet satisfaction of being part of making things happen even though you know that the contribution may never be told. . . . I think it is being recognized for having that expertise . . . when people call you as perhaps their first source to help them out in their situation. . . . Ultimately, if you are fortunate, you do an economic development project or a new business opening. . . . There are 25 people at work there or 50 people at work there, and I played a pretty good part in helping bring that about. It's a sense of community involvement, community accomplishment, what you are doing for other people. And that is a fairly strong incentive, being able to do good while doing what you really want to do and getting paid for it besides. (Chamber 2)

However, new plants and major accomplishments are somewhat unusual and, as indicated, might be attributable to factors over which the economic development practitioner has little if any control. Under such circumstances the professional shifts his or her self-evaluation to the quality of the procedures that would facilitate economic development. One person showed me an updated data base that tracked every available site in his rapidly growing community. Another described the strategy of sending out customized packages—rather than simply publicity brochures—of descriptive material to companies that inquired about the city. They were attempting to act in a professional fashion.

The contrast between professional action, a controllable activity, and uncontrollable outcomes was described by an experienced Chamber official, when I asked him about what success meant to him.

> Success . . . depends on whether you are results oriented or program oriented. I think when you have done your program well . . . we can't promise you results because we don't make the decisions. The best thing we can do is to develop and implement and maintain a professional economic development program, and the results are beyond us. . . . But success, in my term of reference, is when you have a fully developed, implemented, professionally operated program. I call that success. (Chamber 1)

Working with Business and Government

The professional obligations of the economic development practitioners might be to the overall community, to expand its tax base, or to provide increases in the amount of employment. However, practitioners felt that government and developmental boards encouraged formalistic actions that would have little effect in improving the local economy. In contrast, working with and for businesses might provide a more effective way of accomplishing something. Their effort to satisfy business demands might associate their actions with resulting economic changes, no matter whether such changes would have occurred anyway. Working for business provided a way of achieving credit claiming for the practitioner.

In response to a direct question on whom do they really work for, and after some hemming and hawing, the head of a multicity agency responded:

> I guess our philosophy is, when we are working with a company regarding a location, we are working for that company. That company is directing us. They are the boss. Okay. They are basically what we call our prospects. . . .

> Oftentimes they want something from the city. . . . I think the cities have grown to understand that, while . . . we are, over the long pull, working for the community . . . we are, in fact, working for the company, and we are trying to get the best deal we can for the company. (Development Group 2)

However, what might be good for a new company (and by implication the city) need not be to the advantage of other local firms.

> So, really, what you provide is what the prospect asks [for]. . . . And, this is what a lot of our [local business] people don't understand. They say can you tell us who you are talking to and you say, you know you can't. . . . The prospect quarterbacks the visit. Now, if he says, okay, I'm open to your suggestions, then we have a plan; we take him, show him the city, and show him our industries. Give him every plus feature we can. But here again we are also honest with him. We don't try to misrepresent. There is nothing worse than to locate somebody on the basis of misrepresentation. They're mad and so is everybody else. So we're honest with them. (Chamber 1)

You bargain first with the company and then hope that the city is able to match the needs of the company. To illustrate this point, in a minidrama, an economic development practitioner symbolically addressed the city manager:

> Well, you pay me to serve this client, and I am looking out for the best possible deal for this client. . . . The client is the prospective business. I may bring you a deal that you don't think is in the best interest of the city. Your job is to turn it down. . . . I'm still being paid by the city to work for the client. . . . So, again, our job is to help these people with the project. . . . Number one they are coming into a strange community . . . and we feel that once they show an interest in the community, it is our responsibility not only to tell them the positives but the negative things. (Development Group 1)

It seemed easier for practitioners to adopt a probusiness attitude. Throughout the interviews practitioners with private sector experience wondered about the effectiveness of people training in planning or community development in working with business.

> I don't think there is anything as valuable as learning how business approaches things. I don't think that a lot of people coming out of a planning background [can do the job effectively]; they might have information about municipal business needs, but I think that, number one, a business person has a much different perception than a government employee. A business person has to live through those problems that a business is likely to encounter. . . . Elected officials, they have this urgency they have to get something done before the next election, and this is not a quick fix kind of business we are in; you have to take the long view. (Development Group 1)

The flexibility needed in business in contrast to the rule-bound behavior in government was shown by one of the more senior people in the economic development field. His complaint seemed to indicate the unwillingness of public officials to take risks in packaging economic development deals.

> The private sector is easier to work with. Because I found a lot of public sector people are just protecting a little niche . . . they don't really want to stir anybody up, they can't be outspoken or they are gone, so they literally do what they have to do . . . does everything by the book, has no imagination, doesn't ask. He doesn't ask, "It doesn't say we can't do it, let's try it." That the kind of thinking I like, especially on financing projects. Those are the kinds of things you have to do. If you are wrong, they will tell you. What can they do to you? (Development Group 4)

In this proclient view of economic development work, local government should provide the transfers that business cannot afford or lacks the legal powers to accomplish.

> The proper role of government is financing and incentives and to bridge the gap between a project that won't be economically feasible without it or to remove obstacles that an indus-

> try cannot remove for itself such as land assembly. (City ED 1)

Even more succinctly an official from a larger city claimed:

> Local government basically has the power to hand out money in different ways and that is basic, you know. That is basically what we do. We give out a tax abatement. We give out an IRB. We'll put in a road. We expend funds. (Planner 3)

It is even just for a city to expend the funds, since local government will receive the tax profits from such growth. The director of GEDCO; a not-for-profit development group, made a concise argument for government subsidizing risk rather than business.

> Government will provide the facilities, and I think in terms of an industrial park; they have the financing, and, if GEDCO owned the industrial park, we would have to pay taxes on it, and then we would have to have an engineer and we don't have that kind of money. . . . If sales don't go the way you project, you're in trouble; but the city, they have financial resources. They can hold onto the land easier than a small private development corporation, they have engineers on their staff, they have planners . . . so I think the cities should invest in industrial parks, besides which, they are going to get the tax revenue. (Development Group 1)

Personal Responses to Working with Business

Economic development practitioners gained prestige through working with leading business people. This sense of prestige involves working with the movers and shakers while being part of the action. This theme was reflected by a Chamber's economic development official:

> But talking about the ego gratification and the prestige, I am dealing with business owners and managers of large corporations, I have access to these people on a more regular and different basis than the people who work for them. . . . I do get to socialize with people who, from an income, from a social level, can buy and sell me ten times over before lunch. Obviously, you don't compete with them in the country club level. . . . But, again, there is the prestige of being included among those people for art openings, or other kinds of things. You are thought of as being a community leader in that respect even though . . . I drive a Datsun and they drive a you-know-what. (Chamber 2)

Another person responded similarly:

> My picture is in the paper so often that, a lot of time, there is instant recognition. . . . There is a lot of self-gratification. A lot. Again, because I'm exposed and in a position of leadership with the movers and shakers. I mean I have lunch with the publisher of the paper three times a month. (Development Group 3)

It is not simply working with such people but being part of the actions that these movers and shakers bring about. The head of a development group argued, "By and large, the people we deal with want to be the movers and shakers. They want to he the ones who are successful. So there is an electricity there" (Development Group 1). Or as another economic development person in a smaller city stated when asked about the positive aspects of his job:

> I like being a part of virtually everything major that happens. And I don't mean the Chamber does it all, but there aren't many things happening in [city] that we don't have some role in it or at least have some knowledge of it. . . . I like being part of what goes on. (Chamber 1)

Speculations and Conclusions

By themselves, these interviews portray the frustrations of the economic development practitioners in attempting to bring about local economic improvements. They feel that they lack control over development, are often misunderstood at their jobs, and, against their better judgment, participate in an environment that encourages "shooting anything that flies and claiming anything that falls."

More broadly, the interviews suggest how public administrators or quasi-public administrators attempt to cope with uncertainty in their work. My evidence is based entirely on economic development practitioners. I suspect analogous processes hold for any public official working in areas of both high visibility and high uncertainty with an unpredictable technology.

Such practitioners attempt to find procedures and routines that provide a checklist of activities that they can accomplish. Many perform routine tasks, acting as liaison between business and local government, facilitating obtaining permits, or reconciling conflicting demands from separate parts of government. In general, they act as ombudsman for the business community. Others define their successes by providing information about the community, maintaining lists of sites for development, keeping on top of sources of local funding, and similar tasks.

Still, both ego and, perhaps, maintaining their jobs require visible successes, successes that are obtainable only with the cooperation of businesses, especially those about to relocate or expand. Accordingly, these practitioners require ways of claiming credit for visible outcomes. They can receive public thanks from businesses by establishing "one-stop" clearance points for obtaining permits. Such attempts to have available a set of activities helpful to the business community leads economic development practitioners to act as a pressure group to provide incentives to the business community, incentives and concessions that some authors claim are public-to-private transfers.[13]

Whether we are talking about small accommodations or massive relocations for industrial development, the tilt seems to be toward the business community. It is through the day-in-day-out relationship that emerges as the practitioner tries to survive in a difficult work environment, that the public sector's bias toward business needs is increased. A set of business demands (or requests) becomes a checklist that changes an undefinable task into one with a set of concrete sequential steps. Certainty (at least of what to do) replaces the uncertainty that permeates the profession.

What is implied is that economic development practitioners will push for localities to make concessions so that they can show some progress in their work. The bias toward business emerges because it makes the practitioner appear as if he or she is accomplishing something. Such mechanisms are consistent with ideas presented by Stone in both his theory of system bias and his later elaborations of models of "ecological power" in which "the power-wielder is effective without having to engage in direct action to prevail over opposition or to prevent opposition from developing."[14] Stone indicates how business gains an advantage merely because it is simpler and more bureaucratically advantageous for public officials to work with business. As he explains:

> A favorable predisposition might come about in any of three ways. . . . First, and most obviously, a group could achieve a positional advantage through the selection of spokesmen to serve its interest in strategically vital offices. . . . Second, a group might enjoy a positional advantage by virtue of the fact that the occupants of important places in the governmental system share the group's perspective on leading policy questions. Third, a group might possess an advantage in that their leaders have informal ties to and enjoy the respect of major officeholders. Since their views are expressed in a relationship of mutual trust, the group leaders can expect a sympathetic hearing.[15]

The position and network linkages of the economic development practitioner satisfy all three conditions for businesses' obtaining a positional advantage.

Equally important, this "bias" is shown in a low key fashion. It is the sum of many small decisions rather than one or two dramatic capitulations through which the public sector tilts toward the business community. As Stone argues, "Lower visibility provided a protective cover under which advocacy could take place relatively free from the sanctions of unfavored groups.[16] For the economic development practitioners, business people dominate supervising boards, the practitioners see the greater efficiency of business over government and certainly indicate respect for successful business people.

Speculating more generally, it appears that the interviews provide suggestions about some mechanisms of public-private cooperation. It seems that public-private cooperation is likely to occur when there is uncertainty about the task to be performed and there is ambivalence on the part of the public sector in terms of either the effectiveness of the technology to perform the task of the values reflected by the use of the technology.

When such circumstances exist, and business people and government people are asked to work together, the bias toward business values is reinforced. The mere fact that the public sector has asked for cooperation communicates that public officials feel the task is beyond them. Yet the business people consulted are those who have been successful in their respective careers. As such, they bring to the public-private board strong beliefs about what values should be reflected in decisions. They are able to impress these beliefs upon their public sector counterparts, not because of any superiority in talent or resources, but simply because they seem to know, while the public sector counterparts only get involved in such cooperative arrangements because of an uncertainty about what to do. The private sector makes suggestions, suggestions naturally enough in its own self-interest, and in lieu of other viable alternatives, such suggestions are adopted.

Clearly, the interview material presented on economic development practitioners does not prove this model. However, the material is consistent with it. The practitioners find the tasks they confront and the technologies to solve the problems uncertain and that business seems to know what it wants more so than does government. Though realistically cynical about the efficacy of their activities in bringing about local economic development, they undertake probusiness activities anyway. And, to make such activities psychologically easier, either as cause or consequence, they seem to identify with the leaders of the business community.

In their efforts to cope with an uncertain and difficult job, the economic development practitioners, as reasonable people facing administrative uncertainty, make decisions that tilt the system toward the business community.

Notes

1. The interviews were conducted while preparing a survey on economic development practitioners. A total of 20 practitioners working within one day's driving distance of DeKalb, Illinois, were chosen to represent different types of economic developers. Of the 20 people, several were from utility companies and banks, and one was from a consulting firm. This article focuses only on the dozen individuals who work either for a city, a city-based Chamber of Commerce, or urban-based public-private economic development organizations. A total of three are city planners working in economic development, three are city economic development specialists, two work for a chamber or an economic development organization housed in a chamber, and four work for a public-private economic development organization. Their names and locations have been disguised. I followed a loose protocol in formulating the vocabulary of the survey questionnaire, but, more often, I let the respondents discuss the issues they thought would be most useful to my survey. As the work progressed, it became clear that these practitioners were presenting far richer material than needed to design a survey. Much of what they said might constitute a chapter on economic development practice for a book like Studs Terkel's *Working* (New York: Random House, 1972). All the interviews were examined using the logic of analytical induction as described in Barney Glaser and Anselm Strauss, *The Discovery of Grounded Theory* (New York: Aldine, 1967). No themes are included in this article unless they were mentioned by at least three of the practitioners. Though the quotes are chosen selectively for clarity and color, unless indicated in the body of the article, the nonquoted respondents expressed similar ideas to those whose words are presented here.

2. These suggestions are part of the broader question of public-private cooperation in economic development. It is generally assumed that such cooperation is necessary, a point I am willing to concede, as well as good for the common weal, a point I am not willing to concede. Contrast the panegyric to public-private cooperation found in R. Scott Fosler and Renee A. Berger, eds., *Public-Private Partnership in American Cities: Seven Case Studies* (Lexington, MA: D. C. Heath, 1982); Cheryl A. Farr, ed., *Shaping the Local Economy: Current Perspectives on Economic Development* (Washington, DC: International City Management Association, 1984); P. Porter and D. Sweet, eds., *Rebuilding America's Cities: Roads to Recovery* (New Brunswick, NJ: The Center for Urban Policy Research, 1984); or Howard J. Grossman, "Regional Public-Private Partnerships: Forging the Future," *Economic Development Quarterly* 1 (1987): 52-59 with the more cynical interpretations contained in such works as L. Thompson, "The Politics of Economic Development: A Qualitative Case Study," *Urban Resource* 1 (1983): 33-40; Byran D. Jones and Lynn W. Bachelor with Carter Wilson, *The Sustaining Hand: Community Leadership and*

Corporate Power (Lawrence, KS: University of Kansas Press, 1986); Susan Fainstein, Norman Fainstein, Richard Hill, Dennis Judd, and Michael Smith, *Restructuring the City: The Political Economy of Urban Redevelopment* (New York: Longman, 1983); David Fasenfast, "Community Politics and Urban Redevelopment: Poletown, Detroit and General Motors," *Urban Affairs Quarterly* 22 (1986): 101-123; and John Mollenkopf, *The Contested City* (Princeton, NJ: Princeton University Press, 1983). The former praise the accomplishments of public-private partnerships, the latter question, who benefits?

3. Clarence N. Stone, *Economic Growth and Neighborhood Discontent: System Bias in the Urban Renewal Program of Atlanta* (Chapel Hill: University of North Carolina, 1976) and Clarence N. Stone, "Power and Social Complexity," in *Community Power: Directions for Future Research,* ed. Robert J. Waste (Beverly Hills, CA: Sage, 1986), pp. 77-113. I would like to thank Richard Feiock for the suggestion of interpreting my material in light of Stone's theory.

4. Stone, *Economic Growth and Neighborhood Discontent,* p. 189.

5. Stone, "Power and Social Complexity," p. 96.

6. Terry F. Buss and F. Steven Redburn, "The Politics of Revitalization: Public Subsidies and Private Interests," in *The Future of Winter Cities,* ed. Gary Gappert (Beverly Hills, CA: Sage, 1987), p. 292.

7. By a "checklist," I mean a set of definable and achievable tasks. As literature on evaluation points out, there is a tendency for people to want to be evaluated on process measures, which are potentially controllable, rather than less controllable outcome measures. A checklist is simply a list of processes to be accomplished.

8. Though already somewhat dated, Kenneth M. Dolbeare, *Democracy at Risk: The Politics of Economic Renewal* (Chatham, NJ: Chatham House, 1984) provides a good summary of the themes that enter into this debate.

9. Later, I discuss some of their criticisms of local government. I suspect their criticisms of business are primarily targeted at the smaller, local businesses.

10. For a typology of different types of linkage functions in the public sector, see Herbert J. Rubin, "The Meshing Organization as a Catalyst for Municipal Coordination," *Administration and Society* 16 (1984): 215-238.

11. For another discussion of handling uncertainty in the public sector, see Robert McGowan and John Stevens, "Local Governments' Initiatives in a Climate of Uncertainty," *Public Administration Review* 43 (1983): 127-136.

12. For summaries of the "tools" and concessions that practitioners have at their disposal, see Ernest Steinberg, "A Practitioner's Classification of Economic Development Policy Instruments, with Some Inspiration from Political Economy," *Economic Development Quarterly* 1 (1987): 149-161. John P. Blair and Robert Premus, "Major Factors in Industrial Location: A Review," *Economic Development Quarterly* 1 (1987): 72-85, present summaries that indicate concessions have little affect upon business location decisions. Recently Herbert J. Rubin and Irene S. Rubin have indicated that concession making is most likely in cities least able to afford them. See for example, Herbert J. Rubin, "Local Economic Development Organizations and the Activities of Small Cities in Encouraging Economic Growth," *Policy Studies Journal* 14 (1986): 363-388; and Irene S. Rubin and Herbert J. Rubin, "Structural Theories and Urban Fiscal Stress," in *Cities in Stress,* ed. M. Gottdiener (Beverly Hills: Sage, 1986): 177-198. Edward J. Ottensmeyer, Craig R. Humphrey, and Rodney A. Erickson, "Local Industrial Development Groups: Perspectives and Profiles," *Policy Studies Review* 6 (1987): 569-583 indicate that the directors of local economic development groups do not see a direct connection between their actions and economic development progress.

13. For examples, see Todd Swanstrom, *The Crisis of Growth Politics: Cleveland, Kucinich and the Challenge of Urban Populism* (Philadelphia: Temple University Press, 1985). Also, see Jones and Bachelor with Carter Wilson, *The Sustaining Hand*; and Fasenfast, "Community Politics and Urban Development," pp. 102-123.

14. Stone, "Power and Social Complexity," p. 105.

15. Stone, *Economic Growth and Neighborhood Discontent,* pp. 18-19.

16. Ibid., p. 197.

Chapter 19

What Works Best?

Values and the Evaluation of Local Economic Development Policy

Laura A. Reese and David Fasenfest

❋❋❋

The title of this article emanates from a seemingly simple question asked by a student in an urban studies class. During a discussion of various alternative local government structural arrangements and their relative advantages and disadvantages, an exasperated student wanted to "cut to the chase. . . . Just tell us which works best," he said. After an examination of the existing local economic development policy literature, practitioners and academics alike might ask a similar question. What works best? But *best,* of course, lies in the eye of the beholder and is inherently a function of values. Indeed, the notion of evaluation itself is intricately tied to values, however much policy analysts would like to believe otherwise. Regardless of efforts to portray evaluation as a systematic research process, it is, at its root, a method for imputing value. As the dictionary makes clear, the purpose of evaluation is "to find the amount, worth of, or to judge the value of."

That policy evaluation is inherently normative has long been understood in the evaluation literature, but it has not always been explicitly acknowledged in evaluating local economic development policy (Gunnell, 1968; Rein, 1976; Tribe, 1972). There are numerous studies pointing to the success of policies and programs based on aggregate measures, like job creation or growth in tax base within a community; at the same time, other research points to the human consequences and costs of these development programs. If the purpose of evaluation is to determine whether programs are meeting goals, the question of which goals or whose goals becomes critical. At issue is whether economic development efforts have been successful, under what conditions, and for whose benefit. That the answers are related to value needs to be stressed more clearly.

Efforts to evaluate local economic development policy have become "a quagmire of good intentions and bad measures" (Clarke & Gaile, 1992, p. 193). A basic lack of agreement on measurement techniques and policy goals is compounded by the fact that good performance

Reprinted from *Economic Development Quarterly,* Vol. 11, No. 3, August 1997.

on traditional outcome measures, like jobs created, may not readily correspond with community development as distinct from economic growth (Eisenschitz, 1993). This lack of theory regarding local economic development opens the door for "misspecification" of the dependent variable (Warner, 1987). Thus, when evaluations fail to show results, is it because evaluators looked at industrial location rather than economic growth, or increased tax base valuation rather than improvement in business sales? Acceptance of standard measures of growth as success has led to an avoidance of critical assessment—that is, of course, economic development has to be "good." This "uncritical admiration, however, stifles a probing of the constitutive rules that structure our understandings of what economic development means and that delineate what actions and consequences qualify as valid and appropriate" (Beauregard, 1994, p. 267).

This article addresses several issues related to values and the evaluation of local economic development policies. First, it focuses on the issue of goals in relation to the evaluation of local economic development policies: What is one trying to achieve with local economic development policies? What are the implications of the existence of different goal frameworks for policy evaluation? Does the use of differing goal frameworks have implications for what one expects from the expenditure of public dollars on development? Second, it addresses connections between how goals are defined, what analysts choose to evaluate, and the results of the evaluation. In short, it addresses the reality that often what one expects or desires to find becomes a self-fulfilling prophecy, because it frames the design of the evaluation itself. Where such values and desires come from and how they are created are beyond the scope of the current discussion (Fasenfest, Ciancanelli, & Reese, in press). Again, the focus is on the interplay between values and the evaluation of local economic development policies. Finally, the article suggests an alternative approach to evaluating local development policies that rests on an explicit identification of values within the evaluation process.

Identifying Economic Development Goals

What are the goals of economic development? Policy evaluation assumes that goals drive the selection of economic development techniques or mechanisms. In other words, the presumption is that policy makers agree on a goal or set of goals, choose policies accordingly, and then implement those policies. The task of the analyst is to determine the extent to which policies are meeting stated goals. Indeed, recent work by Pagano and Bowman (1995) suggests that local leaders establish a "vision" of the community based on "its history, its place in its hierarchy of cities, and its aspirations to change" (pp. 3-4). This vision, then, serves as a basis for selecting among economic development incentives or, indeed, whether economic development should be pursued at all. However, other research has suggested that the task of goal identification is much more problematic from the start. Beaumont and Hovey (1985) have maintained that the lack of a formal theory of economic development results in a situation whereby "state and local economic development strategies evolve incrementally without [any] underlying economic theory except that more jobs are good and less jobs are bad" (p. 328). Bingham and Blair (1984) have suggested that much urban economic development policy has been "piecemeal," reducing the impact and limiting the attainment of stated goals. And Kirby (1995) has pointed to a general absence of theoretical frameworks in the community development literature guiding local policy choices. Such research seems to suggest that, even given an overriding vision, absent a general theory of economic development, any connection between policy and goals is going to be accidental at best and unlikely at worst.

What works best for local economic development rests first on a specification of the term *works*. What is the locality trying to achieve? The term *economic development* has always been somewhat problematic, because scholars and practitioners have used it to mean different things: economic processes, development activities, and an economic outcome. However, the *economic* part of the concept has traditionally been used in a private, capital-investment, business-growth sense. There has also been an implicit understanding that

> the modifier *economic* is an ideological statement meant to deflect attention from the inherently political nature of economic development (regardless of whether government is actually involved) and to act as a buffer (available when needed) between key investors and elected officials and government bureaucrats who might introduce the scrutiny and accountability of a democratic society. (Beauregard, 1994, p. 269)

The meaning of the term *development* is also open to interpretation: Does it simply imply economic growth, or does it require some larger systematic change—that is, development? Herrick and Kindleberger (1983) employ an analogy to the human organism to describe the difference between growth and development: "Growth involves changes in overall aggregates such as height or weight, while development includes changes in functional capacities—physical coordination, learning capacity, or ability to adapt to changing circumstances" (p. 21). In a similar vein, Cernea (1991) argues that *development* is too often defined in terms of planning and market efficiency, at the expense of the more generalized improvement or development of community residents. Although the former definition is drawn from the international development literature, the purpose here is not to restate ongoing debates. However, it should be noted that concepts of international and Third World development have evolved over the years, from a narrow conception of growth to broader notions including equity and quality-of-life concerns.

In the local development context, the goal of economic *growth* typically has meant how to get the local community back to work and build the local tax base at the lowest price. Such a definition of goals has been played out in communities across the nation. Because job creation and tax base growth usually mean votes, and citizens expect governmental officials to attract business, emphasis has been directed toward incentive packages to attract or retain employers (Rubin & Zorn, 1985). Further, because fiscal retrenchment and funding cutbacks force communities to compete with each other in increasingly intense and destructive local and regional rivalries, the resulting incentive packages become remarkably similar (Markusen, 1987). They stress "smokestack chasing" to replace lost jobs and include a variety of supply-side incentives, such as tax and financial inducements, infrastructure improvement, and land assembly and development (Eisinger, 1988; Reese, 1992; Wassall & Hellman, 1985). In the final analysis, local development policy is deemed successful if it appears to have positive effects on the business climate (Bartik, 1991), improves the economic base in terms of changes in per capita income or employment (Clarke & Gaile, 1992), or, in some cases, if projects are completed at all (Friedan & Sagalyn, 1989).

If economic *development,* in the sense described by Herrick and Kindleberger (1983), defines the valued result of local development policies, the goals are much different, to wit: (a) increase the stability as well as the gross levels of income for the population; (b) increase local control over both market and government operations, particularly in those aspects that affect citizens in poverty; and (c) increase economic and political empowerment of all sectors of a community, including individual citizens (Brown & Warner, 1991, pp. 37-38). Thus, a shift in goals from a narrow conception of growth to a broader notion of development increases the requirements for success in local economic development. Policies that work best are no longer those that merely increase the number of jobs or businesses. Instead, policies that work must foster "structural and institutional changes which promote a more equitable distribution of new jobs and income generated by growth" (p. 29) and enhance a locality's "capacity to act and innovate" (Beauregard, 1994, p. 271).

Is development too ambitious or unrealistic for local economic development policies? Is it too much to expect that such policies should achieve results beyond job creation and tax base growth? If society is spending public dollars on policies, strategies, and incentives to attract or create capital investment, should not the goals enhance development for the community at large? Loftman and Nevin (cited in Loftman, 1995), in evaluating local policies, conclude that "if local authorities in attempting to regenerate the economies of declining urban areas, fail to ensure that benefits are distributed equitably, and indeed compound inequality, questions have to be asked as to why the public sector should become involved in such activity at all" (p. 21). In a similar vein, it has been suggested that economic development policies such as tax abatements face greater public scrutiny, including referenda, because such public expenditures represent "debt issues" and differential cost burdens and service benefits (Nunn, 1994). Issues related to equity are pivotal to the enterprise of local economic development. Indeed, in their presentation of seven "metaphors," or ways of understanding or conceptualizing economic development, Mier and Bingham (1994) included economic development as a "quest for social justice."

Current evaluative research certainly suggests that this is too much to expect. However, even the more narrow measures of goal attainment are problematic. The most frequently used outcome measures revolve around economic health and performance indicators, including employment levels, numbers of or growth in businesses, growth in jobs, change in occupational/industrial categories, income measures, and various indicators of value added by firms (see, for example, Beauregard, 1994; Burchell et al., 1984; Garn & Ledebur, 1980; Ladd & Yinger, 1989). More narrowly, many evaluations have focused almost exclusively on measures of change in local per capita income and job growth (Clarke & Gaile, 1992). On a national scale, economic development has traditionally been examined by focusing on gross national product, gross domestic product, net national product, total income, per capita income, average income, productivity, and employment change (Herrick & Kindleberger, 1983; Ross & Usher, 1986). However, such growth-based measures have certain common flaws: a tendency to reflect only certain types of economic change, to overstate positive outcomes, to devalue effects on social factors in the community, and to ignore attendant negative social externalities, such as crime and pollution.

Measures such as local per capita income, income growth, and employment or job growth can also be misleading. Research has often confused "city economic health" with the "health of city residents." The former, measured by "the number of private jobs per resident, is closely linked to the wages and salaries generated in the city per resident, but is not the same as the economic health of city residents which is measured by their per capita income" (Ladd & Yinger, 1989, p. 29). Definition of goals matters. Even measures of local per capita income can be problematic, because that may be outside the control of local governments, and local efforts may be an artifact of jurisdictional boundaries or commuting patterns. In a similar manner, measures of job growth may be distorted by income inequalities and increased service costs associated with higher levels of inbound workers (Clarke & Gaile, 1992). Thus, at the root, commonly used measures of economic growth fail to make a connection between changes in individual wealth and local fiscal health. In other words, "a city can be healthy in the sense of generating many jobs per resident at the same time that its residents remain impoverished" (Ladd & Yinger, 1989, p. 17).

If the narrow indicators of economic growth are problematic, perhaps a shift in focus toward development might not be so dramatic. Indeed, to the extent that evaluations are intended to identify program improvements, a broader examination of policy outcomes and impacts, opportunity costs, distributional implications, and differential social and economic effects is clearly called for (Loftman, 1995). Defining the goal of local economic development as development rather than simply growth directly affects the type of measures and indicators incorporated in a policy evaluation and avoids some of the pitfalls noted previously. The next section examines how this change in goals could be re-

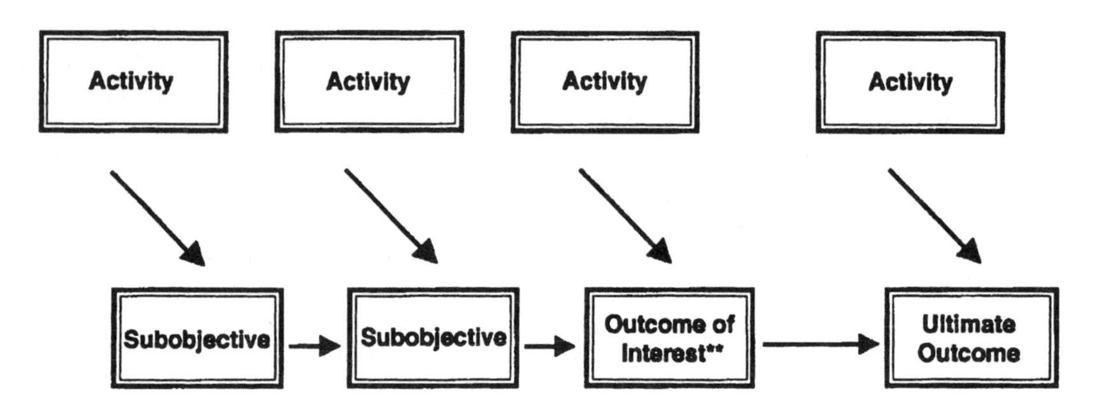

Figure 19.1. The Outcome Line

SOURCE: From Mohr (1995, p. 16). Copyright 1995 by Sage Publications. Reprinted with permission.
**Universal: Must appear in all evaluations.

flected in the design of a policy or program evaluation.

Conceptualizing Goals: Inputs, Outputs, Outcomes, and Impacts

Various conceptions of program results have been suggested in the evaluation literature, including inputs, outputs, outcomes, and impacts (Fischer, 1995; Mohr, 1988). Inputs are widely understood to be the resources and efforts directed at accomplishing a goal, and outputs are seen as the specific activities that directly result from inputs. Outcomes and impacts, however, are conceptually more problematic and implicitly more value laden. Outcomes refer to direct consequences or results that follow from an activity or process; impacts, on the other hand, reflect those implications for society as a whole. This distinction is important: Outcomes are the direct results of a program in the short term, whereas impacts are the longer range, broader societal consequences, more tied to and defined by the values implicit in them.

In economic development, inputs are the resources (budget, staff time, personnel) allocated toward economic development. Outputs are the activities that result: the creation of a Downtown Development Authority, visits to local businesses, the production of a brochure, the development of new water lines. It is the relationship between these inputs and outputs that is considered in assessing the efficiency of development efforts. Development outcomes are the resulting change in conditions; for example, businesses move to the community, foot traffic increases in the downtown, new firms are created. If effectiveness is the major criterion for assessing what works, inputs must be related to outcomes rather than outputs. Finally, the impacts of local development are long-range improvements in attractiveness and, ultimately, the quality of life in the community.

Outcome Line

These basic distinctions are represented in the heuristic of an outcome line (see Figure 19.1). As employed by Mohr (1988), the outcome line identifies subobjectives, the outcome of interest, and the ultimate outcome or benefits of programs or policies. The outcome to be evaluated is the first outcome that is "inherently valued"—that is, its attainment is a valued result in and of itself. That other goals are achieved is fine but not essential; valued outcomes can be assumed to lead to further goals along the line. The stipulations for defining the inherently valued outcome are as follows:

> Outcome Y is said to be inherently valued if either one of two implications for the outcomes

> to its right is valid: (a) if Y is attained, the attainment of outcomes to the right is immaterial—one doesn't particularly care to learn whether they occurred or even what they are; or (b) if Y is attained, one is willing to *assume* that the specified outcomes further to the right will also be attained at a satisfactory level. (p. 15)

Designation of the outcome of interest is inherently subjective or intersubjective, resting on an understanding about what is important and about the nature of causal relationships—but, ultimately, on decisions about goals. In short, identification of the outcome of interest is value driven; the desired result of local economic development policies defines the point along the line that represents inherent value. Further, it defines the relationship between the problem to be solved and various measures of outcome.

The implications of the outcome line for the evaluation of local economic development policy can be illustrated through a simple example: the case of a local property tax abatement. The policy is the enactment of a property tax abatement for a period of years. Reasonable subobjectives of such a policy are as follows: (a) a firm moves into the community, (b) additional jobs are created, (c) local workers are hired for the jobs, (d) unemployment in the community goes down, (e) the purchasing power of local residents increases, (f) demand for local goods increases, (g) sales at other local firms increase, (h) the tax base of the community increases, (i) local service quantity/quality is improved, and (j) the quality of life in the community is enhanced. The critical issue becomes identifying the outcome of value.

If a firm moves into the locality but no additional jobs are created, is the abatement a success? Certainly not. If new jobs are created but local workers do not get them, is the policy a success? Probably not. If local residents get jobs but the unemployment rate does not go down—because insufficient new jobs are created to affect the unemployment rate or because other employers are pushed out of the market, reducing net employment—is the tax abatement a success? Again, probably not. If unemployment goes down but the purchasing power of local residents does not increase—because wages are not high enough to affect purchases or employees are not working enough hours—is the policy a success? Probably not. If purchasing power increases but demand for local goods does not increase—because residents buy goods elsewhere or benefits have not been widely enough distributed for incomes to rise overall—has it been successful? Probably not. If demand increases but sales to other local firms do not increase—again, because goods or services are purchased elsewhere—has the tax abatement been a success? Not for the city. If sales at other firms increase but the tax base does not rise in value and the city gains no additional revenues, is it a success? Not to the extent that revenue enhancement is a goal of local economic development policy. Finally, if the tax base increases but no additional services are provided, is the abatement a success? Probably so, because the provision/production of additional services is a political decision, and the increased tax base could well be used for a variety of purposes. However, if this is the case because the new firm required local services, thus shifting services from other areas or recipients, perhaps improvement in quality of life as the outcome of interest is not unreasonable.

This scenario is admittedly oversimplified; clearly, alternative scenarios could take place with varying chains of results. For example, the tax abatement could have been offered to retain a firm being lured by other cities. Thus, additional employment, increased demand for local goods, and so on would not necessarily be relevant. Or the incentive might be designed to develop local entrepreneurship. It also does not account for regional effects, such as the relocation of workers to surrounding communities. Finally, the value of each subobjective is fairly arbitrary. Indeed, even if no new jobs are created, the tax abatement may lead to other desirable or valued outcomes, such as increased corporate tax revenues, greater use of private services, or enhanced training for those employed by the firm. These caveats aside, the discussion serves to illustrate the value of the outcome line as a heuristic. By fostering identification of objectives and promoting explicit consideration of values, it focuses attention on such issues and stimulates debate.

The outcome of interest in most evaluation studies to date relates to whether the firm arrives in town or whether additional jobs are created—or perhaps, whether overall unemployment goes down. This is evidenced in a recent manual on outcomes monitoring published by the Urban Institute (Hatry, Fall, Singer, & Liner, 1990). Although the manual provides measures at three levels—service quality (an activity measure), intermediate outcome (subobjective), and end outcome—for six types of economic development policies, even the end outcome measures revolve around business starts, sales increases, job growth, and tourist dollars spent. This is even the case for measures of community economic development assistance programs.

This seems to stop short in determining the outcome of interest or the first inherently valued outcome. Evaluations that focus on employment growth, increase in per capita income, job generation, and business starts identify inherent good earlier (or further to the left on the outcome line) than appropriate. If public dollars are spent on economic development, expectations for success should be shifted further down the line: Tax base should increase, people should have salaries sufficient to increase living standards, other local firms should benefit in increased sales, and at least the opportunity for improved services should be the payoff. The criteria typically used in evaluating local economic development policies might well be viewed as means or subobjectives, which produce a further set of inherently valued outcomes (in the language of Mohr, 1988) or ultimately long-term impacts. They reflect a premature conceptualization of outcomes of interest and ignore fundamental questions about the root problems the programs are intended to address. As Mohr (1988) notes, "it is essential to supplement outcome-line construction and analysis with the question, What is the problem?" (p. 18). The techniques and activities selected reflect goal definition and problem identification, and "economic measures of success, including the provision of jobs, new start-ups, reduction in unemployment and an increase in the median income, have become the de facto objective for many [state] economic development plans" (Spindler & Forrester, 1993, p. 2).

This outcome displacement has several implications. First, the outcome—be it more jobs or higher incomes—shifts attention from other problems, such as the maldistribution of jobs and resources, which might create even larger problems of persistent poverty. Second, although outcomes tend to be singular and discrete, the problems they are intended to address are multiple and embedded in systemic patterns. Problems of income inequities, poverty, crime, chronic unemployment, and the like necessitate a broader array of outcome measures. Third, confusing problem with outcome also leads to measurement of the wrong outcomes. Finally, the nature of the outcomes determines whose problems are addressed.

Practical Deliberation

Another methodology for reconceptualizing the evaluation of local economic development policies can be drawn from Fischer's (1995) practical deliberation evaluation. The purpose of a practical deliberation evaluation is to explicitly address issues of value within the evaluation methodology. To again use the example of local property tax abatement, the first step would be *program verification*: Does the abatement create jobs, improve the tax base, have spin-off effects to other firms, or reduce unemployment? Do these results occur at a lower cost than for alternative development strategies, such as bonds/loans, infrastructure development, training, or downtown development authorities?

The second stage establishes *situational validation*: Are increased jobs, reduced unemployment, and increased tax base related to the problem of economic underdevelopment? Are these outcomes appropriate remedies for the social and economic ills facing the community? Are they sufficient to deal with the problems associated with slowing or negative economic development? Are there some groups of residents or areas of the city for which the program has not been successful?

The third stage of the evaluation would consider *societal vindication*: Does the tax abatement produce outcomes that contribute to society as a whole? Does it reduce inequality in opportunity or outcome? Does it allow marginal groups increased participation? Does it increase choices for residents? Does it favor certain segments or interests—private versus public, for example—over others? Does it have negative societal by-products—increased pollution or congestion, for example? Finally, the evaluation would consider *social choice*: Are outcomes compatible with the overall ideology or societal order in terms of equality, individual initiative and responsibility, and appropriate processes for systemic change?

It is likely (and reasonably) beyond the scope of most economic development policy evaluations, particularly those commissioned or conducted by governmental entities, to consider impacts on social ideology. However, considerations beyond mere program validation are pivotal, in that they promote consideration of the real goals of programs and examination of the distribution of program benefits. Careful consideration of outcomes of interest does the same. Both force attention to the radical question (using *radical* in the true sense of the word—i.e., getting to the root of), What does the community hope to achieve through the use of public money for local economic development? What is expected of economic development policies, given the competing uses of public dollars to deal with urban ills? Are programs such as tax abatements, infrastructure investment, and loans the best ways to achieve social, economic, and political goals? Are there other ways of using resources that might achieve such ends more efficiently or effectively?

These are questions raised by the outcome line and deliberative evaluations. They are questions that require making values explicit within evaluations. The next section moves from the conceptualization of the evaluation itself, and the goals inherent in it, to more specific measurement issues. If values are to be explicitly addressed in assessing economic development policies, the indicators used to measure success must also be carefully addressed. Although Beauregard (1994) suggests that because policy is inherently normative, one must "go beyond the data" to decide what policies should be pursued, it seems clear that such normative decisions can be better informed if measures are drawn from further along the outcome line, indicating impacts on broader social and economic issues.

Developing Indicators

Evaluating the broader impacts of local economic development policy is more difficult than assessing growth. Current indicators are far from perfect in what they measure—and are particularly noteworthy for what they miss (Khan, 1991; Myers, 1987). Traditional growth measures tend to overlook the "variegated composition of the economy, the differences in sectoral responses, and the reactions of individuals within a setting of poverty" (Herrick & Kindleberger, 1983, p. 22). As Beauregard (1994) notes, "absent from mainstream economic development are indigenous and informal economic activities, the needs and desires of workers, the democratizing influences of the state, a sense of developmental capacities, collective memory, and an understanding of space as socially negotiated and dialectically uneven" (p. 275). Measurement models that include such complexities, however, may be "less purely scientific" and "considerably more suggestive," making identification of indicators much more challenging (Herrick & Kindleberger, 1983).

Indeed, Molotch (1991) explicitly calls for more social criteria in evaluating economic development policies, highlighting the shortcomings of approaches that focus solely on job creation. He suggests that additional outcomes be examined, including

> the quality of the work experience that will be used in constructing the project and in the labor it would house (e.g., craft work v. tedious assembly, safe v. dangerous work, well-paid v. poorly paid); the impact of the project on the external environment, social as well as physical, including the capacity of the project to contribute to a self-sustaining local region; the

> use value of the product that would result from the project (e.g., production of bread v. production of cigarettes). (p. 49)

The concept of livelihood is useful in developing outcome measures that account for broader changes in social, economic, and political spheres. Livelihood implies a focus on economic units such as households, neighborhoods, or communities and is broader than "politically defined" standard-of-living measures or income-based definitions of well-being, which stress the "physical quality of life" (Howes & Markusen, 1981, p. 438). Such measures go beyond employment and economic growth to include economic empowerment, sustainable improvements in income levels, income stability for large numbers of people, and redirection of resources to the poorest segments of the population (Tendler, 1987).

Political empowerment or voice might also be included as an indicator, because economic development has not occurred evenly across regions and within individual cities because of "a failure to articulate the economic interests of a broader spectrum of society in development discussions" (Brown & Warner, 1991, p. 35). Indeed, Stone (1989) has suggested that, to govern effectively, local officials must engage in a process of "social learning." This implies that, for effective and equitable economic development policies to be implemented, a wide array of alternatives must be brought before the governing regime, and "to the extent that urban regimes safeguard special privileges at the expense of social learning, democracy is weakened" (p. 244). Community development initiatives, in particular, appear to depend on the extent of neighborhood input or control in the policy-making and implementation process (Betancur & Gills, 1994). Thus, mobilizing political power is an integral part of what community development means (Wiewel, Teitz, & Giloth, 1994). Assessments of local economic development outcomes must include not only indicators of employment and economic growth but empowerment and systemic change. The former represent whether populations in need have a role in collective action; the latter indicate whether power and resource patterns are altered in a sustainable manner (McKee, 1989).

In discussing appropriate social indicators of development, Herrick and Kindleberger (1983) suggest a focus on life expectancy, health and nutritional status, educational attainment, housing quality, and social insurance program coverage. Such indicators lend themselves to empirical assessment, are less sensitive to transitory changes or outlying extremes (as is the case with many other aggregate measures), and do not require extensive assumptions about valuation. Other social indicators, such as infant mortality and illiteracy, could easily be added to the list, along with a variety of measures of income disparity or inequality, including social minimums, participation or integration into the mainstream economy, income share analysis, comparisons of pay to work performed, and analysis of interoccupational and intergenerational disparities.

Figure 19.2 provides a direct comparison of indicators traditionally used to measure the effectiveness of local economic development policies and logical indicators for evaluations that are conducted further to the right on the outcome line. The difference between the indicators in the two columns lies not only in that alternative measures will be located further out on the line, encompassing both outcome and impact measures, but also in their inherent link to value judgments. For example, traditional evaluations employ indicators of employment growth, job generation, and jobs per resident; valuing outcomes further to the left on the outcome line would require attention to the quality and distribution of jobs generated as well. Are the jobs in sectors appropriate to the local economy? Is diversification of the economic base enhanced? Are the jobs safe? Do they provide long-term security and incomes sufficient for sustainable households? Indeed, such measures reflect more than just a call for broader notions of economic development; they also have particular implications for understanding the full impact of economic development policies. For example, although a recent evaluation of "model" local economic development programs indicates that job generation has occurred, be-

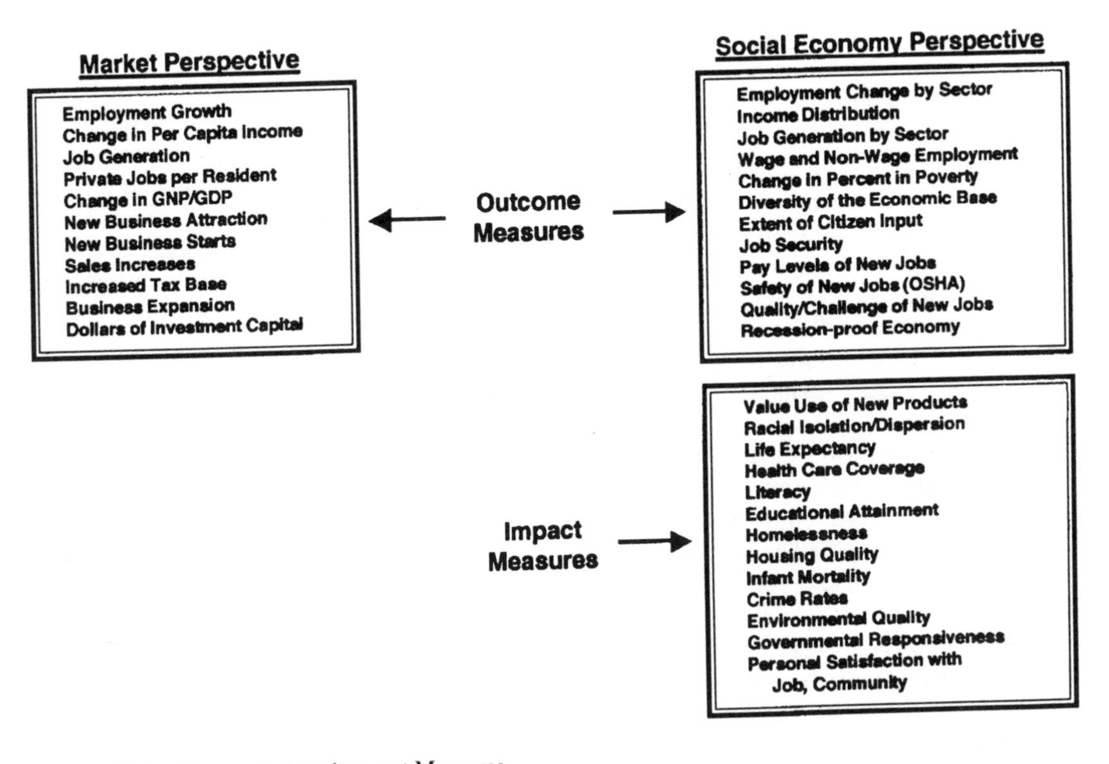

Figure 19.2. Economic Development Measures
NOTE: GNP = gross national product; GDP = gross domestic product; OSHA = Occupational Health and Safety Administration.

cause many of the jobs are part-time or at low wages, the net local effect has been to increase the number of working poor (Schwarz & Volgy, 1992).

Concluding Discussion

The conceptual framework suggested above may be intuitively appealing, but empirical applications are largely absent and critically needed. Future evaluations of local economic development policies should apply methods and goal structures further to the left on the outcome line. Indeed, assessments that employed both traditional outcome measures, such as job creation and tax base growth, and nontraditional indicators would be particularly instructive. Policy makers and analysts would then be able to see how policies deemed successful on traditional measures may fall short on other criteria.

It is also clear that there are some important problems associated with designing and implementing evaluations that incorporate broader values. From a purely technical standpoint, data on the required indicators are often difficult to collect. Although changes in tax base valuation are fairly concrete, measures of qualitative improvements in service, political empowerment, and job quality are problematic. Proxy measures of somewhat amorphous concepts are needed; they require greater skill on the part of evaluators, demand intersubjective validation of operationalizations, and offer greater challenges in collecting necessary data. Just because they are more difficult, however, furnishes no reason to discard such evaluative approaches. Some specific circumstances in which a broader approach might be feasible come to mind. Impracticalities abound in evaluating a community's completed economic development effort, but application to specific and more limited efforts may offer an opportunity to begin. For example, evaluating a business incubator, a new training effort, or a particular capital development strategy, such as a community development corporation, might prove more feasible.

A cooperative community development effort by faith-based institutions and state agencies in Michigan provides a real-life evaluation model. This case highlights several features that increase feasibility: innovative and relatively focused economic development efforts, community-based efforts conducted by or with the participation of groups sympathetic to a broader range of value considerations, and evaluation requirements associated with many state or federal grants—which ensure not only that evaluation takes place but that it be considered and planned before the project begins. To the extent that evaluators enter the process at this stage, plans can be made for collecting necessary data.

Political issues also surface in efforts to include alternative measures in economic development policy evaluation. First, practitioners tend to be judged on outputs by both political actors and the public; judging their efforts on outcomes is likely to be threatening, unwelcome, and a change in orientation from past practice. For elected officials, with a short-term perspective dictated by an election cycle, consideration of policy outcomes and impacts may lie beyond their frame of reference. Although the latter problem is intractable, academic evaluations are clearly not as bound by such political considerations and could more easily move toward broader conceptualizations of valued outcomes.

How to get such evaluations used is another question and beyond the scope of the current article. However, local legislation could incorporate requirements for evaluations, as could state enabling legislation; greater public involvement might well increase the demands for evaluation, and job expectations for economic development professionals could be redefined, to name a few possibilities. Indeed, citizen pressure, current calls for reinventing government, and performance-based resource allocation may coalesce to foster political environments in which local leaders become much more interested in broadly based evaluations, both to meet constituent concerns and, admittedly, to support desired programs. Providing assistance on a consultative basis for communities establishing performance-based systems may again afford academic evaluators the opportunity to affect the types of data collected, evaluation designs used, and operationalization of indicators.

Yet another political issue deals with weighting or prioritizing various policy objectives. All policy decisions and public-sector investment decisions involve (whether explicitly or implicitly) trade-offs between competing projects under conditions of limited resources. Further, business relocations that increase jobs and the tax base and even equalize incomes may also increase pollution and traffic congestion, perhaps thereby decreasing overall quality of life. No effort has been made here to weigh alternative outcomes; this ultimately must be done in reality. Shadow-pricing schemes are commonly used to "value" intangibles and foregone projects in cost/benefit analyses. Although such calculations are cumbersome and expensive, local governments can undertake something akin to what the business world calls a *sensitivity analysis* of prospective projects. This provides a range of circumstances and situations under which expected revenue and outcome streams will be forthcoming—and thereby serves as a guideline for conditions to impose so as to ensure that public investments provide public benefits (Fasenfest, 1986; Fasenfest & Ciancanelli, 1988). Furthermore, an analysis that weighs and monatizes outcomes permits the imposition of social values and community concerns similar to shadow pricing. The inclusion of a broader array of indicators and more normative goals, plus evaluation requirements to identify successive outcomes, enhances the identification and assessment of differential policy effects. Again, how these are weighted will likely and most appropriately be determined politically. However, the value of broader evaluations is that they facilitate and foster examination of the spectrum of policy effects.

Another central concern is the extent of effect local governments have—or can be expected to have—on some of the indicators contained within these broader evaluation approaches. The point could be made, for example, that income distribution, poverty, job security, safety and quality, and even economic-base diversity are beyond the control of local entities. Although this is largely the case, local policies do have some effect; for example, local property tax abatements could be given to firms producing cigarettes or baking bread, firms that are unionized or not, firms that have safe working conditions or those with problems in the past, firms that have comprehensive health care plans or firms that do not, firms that pollute the environment or firms that do not, footloose multinationals or local entrepreneurs. Cities obviously have choices in distributing benefits to enterprises that either meet social goals or do not. By allocating incentives to employers based on wage rates, health care plans, job safety, or quality of employment, cities can have some effect on the distribution of income for residents and, ultimately, the quality of life. Thus, although there are many external factors and dynamics that affect these larger social and economic issues, local economic development policies can be used to either mitigate or exacerbate such trends and should be judged accordingly.

Related to the issue of sphere of influence of local efforts is a more theoretical concern regarding causality: Specifically, can it be logically posited that economic development incentives should cause changes in infant mortality, illiteracy, and crime as implied in Figure 19.2? On one hand, the complexities inherent in the development process, uncertainty regarding causal relationships, and constraints on data availability and evaluation design technology, as noted previously, suggest that the answer is no. Further, a positive response would fly in the face of everything that is known about realistic limits of public policy.

Still, as suggested above, this need not be a reason to discard such evaluative approaches. To the extent that such causal connections are explored through incorporation into economic development policy evaluations and introduced into debates about policy alternatives, the ties between public policies and long-term impact can begin to be explored. If the most pressing urban problem is homelessness, for example, such assessments can begin to address issues of trade-offs: Should we spend public dollars on business incentives or shelters? To avoid exploring these issues is akin to the political response to former Surgeon General Elders's call for research on legalizing drugs—to wit, it is better to remain ignorant altogether than study the issue.

Still, another issue becomes problematic for program evaluation. In most cases, economic development efforts are undertaken by individual local governments; the effect of activities, however, might well extend beyond the boundaries of a city. Given the fragmented structures of urban government in the United States, it is just as likely that the effect, both positive and negative, of local economic development efforts is felt by citizens in other local governments. The tax abatement example illustrates the problem: If an abatement is awarded and a firm locates within the city but the bulk of new jobs go to residents in a neighboring community, is the abatement a success? If purchasing power goes up but residents buy goods and services in a neighboring community—perhaps because the sales tax is lower—is the abatement a success? If the demand for local services, for schools to accommodate families of new workers, increases in neighboring communities without the concomitant growth in tax base, has the abatement been a success? These questions make explicit the interjurisdictional consequences of most local economic development activity. Evaluation models and measures must incorporate attention to the distributional consequences of policy action, among the residents within a community and between communities within an urban area. Valued outcomes cannot be defined solely in terms of a single community.

In the final analysis, explicit attention to values and broader notions of outcomes may say more to the policy-making process than to evaluation. The debate between perspectives and the alternative frameworks provided is critical in informing the decision-making process. The problems identified through a broader perspective can only serve to challenge the economic development enterprise, forcing reconsideration and possible reinvention of strategies and action.

References

Bartik, T. J. (1991). *Who benefits from state and local development policies?* Kalamazoo, MI: Upjohn Institute.

Beaumont, E. F., & Hovey, H. A. (1985). State, local and federal economic development policies: New federal patterns, chaos or what? *Public Administration Review, 45,* 327-332.

Beauregard, R. A. (1994). Constituting economic development. In R. D. Bingham & R. Mier (Eds.), *Theories of local economic development* (pp. 267-283). Thousand Oaks, CA: Sage.

Betancur, J. J., & Gills, D. C. (1994). Race and class in local economic development. In R. D. Bingham & R. Mier (Eds.), *Theories of local economic development* (pp. 191-209). Thousand Oaks, CA: Sage.

Bingham, R. D., & Blair, J. P. (1984). *Urban economic development* (Urban Affairs Annual Reviews 27). Beverly Hills, CA: Sage.

Brown, D. L., & Warner, M. E. (1991). Persistent low-income nonmetropolitan areas in the United States: Some conceptual challenges for development policy. *Policy Studies Journal, 19,* 22-41.

Burchell, R. W., Carr, J. H., Florida, R. L., Nemeth, J., Pawlik, M., & Barreto, F. R. (1984). *The new reality of municipal finance: The rise and fall of the intergovernmental city.* New Brunswick, NJ: Center for Urban Policy Research, Rutgers University.

Cernea, M. (1991). Knowledge from social science for development policies and projects. In M. Cernea (Ed.), *Putting people first: Sociological variables in rural development* (pp. 1-42). Oxford: Oxford University Press.

Clarke, S. E., & Gaile, G. L. (1992). The next wave: Postfederal local economic development strategies. *Economic Development Quarterly, 6,* 187-198.

Eisenschitz, A. (1993). Business involvement in community: Counting the spoons or economic renewal? In D. Fasenfest (Ed.), *Community economic development: Policy formation in the US and UK* (pp. 141-156). London: Macmillan; New York: St. Martin's.

Eisinger, P. K. (1988). *The rise of the entrepreneurial state.* Madison: University of Wisconsin Press.

Fasenfest, D. (1986). Community politics and urban redevelopment: Poletown, Detroit and General Motors *Urban Affairs Quarterly, 22,* 101-123.

Fasenfest, D., & Ciancanelli, P. (1988). Public costs and private benefits: The pitfalls of capital budgeting for reindustrialization. *Journal of Urban Affairs, 10,* 291-307.

Fasenfest, D., Ciancanelli, P., & Reese, L. A. (in press). Value, exchange and the social economy: Framework and paradigm shift in community development policy making. *International Journal of Urban and Regional Research.*

Fischer, F. (1995). *Evaluating public policy.* Chicago: Nelson-Hall.

Friedan, B. J., & Sagalyn, L. (1989). *Downtown, Inc.* Cambridge, MA: MIT Press.

Garn, H. A., & Ledebur, L. C. (1980). The economic performance and prospects of cities. In A. Solomon (Ed.), *The prospective city* (pp. 204-251). Cambridge, MA: MIT Press.

Gunnell, J. G. (1968). Social science and political reality: The problem of explanation. *Social Research, 35,* 187-200.

Hatry, H. P., Fall, M., Singer, T. O., & Liner, E. B. (1990). *A manual for monitoring the outcomes of economic development programs.* Washington, DC: Urban Institute Press.

Herrick, B., & Kindleberger, C. P. (1983). *Economic development.* New York: McGraw-Hill.

Howes, C., & Markusen, A. R. (1981). Poverty: A regional political economy perspective. In A. H. Hawley & S. Mills (Eds.), *Nonmetropolitan America in transition* (pp. 437-463). Chapel Hill: University of North Carolina Press.

Khan, H. (1991). Measurement and determinants of socioeconomic development: A critical conspectus. *Social Indicator Research, 24,* 153-175.

Kirby, A. (1995). Nine fallacies of local economic change. *Urban Affairs Quarterly, 21,* 207-220.

Ladd, H. F., & Yinger, J. (1989). *America's ailing cities' fiscal health and the design of urban policy.* Baltimore: Johns Hopkins University Press.

Loftman, P. (1995, May). *The politics of evaluation research: A case study of Birmingham's prestige projects.* Paper presented at the annual meeting of the Urban Affairs Association, Portland, OR.

Markusen, A. R. (1987). *Regions: The economics and politics of territory.* Totowa, NJ: Rowman and Littlefield.

McKee, K. (1989). Microlevel strategies for supporting livelihoods, employment, and income generation of poor women in the Third World: The challenge of significance. *World Development, 17,* 993-1006.

Mier, R., & Bingham, R. D. (1994). Metaphors of economic development. In R. D. Bingham & R. Mier (Eds.), *Theories of local economic development* (pp. 284-304). Thousand Oaks, CA: Sage.

Mohr, L. B. (1995). *Impact analysis for program evaluation.* (10th ed.). (Figure 2.1, p. 16). Thousand Oaks, CA: Sage.

Molotch, H. (1991). The political economy of growth machines. *Journal of Urban Affairs, 15,* 29-53.

Myers, D. (1987). Community-relevant measurement of quality of life: A focus on local trends. *Urban Affairs Quarterly, 23,* 108-125.

Nunn, S. (1994). Regulating local tax abatement policies: Arguments and alternative policies for urban planners and administrators. *Policy Studies Journal, 22,* 574-588.

Pagano, M. A., & Bowman, A. O'M. (1995). *The politics of urban development.* Baltimore: Johns Hopkins University Press.

Reese, L. A. (1992). Local economic development in Michigan: A reliance on the supply-side. *Economic Development Quarterly, 6,* 383-393.

Rein, M. (1976). *Social science and public policy.* New York: Penguin.

Ross, D. P., & Usher, P. J. (1986). *From the roots up economic development as if community mattered.* Croton-on-Hudson, NY: Bootstrap.

Rubin, B. M., & Zorn, C. K. (1985). Sensible state and local economic development. *Public Administration Review, 45,* 333-340.

Schwarz, J. E., & Volgy, T. J. (1992, September). *The impacts of economic development strategies on wages: Exploring the effect on public policy at the local level.* Paper presented at the annual meeting of the American Political Science Association, Chicago, IL.

Spindler, C. J., & Forrester, J. P. (1993, April). *The state of economic development policy: Is it time for reform?* Paper presented at the annual meeting of the Urban Affairs Association, Indianapolis, IN.

Stone, C. N. (1989). *Regime politics governing Atlanta 1946-1988.* Lawrence: University Press of Kansas.

Tendler, J. (1987). *What ever happened to poverty alleviation?* New York: Ford Foundation.

Tribe, L. J. (1972). Policy science: Analysis or ideology? *Philosophy and Public Affairs, 2,* 616-110.

Warner, P. D. (1987). Business climate, taxes, and economic development. *Economic Development Quarterly, 1,* 383-390.

Wassall, G., & Hellman, D. (1985). Financial incentives to industry and urban economic development. *Policy Studies Review, 4,* 626-640.

Wiewel, W., Teitz, M., & Giloth, R. (1994). The economic development of neighborhoods and localities. In R. D. Bingham & R. Mier (Eds.), *Theories of local economic development* (pp. 80-99). Thousand Oaks, CA: Sage.

Chapter 20

The Economic Impact of Development

Honda in Ohio

Mary K. Marvel and William J. Shkurti

❀❀❀

Honda's investment in Ohio and Ohio's investment in Honda began on October 11, 1977, when Honda officials joined Governor James A. Rhodes to announce construction of a $35 million motorcycle assembly plant in a soybean field near Marysville, Ohio. In conjunction with the site announcement, state officials announced a series of state-financed economic incentives as part of the package to bring Honda to Ohio. At the time it was the largest incentive package ever provided by the state of Ohio and the first ever by any state to a Japanese auto firm. In the years that followed, Ohio and other states would offer larger and more complex incentive packages to a growing list of foreign and domestic companies.

Although states maintain an active, continually expanding portfolio of tax incentives, technological assistance, subsidized loans, tax exempt bonds, and direct loans to pursue economic development, the effectiveness, efficiency, and equity of these efforts have not been clearly established. On one hand, the Council of State governments (CSG) reported that its 1988 survey of state economic development agencies revealed that 68% of those state agencies surveyed felt that incentives had a significant effect on new business investment and creation.[1] On the other hand, CSG reported that "nine out of 10 studies have documented that business incentives have minimal effects or no effect whatsoever on business location decisions. But we also know that most, if not all, of these studies were based on information and data that were neither reliable nor comprehensive from the perspective of state policy-makers."[2]

After a review of development activity in a number of states, Scott Fosler, Vice President

AUTHORS' NOTE: We would like to thank Bill Burns, Don Mucha, Eloise Dowell, and Howard Wise for making data available. Lee Walker provided valuable comments on an earlier draft. Special thanks go to John Downs for excellent research assistance.

Reprinted from *Economic Development Quarterly,* Vol. 7, No. 1, February 1993.

and Director of Government Studies for the Committee on Economic Development identified what he called a "knowledge gap" in evaluating the effectiveness of these programs. This knowledge gap reflects in part the relative newness of these programs, their complexity, the reluctance of policymakers to disturb the status quo and the paucity of data available to make judgments. The CSG report on economic development indicated that most states identified lack of good information as the biggest obstacle to their effort to encourage business to locate or expand in their states.[3]

The purpose of this article is to help address that knowledge gap. The costs and benefits of the state of Ohio's investment in Honda are calculated to provide an assessment of that seminal initiative. To date, no such assessment of Honda in Ohio has been undertaken. In addition to providing the first accounting of the costs and benefits of that investment, the analysis considers the portability of the findings to other economic development programs.

The plan of the article is as follows. Section 1 briefly reviews a segment of the cost-benefit portion of the economic development literature dealing with recent auto investment in the states. Section 2 provides the context for the incentive package provided by the state of Ohio for Honda. Section 3 presents the research design and the data employed to execute the design. Direct costs and benefits are reported in Section 4. Indirect costs and benefits of the project and data on Honda's supplier network are considered in Section 5. Section 6 concludes briefly with suggestions for further research.

1. Cost-Benefit Studies

A voluminous body of research is focused on whether policy incentives do, in fact, alter business decisions. That is, do firms merely extract concessions for actions that would have been taken in any case, or do the concessions alter those actions in a significant manner? A smaller number of studies consider the specific, as opposed to anticipated, benefits associated with industrial expansion and compare them with the costs incurred to achieve them. This latter issue constitutes the core question motivating this research. These benefits need to be measured whether or not that expansion effectively was facilitated by government policy. Such benefits will overstate the benefits of government development policy because they might have been obtained without government intervention. Moreover, the costs directly associated with successful development efforts will understate expected costs due to the selection bias associated with successful projects. The costs of unsuccessful overtures would need to be considered in an overall evaluation of development programs. Nevertheless, measurement of benefits achieved is a necessary precondition for the evaluation of any development program. A feasible first step in an evaluation is an assessment of the benefits of development that appear to have been fostered by state and local efforts. Successful development efforts require that, at a minimum, the benefits actually obtained exceed the direct and indirect costs of concessions. Only programs that meet this test need to be evaluated further to determine that state and local efforts were necessary for the supported projects to be undertaken and their benefits obtained.

Milward and Newman provide an excellent overview of the incentive packages put together by the states in their attempts to create a "Japanese Auto Alley" in the Midwest.[4] The costs contained in six of those packages are examined. They estimate the financial incentive cost per plant employee to be as follows: Nissan (Smyrna, Tennessee), $11,000; Mazda (Flat Rock, Michigan), $13,857; Saturn (Spring Hill, Tennessee), $26,667 (based on only Phase 1), $13,333 (based on Phase 2); Diamond-Star Motors (Bloomington-Normal, Illinois), $33,320 (based on 2,500 workers), $28,724 (based on 2,900 workers); Toyota (Georgetown, Kentucky), $49,900; and Fuji-Isuzu (Lafayette, Indiana), $50,588 (based on Phase 1), $34,688 (based on Phase 2).

Milward and Newman deliberately excluded the Honda plant because they felt the auto assembly plant in Marysville was an expansion rather than a commitment to a new location.

They believed an analysis of the costs associated with that endeavor would prove to be more misleading than informative. However, it can also be argued that the totality of the Honda investment in Ohio should be included in any assessment of the impact of that investment. It is instructive to note that the above transactions are arrayed in chronological order dating from 1980 until 1986. Over that time the price tag for the auto plants has become more expensive. The benefit side of the ledger is more difficult to isolate and quantify. The unfolding of benefits over time, the presence of a host of threats to internal validity and, as previously stated, the paucity of data available to make such judgments contribute to the challenge. A brief review of four such efforts to document the benefits of siting of an auto plant and relate them to the costs of attracting the plant follows.

Fox assessed the economic effects of the Nissan decision to locate a facility in Rutherford County, Tennessee. He concluded that "Rutherford County's growth rates have been good relative to other counties, but have not been spectacular and do not appear to have accelerated in the intervening years since Nissan's location." He further found that the county did not experience fiscal stress as a result of the plant location and "no significant negative noneconomic influences appear to have resulted because of the facility." Because of the expansion of the facility, Fox concludes that "the chances are even higher that inducement activities were a sound investment."[5]

Bachelor in her assessment of the Mazda impact in Michigan found that the unit of analysis employed was important in determining if benefits outweighed the costs. The benefits to the city of Flat Rock, the location of the facility, were less clear than those accruing to the state of Michigan. "The absence of spin-off development, the continuing presence of the tax abatement, and the small number of residents employed at the plant yield a somewhat negative response, but at least two of these concerns [tax abatement and spin-off] may be alleviated over time." The imprecision of the extent of indirect employment effects and the absence of an accounting of the opportunity costs associated with the incentive package render a firm conclusion about the benefits garnered by the state unclear. Bachelor reports that the state's economic model projected a positive benefit-to-cost ratio.[6]

In their 1986 assessment of the Diamond-Star transaction, Lind and Elder also found that the unit of analysis is critical in determining if benefits outweighed costs. However, in this case it was the local unit of government (McLean County) that reaped the benefits. "In the long term, those who benefit from this particular combination of strategies are likely to be the residents of the McLean County area who will see new jobs and new economic opportunity largely underwritten by the State of Illinois."[7] They also point out that regional effects were likely as the plant was located near the state border and was supplied by a number of out-of-state concerns.

Milward evaluated the effects of the Toyota plant on Kentucky's economic health. Using the Regicnal Input-Output Modeling System [RIMS II], he estimated that at full capacity production, the state's investment of $325.4 million and Toyota's $800 million would result in an "additional $3,792.16 million in output and $768.48 million in earnings annually and 35,520 jobs created."[8] In addition an internal rate of return analysis was performed indicating a 25% rate of return to the state from the Toyota initiative.

The above review of the literature reveals that unambiguous conclusions regarding the economic effects of state and local attempts to attract Japanese investment are not yet in hand. Although benefit determination has been positive in most instances, some issues remain unresolved. Benefits and costs have been identified but more detailed information on their distribution is needed. Assignment of benefits to a specific project has been attempted but work remains on disentangling those benefits from those that would have occurred in the absence of the project. With those issues in mind the next section addresses the economic effects of Honda in Ohio.

2. Honda and Ohio

Honda's initial foray into Ohio was announced in 1977 with the unveiling of plans to construct a motorcycle assembly plant in Marysville, Ohio. The announcement indicated that the motorcycle facility would be designed to produce 60,000 motorcycles annually and employ 400 workers and that it may later be expanded to produce automobiles. Honda officials stated that the reason the Ohio site was chosen over the other states vying for the plant was market location, outstanding transportation system, its supply of good labor, supply of parts, and its industrial environment. The state of Ohio also provided Honda officials with a package of economic incentives including $3.6 million in state tax revenue for site improvements and $11 million in improvements in highways near the plant, a limited property tax abatement, and an option to purchase 260 acres of state-owned property next to the site.

The motorcycle plant began production on September 10, 1979. The following year, Honda announced plans to purchase the adjoining land for a $250 million automobile plant projected to employ up to 2,000 persons. The incentive package for this plant included $1.7 million in one-time site improvements, $35 million in improvements to nearby highways, $8 million in improvements to local sewer and water facilities, and a partial property tax abatement.

Production of Honda Accords began at the Marysville plant in November 1982. By 1989 the Honda Accord became the best selling car model in America. That year Honda sold nearly 800,000 units in the United States, with about 400,000 of those produced in Marysville.

Honda's investment in Ohio and Ohio's investment in Honda increased as well. In 1984, Honda officials announced plans to spend $240 million on an expansion of the Marysville auto plant, to build a plastics plant in Marysville, and to build an engine plant in Anna, Ohio. The next year, the Honda Research and Development branch was established in Marysville. That was followed in 1987 with the expansion of the Anna engine plant, expansion of Honda Research and Development, and the announcement of a second automobile plant in East Liberty, Ohio.

By 1990 Honda's investment in Ohio exceeded $2 billion. The company employed 8,000 associates (Honda's term for its employees) in its own plants. Hundreds more were employed by foreign and domestic auto suppliers located in Ohio. Economic incentives from the state of Ohio included $25 million in direct economic assistance and $65 million in highway work, as well as additional assistance from local governments in the form of property tax abatements and expanded sewer and water facilities.

3. Research Design

The economic impact portion of this study is divided into two distinct components. The first component measures direct costs and direct benefits. Direct costs are defined as state general fund tax dollars spent on Honda's behalf for site preparation and worker training, and local tax dollars spent for sewer and water expansion. These data were gathered from unpublished records at the Ohio Department of Development, Honda of America, contemporary press reports, and local officials.

Direct benefits are defined as property taxes paid by the auto plant and state income taxes paid by Honda associates. The property tax figures were provided by the Union County Auditor. State income tax estimates were obtained from the Ohio Department of Taxation based on estimates of the number and average wage of Honda associates. It should be noted that this definition of direct benefits does not include local sales taxes, federal income taxes, or corporate income taxes, which were not available.

Indirect costs included state and federal expenditures for highways near the Honda plant but not on the premises, federal sewer and water grants used to match local dollars, and the increased state tax dollars spent in support of do-

TABLE 20.1 Costs and Benefits of Honda in Ohio

Direct
To the state
Cost of incentive packages
Benefits of additional income taxes paid by Honda associates
To local governments
Cost of local share of sewer and water grant
Benefits of additional property taxes
Cost and benefits of property tax abatements
Indirect
To the state
Cost of the loss of jobs at existing auto plants
Cost of additional state assistance to domestic automakers
Cost of incentive packages for suppliers
Benefits of additional jobs created by suppliers
To local governments
Cost of additional public services
Benefits of reduced welfare costs
Benefits of multiplier effect on regional growth
Costs or benefits not directly addressed in this study
To the state
Benefits of other taxes (e.g., corporate income)
Benefits of multiplier effect statewide
To local governments
Impact on quality of public education
Impact on crime
Impact on transportation system
Impact on environment
Impact on charitable contributions

mestic auto plants and Honda suppliers. Measuring the indirect benefits was a more difficult challenge. Fortunately, the first Honda plant was located in the middle of an empty field 30 miles from the largest metropolitan area. This provides a relatively pristine atmosphere to measure the indirect economic effects of the plants' operations. The method chosen was to compare Union County, home of the Marysville plant, and four surrounding counties with five matched comparison counties for 5 years before and after the auto plant opened on a variety of dimensions.

The five experimental counties chosen were Union, Champaign, Hardin, Logan, and Madison counties, with the latter four bordering Union County and sharing in the indirect economic effects associated with the plant. Three neighboring counties, Delaware, Marion, and Clark, which may have benefited from Honda, were omitted because of proximity to major metropolitan areas and major manufacturing facilities. The five comparison counties were selected from among the state's remaining 80 counties based on characteristics similar to the five experimental counties including population growth rates prior to 1982, per capita income, unemployment, percentage of work force in manufacturing, urban-rural mix, and distance from a major metropolitan area. This approach constitutes a quasi-experimental design that allows one to pose the counterfactual and offer tentative conclusions regarding what would have happened in the absence of the Honda initiative in Union and surrounding counties.

A variety of measures were chosen for comparison between the experimental and comparison counties. Economic growth variables include population, per capita income, employment, and retail sales. Public service variables highlighted include local and county government spending, including spending on welfare, and education expenditures. Table 20.1 provides an enumeration of the costs and benefits employed in this analysis.

As we will discuss in a later section, the time period chosen for analysis has a great deal of influence on the outcome of any cost benefit study. For purposes of this analysis, the period 1977-1990 was used for the measurement of direct costs and benefits. The first year that incentives were announced was 1977, and the most recent year for which reliable numbers are available is 1990. The 13-year period encompasses 8 years of full operation of the main assembly plant.

For purposes of comparing experimental and comparison data, the period 1976-1981 as the before period and 1982-1987 as the after period were used. The auto plant began production in 1982 allowing for the collection of 5 years of data both before and after the onset of production.

4. Direct Costs and Benefits

Between 1977 and 1990, state and local governments invested nearly $27 million of their own tax dollars in incentive packages to Honda. Honda invested more than $2 billion of its own funds and created more than 8,000 jobs. The purpose of this section is to examine the direct costs to the state and local taxpayers of the Honda incentives and to estimate the direct tax benefits occurring to these same state and local governments from Honda's investment.

Direct Costs and Benefits to the State

The direct costs of the incentive packages to the state of Ohio are Industrial Inducement (412) grants for infrastructure and site improvements and Ohio Industrial Training Program (OITP) grants for job training and retraining. Table 20.2 summarizes the direct governmental incentives provided to Honda. Appropriations for 412 and OITP grants are made out of the state's General Revenue Fund to the Ohio Department of Development, which in turn distributes them to companies who meet the program requirements.

When the Honda incentive packages were announced, state officials consistently pointed to the importance of the benefits derived from the new jobs Honda would bring to Ohio and Marysville and the surrounding communities. In a subsequent section the multiplier effects of those jobs and other indirect benefits of Honda's location and expansion in west-central Ohio will be considered. For purposes of this section, only the number of jobs and the direct tax revenue to state government attributable to those jobs will be examined.

Tax records of individuals and individual corporations are confidential under Ohio law. However, using Honda's employment figures and published information on the average pay of Honda associates, an estimate of the personal income tax paid to the state of Ohio by these associates was derived with the assistance of the Ohio Department of Taxation. These figures are shown in Table 20.3, which also traces the number of Honda associates in Ohio, which increased from only a handful in 1980 to more than 8,000 in 1990.

This analysis probably underestimates the tax revenue generated by Honda and its employees. For example, Ohio also has a corporate income tax. However, these records are confidential and their variability so great that no estimate of corporate income tax revenues can be made at this time. Overall, Ohioans pay about 20 cents in corporate income taxes for every dollar paid in personal income taxes. As a foreign-based company, however, much of Honda's liability for state purposes would depend on how the company chose to allocate taxable resources among various domestic and foreign jurisdictions. In addition, the state of Ohio permits a credit for payment of property taxes on inventory and equipment to be taken against income tax liability. Therefore, a reliable estimate of corporate tax liability is not possible.

State tax records show that during the period of this study, in aggregate, Ohio taxpayers paid roughly one dollar in sales tax revenue for every

TABLE 20.2 Direct Governmental Incentives to Honda, 1977-1990

Project and Nature of Support	*Dollar Amount*	*Funding Source*
Motorcycle plant (1977)[a] inducement grant[b]	3,610,598	State
Marysville auto plant (1980)[c]		
Inducement grant	1,688,909	State
Sewer and water expansion[d]	2,000,000	Local
Subtotal	3,688,909	
Auto and motorcycle plant expansion (1984)		
Job training	205,560	State
Plastics plant (1984)		
Inducement grant	1,500,000	State
Anna engine plant (1984)		
Inducement grant	1,000,000	State
Job training	36,274	State
Subtotal	1,036,274	
Anna engine expansion (1987)		
Inducement grant	3,160,000	State
Job training	500,000	State
Subtotal	3,660,000	
East Liberty auto plant (1988)[e]		
Inducement grant	11,000,000	State
Job training	2,160,000	State
Subtotal	13,160,000	
Totals[f]		
Inducement grants	21,959,507	State
Job training	2,901,834	State
Water and sewer work	2,000,000	Local
Total	26,861,341	

SOURCE: Unpublished files, Business Development Division, Ohio Department of Development.

a. Dates reflect date of announcement, not date of expenditure.

b. Inducement grants are 412 (Industrial Inducement Program) grants used primarily for site improvements. Does not include an estimated $11 million in highway work. 1977 package also included state assistance in applying for foreign trade subzone and option to buy 260 acres of property held by the Transportation Research Center (TRC).

c. Does not include $54 million of work on widening State Route 33 that was originally promised in the 1981 inducement package, but not begun until 1988. Original package included purchase of an additional 400 acres of state-owned land at "fair market value."

d. Excludes $6 million of federal matching money.

e. The 1988 package also included sale of TRC to Honda for $31 million.

f. The 1987 motorcycle plant renovation ($10 million) and 1985 announcement of Honda Engineering and Honda R&D ($27 million) are not included in these figures because no incentives were requested.

dollar they paid in income tax. If this relationship held for Honda employees, the sales tax revenue generated for the state would be equal to the amount of income tax revenue indicated on Table 20.3. However, state tax records do not permit the production of a reliable estimate of the relationship between income tax revenue paid by individual tax payers at a certain income level and sales tax revenue generated by those same individuals. Consequently, we have not included an estimate of sales tax revenue generated by Honda employees for this study. This means the positive benefits to the state are underestimated, but to an unknown amount.

Using the data presented in Tables 20.2 and 20.3, the cost per job and the return per job to the state of Ohio can be estimated. These esti-

TABLE 20.3 Estimated Ohio Personal Income Tax Payment by Honda Employees, 1980–1990

Year	*Number of Employees*[a]	*Estimated Average Income Tax per Employee*[b]	*Total (in millions of dollars)*
1990	8,200	882	7.2
1989	7,475	840	6.3
1988	6,150	840	5.2
1987	4,738	764	3.6
1986	3,715	854	3.2
1985	2,752	908	2.5
1984	2,218	886	2.0
1983	1,451	855	1.2
1982	700	618	0.4
1981	419	554	0.2
1980	227	504	0.1

SOURCES: Number of employees, Honda of America. Tax per employee, Ohio Department of Taxation.

a. Mid-point of January of that year and January of the succeeding year.

b. Assumes an average annual wage of $30,000 in 1988; figures for other years based on annual percentage change for all weekly wages of vehicle and equipment workers in Ohio. Tax figures were provided by Ohio Department of Taxation assuming Honda employees are similar to all taxpayers in that bracket. Figures fluctuate from 1983 to 1987 because of significant changes in Ohio income tax rates.

TABLE 20.4 Cost per Job to the State of Ohio of Honda Incentives, 1980–1990

Component	*Year*	*Direct Cost*[a] *(in millions of dollars)*	*Estimated Jobs to Be Created*	*Anticipated Cost per Job*[b] *(in dollars)*	*Actual Jobs Created*[c]	*Actual Cost per Job*[b] *(in dollars)*
Motorcycle plant	1977	3.6	400	9,000	380	9,474
Auto plant	1980	1.7	2,000	850		
Auto expansion	1984	0.2	500	400		
Auto plant and expansion	—	1.9[d]			6,200	307
Plastics plant	1984	1.5	140	10,714	140	10,714
Engine plant	1984	1.0	200	5,000		
Engine expansion	1987	3.7	800	4,625		
Engine plant and expansion	—	4.7[d]			1,700	2,764
Auto plant 2	1988	13.2	1,800	17,778	1,800	7,333
Total		24.9	5,840	4,263	10,220	2,436

SOURCE: Ohio Department of Development and Honda of America.

a. State costs only, does not include local.

b. To the state.

c. As of mid-1990, includes all jobs created for Auto Plant 2, even though some of these jobs were created after mid-1990.

d. Total of two preceding lines.

mates are presented in Table 20.4. Table 20.4 indicates that after 10 years the cost per job to the state General Fund was about $2,500, and based on current income tax rates, the state of Ohio should be expected to recover that cost in income tax revenue alone in less than 3 years.

Direct Costs and Benefits to Local Governments

As part of the 1980 package for the Marysville automobile plant, the city of Marysville agreed to provide $2 million in order to obtain a federal grant of $6 million for upgraded sewer

TABLE 20.5 Property Tax Payments and Abatements: Honda of America in Union County, 1980–1990 (in thousands of dollars)

Year	*Taxes Paid*	*Taxes Abated*[a]	*Ratio*[b]	*Honda Tax Share*[c]
1990	4,877.0	1,708.7[d]	2.8	63.2
1989	4,472.0	1,512.3	2.8	65.2
1988	4,761.4	1,189.6	4.0	67.4
1987	4,242.5	1,149.5	3.7	67.7
1986	3,174.1	730.5	4.4	61.4
1985	1,526.8	692.7	2.2	43.2
1984	1,765.8	692.7	2.6	46.8
1983	1,878.8	692.7	2.7	41.9
1982	164.6	41.9	3.9	6.3
1981	93.2	43.9	2.1	3.6
1980	100.2	42.5	2.4	4.5
Total	27,056.4	8,497.0	3.2	

SOURCE: Union County Auditor, Eloise Dowell. Date are for Union County only. Honda also paid an estimated $4.1 million in property taxes to Shelby County between 1986 and 1990.

a. Property taxes Honda would have paid to Union County had tax abated property been taxed at rate of all other personal property.

b. Ratio of taxes paid to taxes abated.

c. Honda taxes as a percentage of total Union County tax collection.

d. Excludes $925,000 abatement for installation of pollution control device.

and water service for the Marysville plant. Table 20.5 presents Union County personal property tax collections for the past 10 years with Honda's portion of the total highlighted, as well as the value of the tax abatement received by Honda for the period. It is clear that the increased property tax paid to Union County exceeded the local share of the sewer and water grant in the first full year of operation for the auto plant and that by the fifth year of operation, the cost of the entire state and federal portions was recovered. However, under Ohio's tax structure, the city of Marysville was not the principal recipient of these additional tax dollars. Approximately 68% of the increased taxes went to support the Marysville city school system and most of the remainder to Union County.

The distribution of costs and benefits to local government in this case illustrates the importance of political boundaries in analyses of this nature. For example, the city of Marysville bore the direct cost of the $2 million sewer and water expansion, but Marysville city schools and Union County, both of which are political subdivisions entirely separate from the city of Marysville under Ohio law, gained almost all of the additional property tax revenue.

It is clear that Honda is important to Union County, constituting 63% of total personal property tax collections in 1990, up from a total of 4.5% in 1980. A list of the largest employers was developed for the other counties included in the analysis. For those counties for which data were available, no other county reported reliance on a single employer for such a large share of property tax collections. The closest found was 38%. As long as Honda continues to do well, Union County will continue to benefit, but this arrangement makes Union County and the Marysville schools vulnerable should Honda's fortunes be reversed in the future.

The value of property tax abatement was and continues to be the most controversial aspect of state and local economic development incentives. The Honda move into Marysville marked the first major use of tax abatement in Ohio; in fact, Ohio law was changed in 1977 specifically to provide for tax abatement in this case. The 1977 law provided for full abatement of local property taxes on the increased value of land and buildings for a period of up to 15 years. Under terms of the law, Honda would still pay property taxes on the original value of the land and on all machinery and equipment.

Table 20.5 matches the amount of taxes paid by Honda to Union County from 1980 to 1990 with the value of the tax abatement received by Honda in that same period. It shows that Honda paid a total of $27 million in property taxes and received $8.5 million in property tax abatement.

5. Indirect Costs and Benefits

Indirect Costs and Benefits to the State

A more complete accounting of the economic effects of the Honda initiative in Ohio requires an analysis of the network of domestic and Japanese suppliers spawned by the original investment. The Office for the Study of Automotive Transportation reports that at least 49 Japanese automotive suppliers are located in Ohio. Forty of the 49 were willing to name their customers and 35 listed Honda as a customer. Some 6,800 jobs were reported by the automotive part and component assembly manufacturing facilities and material and capital tool facilities that supply the automotive industry.[9]

According to the Ohio Department of Development's 1991 Loan/Grant report to the Ohio Legislature, at least 28 of Japanese auto supply companies received direct or indirect support from the state of Ohio and local governments. The report covering the 1983-1990 period indicated at slightly over $12 million had been offered to these companies. Approximately $3.5 million in Community Development Block Grants, $5.4 million from the 412 or Business Development Grant program, and $3.2 million from the Ohio Industrial Training Program have been allocated. According to the report, over 4,000 jobs have been created (at a cost to the state of about $2,500 per job), and the affected companies' investment totaled $1.9 billion. However, if pay scales are similar to Honda workers, these 4,000 employees generated $3.5 million in income tax revenues in 1990 alone.

Over the 1977-1990 period, the domestic auto manufacturers also were receiving support from the state of Ohio. Governor Richard Celeste articulated the position that state assistance for foreign firms should also be forthcoming to domestic companies. Ford received $9.8 million, General Motors received $6.8 million, and Chrysler $1.1 million for a total of $17.8 million, or $2,625 for each new job created. This figure understates the total as assistance for domestic auto suppliers could not be discerned from the Loan/Grant report. Thus it could be argued an indirect cost of the commitments made to Honda was the establishment of a precedent that required the state of Ohio in a political sense to provide similar assistance to domestic automakers when that might not have been done otherwise. In addition, it is possible that Honda's success accelerated the decline of domestic automakers who closed plants in Ohio in the 1980s. On the other hand, it is also possible that these plants would have been closed anyway.

Indirect Costs and Benefits to Local Government

Proponents of government support of economic development agree that an important component of investments of this nature is the multiplier effect of a large manufacturing plant in the form of parts supply plants and other related industries. For example, Ohio development officials use a multiplier of three secondary jobs created for every new manufacturing job brought into that state. The argument is made also that more jobs bring in more tax dollars and reduce welfare rolls. In an attempt to quantify these effects, the authors compared five counties near the Honda plant with five matched counties in other parts of the state. The period 1981-1986 was used as it was the period of greatest percentage growth with Honda employees increasing from 419 to 3,715. This period was compared with the preceding 5 years.

As Table 20.6 indicates, the growth rates for population, total standard industrial classification (SIC) employment, and retail sales in

TABLE 20.6 Economic Indicators: Union, Experimental, and Comparison Counties

	Union County		*Experimental Counties*		*Comparison Counties*	
Percentage Change in	*Before (1976-81)*	*After (1982-87)*	*Before (1976-81)*	*After (1982-87)*	*Before (1976-81)*	*After (1982-87)*
Population	9.1	6.6	4.8	3.4	5.2	1.4
Per capita income	54.0	40.1	48.5	38.6	54.1	36.0
Total employment	27.2	46.6	15.1	21.0	13.3	17.1
Retail sales	31.4	43.0	26.6	41.1	25.3	39.5

SOURCES: U.S. Department of Commerce, Bureau of the Census and Bureau of Economic Analysis; Ohio Bureau of Employment Services.

Union County were outpacing those in both the experimental and comparison counties for the 1976-1981 period. Growth in per capita income equaled that of the comparison counties and surpassed that of the experimental counties during that same period. That time interval encompasses the announcement and opening of the motorcycle plant as well as the announcement of the automobile plant.

The after-time interval includes 5 years of auto production and expansion at the Marysville plant, the plastics plant in Marysville, and the announcement of the Anna engine plant. Growth rates in Union County exceeded those in the experimental and comparison counties in all categories. It is interesting to note that the after comparisons between Union County and the experimental counties are smaller than those in the 1976-1981 period in three of the four categories. Only total SIC employment continued to grow at a more rapid rate. The comparison counties present a mirror image. The after-growth rates dominate the before rates in three of the four categories with retail sales representing the only deviation. The presence of Honda suppliers in the experimental counties could account for their growth rates relative to Union and the comparison counties.

The source of the population growth experienced by Union County during both periods under study is interesting. The Honda plant did not attract a flood of job seekers from out of state. Less than 20% of migrants into Union County came from outside Ohio borders during both periods. Migration into Union County came primarily from other Ohio counties, 82% during 1978-1982 and 81% during 1982-1985. Approximately one-quarter of the within-state migration into Union County came from Franklin County, site of the state capital Columbus, some 30 miles from the plant. Madison County, bordering Union County, contributed the next highest percentage with 11%. Less than 1% of the in-migration was from a foreign country.

Table 20.7 presents comparisons on several public-service dimensions. General expenditures in Union County exceeded those for the other two groups in the 1976-1981 period. This is reversed in the 1981-1986 period. A significant portion of the explanation for the reversal lies in the welfare expenditure variable. The rate of gain in Union County welfare expenditures in the after period was reduced greatly when compared with the before period and when compared with the experimental and comparison counties. Local government and welfare expenditures are available for only 1981-1987. General expenditures in Union County increased at a faster rate than in the other two groups of counties. Spending on police and fire increased at a much faster rate in Union County. Increases in local government expenditures on welfare in Union County trailed those in the other two groups.

6. Conclusions

With the relationship between Honda and Ohio now more than a decade old, it is possible to draw some conclusions about what that relationship has produced. This analysis suggests the following conclusions.

TABLE 20.7 Public Sector Comparisons: Union, Experimental, and Comparison Counties

Percentage Change in	*Union County*		*Experimental Counties*		*Comparison Counties*	
	Before (1976-81)	*After (1982-87)*	*Before (1976-81)*	*After (1982-87)*	*Before (1976-81)*	*After (1982-87)*
County government						
General expenditures	109.5	29.3	99.7	38.8	101.8	29.5
Welfare expenditures	143.1	46.3	215.2	51.8	131.2	74.3
Local government						
General expenditures	n.a	48.8	n.a.	45.4	n.a	31.7
Welfare expenditures	n.a.	42.5	n.a	50.8	n.a	77.5
Education						
Property valuation per pupil	58.5 (1980-84)	21.3 (1985-89)	38.2 (1980-84)	9.1 (1985-89)	34.4 (1980-84)	12.8 (1985-89)
Expenditures per pupil	40.6 (1980-84)	56.7 (1985-89)	40.4 (1980-84)	34.8 (1985-89)	46.0 (1980-84)	39.6 (1985-89)

SOURCES: Census of Governments and Ohio School Board Association.
NOTE: n.a. = not available.

Direct Effects

Direct benefits in terms of additional tax revenues to the state of Ohio and to local governments directly attributed to the Honda plants clearly exceed direct investment of tax dollars by these same governments early in the life of the project. This is true whether comparing state income tax revenue directly attributed to the Honda plant or comparing property tax revenues directly attributed to Honda, including revenues of the Marysville city schools. A cautionary signal is contained in the growing dependence on Honda as a source of revenue for local governments in Union County.

How important was tax abatement as part of the incentive package? The data indicate that the value of the abatement accounted for a small proportion of the incentive package, compared to other elements. Would Honda have moved into Marysville or Ohio without tax abatement? This question probably will never be answered with certainty. Officials at Honda and at the Ohio Department of Development, however, indicate that tax abatement served as a key indicator in the mind of Honda officials as a measure of the community's interest in establishing a relationship with Honda. The offering of the tax abatement was seen as more important than the specific amount of the abatement. Some evidence in support of this argument is the relatively small value of tax abatements in the first 10 years (about $8.5 million) compared to the value of other incentives (nearly $27 million).

Indirect Effects

Indirect effects are much more difficult to measure; however, it is clear that both indirect costs and indirect benefits are significant. Indirect benefits include higher economic growth for the entire region surrounding the Honda plant. Evidence of this includes greater growth on a wide range of indicators for Union and surrounding counties compared to the comparison counties. Additional evidence comes from the documented growth of Honda suppliers in Ohio and the migration patterns of new workers into Union County. The growth of Honda seemed to require additional public services but also seemed to have a positive effect on welfare costs compared to the comparison counties. An indirect cost may have been the loss of jobs in domestic auto plants and the cost of additional incentives to domestic auto makers. However, the evidence does not establish clearly a 3:1 ratio of jobs produced in Union County or the four surrounding counties measured in this study. This may be a result of those jobs being created in large metropolitan areas such as Columbus or Springfield (which were not included in this study) or may mean that these effects have been exaggerated. In addition, the share of personal property taxes paid by Honda to Union County has increased substantially over time. If Honda's fortunes should change, the county and school district could find themselves in a vulnerable position.

Generalizability

Although the state of Ohio's relationship with Honda has clear direct benefits in excess of direct costs and indirect benefits in excess of indirect costs, it is not at all clear that these conclusions are portable to other economic incentive packages in Ohio and elsewhere for at least two reasons.

First, the long-term nature of the relationship between Honda and the state of Ohio, and Honda's subsequent additions to the original investment, have resulted in a cost per job that is much lower than the original investment would have portended. The figure is lower than subsequent inducements offered by other states in their pursuit of economic development. Extreme caution, however, needs to be taken in judging the cost-per-job data that accompany any assessment of economic development initiative. The Honda cost per job varies by a factor of three depending on what point in time one chooses to examine. Adjusting for inflation presents another analytical issue. For example, if the cost of the incentives for the motorcycle plant is adjusted to 1989 dollars using the gross national product deflator and to include the cost of highway improvements, the cost becomes $66,795, clearly the most expensive of the auto and motorcycle plant investments. On the other hand, if Ohio's total investment in Honda limited to direct assistance (excluding highway work) is computed in nominal dollars and reflects actual rather than projected employment, the cost is less than $2,500 per job, a bargain in today's environment.

What this means is that public officials and others involved in judging economic value of various incentive proposals need to be very careful to understand the assumptions on jobs created, timing, the time value of money, and the implications of the long-term relationship between the business and the community before drawing conclusions about the relative merit of various investments.

Second, employment and expansion of Honda in Ohio exceeded initial projections in large part because of the success of the Honda Accord, success which cannot be easily be replicated. One need not look far to find instances of products that did not meet with success in the market despite a significant investment by the company and the state (i.e., the VW Rabbit plant in New Stanton, Pennsylvania).

The relationship between Honda and Ohio has entered its second decade. A review of that association reveals the importance of time in a variety of ways. The point in time when one chooses to calculate the cost per job of the incentives offered, the particular configuration of the auto market and the success enjoyed by the Honda Accord in the 1980s, and the unfolding of both benefits and costs over time, mandate that a time series approach be taken in the evaluation of economic development initiatives. A one-shot approach will give a distorted picture of the actual state of affairs. The shape and character of the Honda-Ohio relationship will no doubt continue to evolve with the inevitable changes in market forces and economic and political conditions.

Also, further research is needed to assess Honda's impact on a variety of quality-of-life measures. For example, the data on residential property valuations for Union County represent a puzzle. Between 1980 and 1990, residential taxable property values increased by 63% ranking it eighth out of the 10 counties included in the study in terms of growth in residential property valuation. Union County ranked sixth in absolute value of residential taxable property values in 1990. These are surprisingly modest gains in light of the population growth experienced by Union County.

Concerns regarding education issues become salient when tax abatement enters the development equation and needs to be explored in more depth. Property valuation per pupil increased at a faster rate in Union County than in the other two groups of counties from 1980 to 1984. For the 1984-1989 period the rate of gain abated in Union County but continued to grow faster than in the experimental and comparison counties. Expenditures per pupil, increasing at about the same or lower rate in Union County when compared with the other two groups for the earlier period, increased at a much higher rate in the later period. The impact of these expenditures on the quality of the educational product needs to be considered.

In addition, impact on crime, congestion, environment, and charitable contributions requires additional study to round out a more comprehensive assessment of Honda in Ohio.

Notes

1. Council of State Governments, *The States and Business Incentives: An Inventory of Tax and Financial Incentive Programs,* 1989, p. 15.

2. Council of State Governments, *The Changing Arena: State Strategic Economic Development,* 1989, p. 11.

3. Council of State Governments, *The Changing Arena,* p. 15.

4. H. Brinton Milward and Heidi Hosbach Newman, "State Incentive Packages and the Industrial Location Decision," *Economic Development Quarterly* 39 (1986): 341-48.

5. William F. Fox, "Japanese Investment in Tennessee: The Economic Effects of Nissan's Location in Smyrna," in *The Politics of Industrial Recruitment* (New York: Greenwood, 1990), p. 186.

6. Lynn W. Bachelor, "Michigan, Mazda, and the Factory of the Future: Evaluating Economic Development Incentives," *Economic Development Quarterly* 5 (1991): 123.

7. Nancy S. Lind and Ann H. Elder, "Who Pays? Who Benefits? The Case of the Incentive Package Offered to Diamond-Star Automotive Plant," *Government Finance Review* 2 (December 1986): 23.

8. H. Brinton Milward et al., *The Estimated Economic Impact of Toyota on the State's Economy* (Lexington: University of Kentucky, 1986), p. 4. Report submitted to the Center for Business and Economic Research, December.

9. Office for the Study of Automotive Transportation, *Japanese Automotive Supplier Investment Directory.* 3d ed. (Ann Arbor: University of Michigan, Transportation Research Institute, 1990).

Customization and Macroeconomic Efficiency

John P. Blair and Laura A. Reese

It is apparent from the recent literature on economic development (ED) that the "one size fits all" approach to ED is now obsolete. Planners and ED officials should do more to tailor plans to fit the needs and resources of particular areas. Recent literature, including the articles in this book, have enhanced the ability of local officials to adapt strategies to local environments. The increased capacity to customize ED efforts has improved the efficiency of economic practice locally and helped strengthen the national economy.

Historically, communities have frequently pursued similar strategies at similar times as certain ideas became faddish. The pursuit of similar routes to ED has led to excess capacity and waste in some circumstances. For example, in the mid-1800s, cities excessively subsidized rail lines. As a result, some lines connected so few places as to be inefficient. The idea of encouraging development by improving access to materials and markets was valid. The waste of resources resulted from copying one community's ED activity without considering how the valid theoretical idea should be modified in light of local resources and other characteristics. During the 1960s, many cities spent ED resources on tourism, leading to excess capacity in that activity. The remnants of the tourist development phase can be seen in the numerous small museums (closed most of the year) in many areas. Today, sports-based ED strategies have resulted in too many major league ball parks compared to the number of teams. In each example, the essential reason for the excess capacity was that communities were replicating ED approaches that were successful in some places without adequately considering whether the plan would work in their particular community.

Today, popularity swings still characterize ED strategies but with a difference—increasingly sophisticated analysis is devoted to customizing strategies to reflect the local environment. Industrial parks, incubators, and enterprise zones have been widely implemented throughout the United States. However, each of these initiatives can be modified to fit local circumstances. In some cases, the alterations fit the local situations well; in other cases, the tailoring amounted to rolling up the cuffs. The need to modify existing ideas so as to match local conditions is a challenge to ED officials. The improved capacity to customize ED approaches is one of the key benefits from the increased knowledge regarding ED practice and from the closer relationship between theory and practice that characterize articles in this volume.

Recent literature on local ED suggests several related areas that must be considered when adapting a development idea for a local community. First, differences in the external environment should be assessed. A development program that works during a period of rising prosperity may fail at another time. As noted earlier, after an idea has been replicated numerous times, the market may become saturated, and thus success will be more difficult. Along with the external environment, important local considerations are resources, political circumstances, values, and needs.

Areas differ greatly in their resource base. It is obvious that a community with a very cold climate cannot pursue a winter tourist strategy in the same way that a southern coastal region can. Resources is a very broad term. Included in a community's resources are the skills of the workforce, the social infrastructure (roads, airports, universities, and so forth), other types of business, and reputation as well as the more commonly discussed natural resources. Often, a plan that worked elsewhere may fail when transplanted because the full range of resources needed for success is not recognized. For example, consider a community that seeks to develop by encouraging growth of advanced technology activities. It may have many of the resources that other advanced technology communities have, including good universities, an initial cluster of advanced technology enterprises, and a quality of life that will attract high-tech workers. However, an airport with frequent direct service to major cities may also be a critical, but unrecognized, resource for the success of such a strategy. Lacking that resource, the community's advanced technology strategy may have a much smaller chance of success.

The political atmosphere also varies among communities and should be a consideration in policy design. The state places limits on the ability of local governments to take certain ED actions, and these differences should be acknowledged. A good ED official should also be able to understand the limits placed on potential ED actions by the local political culture. Some community leaders are not risk takers, for example. Political influence is also an important resource. Some communities can get public assistance and resources based on their state or national political clout.

Value differences represent another reason why applying an ED template without careful modifications to account for the region may be inefficient. Concern about the relationship between local values and ED is more evident today than previously. Some communities may value job growth and low taxes, whereas others may value slower job growth and more public amenities, which usually translate into higher taxes. The value differences indicate the need for alternative development strategies. Particularly in a metropolitan environment where families have substantial mobility, individuals with similar tastes tend to live in the same neighborhoods. For instance, families of equal income that value good schools compared to other things will cluster in certain areas whereas families that place a relatively higher value on open spaces choose to locate in other areas.

Enhancing quality of life is emerging as an important ED strategy. Defining a good quality of life is inherently a value-laden task, and cities are likely to reflect differences regarding which amenities are most valued relative to their costs. In determining how to improve the quality of life, planners have had to confront the values issues, develop techniques for aggregating preferences, and reconcile value conflicts between groups. In some instances, value conflicts and inconsistences may emerge within a single group.

Needs are closely related to values, for values define needs. Some observers might consider needs a subset of values. However, because communities with roughly the same values may have vastly different needs, it should stand alone as a reason for differentiation of development strategies. A community with high unemployment or low per-capita income should have a different ED strategy from that used in a community that is subject to high average incomes but has serious pockets of distress.

What are the implications of the tendency toward increased customization for ED research and practice? Almost every application of an existing idea or academic study to a local context will require more than superficial modifications for optimal results. Decision makers who understand the theory of ED processes and the environment in which ED occurs will be more effective planners. Customizing such plans will require that practitioners understand development theory, think creatively, and account for and weigh the aims of important political stakeholders. For instance, two cities may both decide that neighborhood revitalization is an appropriate strategy. Because of differences in the external environment, resources, politics, values, and needs between the two, in one city that goal may best be addressed by attracting middle-class residents, whereas in the other city, local job training in the construction trades may be a preferable solution. Almost by definition, developing customized plans requires innovative analysis.

The argument for customization is not intended to suggest that economic developers should ignore or discount what other places are doing. A laboratory function has been considered one of the advantages of decentralized government. As usually articulated, this laboratory role favors allowing local governments to try new approaches. The ideas that work are then adopted by other places. By this process, the theory suggests, local governments will weed out bad ideas, and the workable ideas will spread. With regard to ED, excessive and uncritical copying will be inefficient for the national economy and probably for the locality doing the copying. Thus, at least when applied to ED, the importance of customization should be an amendment to the popular laboratory concept.

The increasing importance of customization has blurred the distinction between research and practice in local ED. Increasingly, practice requires characteristics traditionally associated with academic research, such as originality and a grounding in established theory. The word "practice" has had the connotation of replicating what had been done elsewhere. As argued above, pure replication is likely to be inefficient. ED officials cannot simply copy what has been done or suggested elsewhere. Applied researchers attempting to make ideas fit local circumstances frequently encounter problems not anticipated in the academic literature. Original analysis is required to overcome such obstacles, and in the process, the knowledge base will expand. Furthermore, applied research should not necessarily play second fiddle to scholarly articles. The distinguished urban geographer Harold Rose once said that if academic researchers make a mistake in an article their colleagues will likely catch the mistake in the review process or when the manuscript is published. However, if a practitioner makes a mistake, it may hurt thousands of people.

When cities pursue very similar plans, they frequently end up competing against one another. Many persons view ED as a zero-sum game for the national economy. One city gains jobs or economic activity at the expense of another. For instance, when cities that are near each other construct nearly indistinguishable industrial parks, the decision of a firm to locate a facility in one area will likely represent a lost opportunity to another city. The same phenomenon can be seen in the zero-sum efforts to attract major league sports teams. However, as customization of plans improves and becomes more widespread, ED efforts can be seen as helping to make cities more efficient production and consumption "machines." Hence, ED efforts can add value to the economy as a whole.

There is reason to believe that local ED efforts make a substantial contribution to the national economy. With the exception of a very brief downturn in the early 1990s, the U.S. economy has experienced a relatively unique, extended period of expansion. Not coincidently, the devolution of ED responsibilities to state and local governments and the rapid expansion in the level of professional ED education occurred about the same time that the era of expansion began.

Undoubtedly, there are multiple causes of the current expansion, so local economic efforts are not the only engines of prosperity. Traditional economists often attribute the decrease in the unemployment rate and the simultaneous reduction in the rate of inflation to structural changes in the economy. Exactly what these structural changes are is often not specified. However, consider some of the positive ways that local ED has contributed to favorable structural change. Local ED subsidies have helped many businesses operate that otherwise could not. Various incentives may have tilted businesses to profitability. As a result, there is less unemployment and more investment. Local programs provided labor-force training that matched area business employment needs. Consequently, the resource base was enhanced, and structural unemployment was mitigated. Local programs also helped stimulate agglomeration economies by encouraging businesses with linkages to locate near one another, thus increasing productivity. Other local development programs have encouraged businesses that are already near one another to develop beneficial partnerships, which increased productivity. Some businesses may have been encouraged to locate in regions that are more efficient due to information and incentives offered under local programs. Local ED efforts have recently discouraged wasteful intergovernmental competition, furthering efficiency. Thus, a very plausible case can be made that local efforts are part of the structural changes that have stimulated the national economy.

As a result of the boost that local ED efforts can give the national economy, they should be considered an important tool for macroeconomic policy, along with fiscal, monetary, supply-side, and trade policies. Adding to their importance, the globalization of the economy is limiting the use of monetary and fiscal policy.

However, ED practice has yet to receive adequate recognition of its importance to the national economy. Some observers have noted that the success of the national economy is largely a function of what is occurring in cities. Yet this success has not been attributed to local ED practice. The lack of recognition may be due in part to the interdisciplinary nature of ED practice. The more traditional macroeconomic tools are generally analyzed by professional economists, so economists more readily recognize and understand the contributions in these areas. Local ED is at best a subspeciality within economics, not at the core of the discipline. Furthermore, traditional economists might not recognize what is being accomplished because the practice of economics at the local level has shifted to planners, political scientists, marketing experts, and individuals trained in other fields. Understandable turf issues may lead some economists to undervalue contributions from allied disciplines.

Recognition of the importance of local decisions is not to argue that the practice of local ED is problem free. Inefficiencies in practice exist, and as they are further reduced, both local economies and the national economy will be enhanced. Although local development efforts are probably not as strong as monetary and fiscal policies, any analysis of the reasons for the current long-lasting expansion would be incomplete without considerations of the salutary effects of ED practice that have been fueled by customization.

Along with illustrating much about the positive nexus between theory and practice, the articles in this volume also say much about the limitations inherent in both the research on and practice of ED at the local level. Local communities still practice many ineffective techniques either because the academic literature has not provided sufficient or sufficiently useful evaluations or because academics have not suggested how effective policies might operate within the real political, economic, and institutional constraints that local decision makers face. Many local development policies have been shown to be important to businesses only at the margins, to be in excess of what businesses actually need, and to favor business interests over community needs. Similarly, the goals driving, and the evaluations of, local policies pay insufficient attention to policy effects beyond jobs created and tax-base enhancement to consider distribution issues, negative externalities, benefits accruing to the community and to the poor in the community more specifically, and ultimately, broader quality-of-life issues.

In short, the articles in this volume reflect both the strengths and weaknesses of current policy practice and our knowledge base about such practice. Many important questions remain unanswered or insufficiently answered, and the articles here point clearly to many of these questions. For example,

- What are the limitations of local markets in stimulating and sustaining ED?
- Are inner city markets really sufficiently robust to sustain ED?

- What is the appropriate role of the various levels of government in addressing market failures?
- Which are more effective, targeted or universal ED policies? What might be an effective mix?
- What should be the proper emphasis for governmental ED programs at all levels—people or place?
- Should job programs emphasize training of local residents and then job creation or enhanced mobility of said residents?
- How might local strategies be improved to ensure local benefits and that any resulting development is actually commensurate with resources spent?
- Which are more effective, traditional supply-side programs, which are still widely practiced, or demand-side efforts to build local capacities, which are still very little evaluated?
- To what extent is development local, regional, or national? How can ED efforts be designed to address neighborhood needs while taking into account the regional and national framework in which they are embedded?
- How might the realities of interlocal competition be addressed in a manner that enhances cooperation? How can local officials be encouraged to cooperate within a political atmosphere that encourages fragmentation and attendant competition?
- How can researchers better understand local political realities in an effort to show practitioners how to work within those realities to select and implement more effective development policies?
- What *are* the more effective local development politics, and how might our notions of "effectiveness" be broadened to include all stakeholders? Just how much should we reasonably expect of our public expenditures to foster economic development?

These questions and many more are addressed in some manner by the articles in this volume. However, our knowledge, and hence ability to come to concrete answers, remains limited. Of course, most of these questions do not lend themselves to universal generalizations. The answers will vary, depending on local circumstances. In providing a sense of the "state of knowledge" in the ED field to date, the work here says as much about what we know as what we don't know. The challenge for the future efforts of both academics and practitioners is to move on and jump into the breach.

Index

About the Contributors

❋❋❋

Timothy J. Bartik is Senior Economist at the W. E. Upjohn Institute for Employment Research, where he is responsible for conducting research on state and local economic development and local labor markets. Recent publications include "Can Economic Development Programs Be Evaluated?" (with Richard Bingham) in *Significant Issues in Urban Economic Development* and *Who Benefits from State and Local Economic Development Policies?* He is currently working on a book titled *Jobs for the Poor: Can Labor Demand Policies Help?*

John J. Betancur is Assistant Professor of Urban Planning in the College of Urban Planning and Policy at the University of Illinois at Chicago. Previously, he was Professor of Sociology and Philosophy at Columbia University. His interests include urban labor markets, corporate structures, and Latin American studies.

Thomas Bier has been Director of the Housing Policy Research Program at the Maxine Goodman Levin College of Urban Affairs at Cleveland State University since 1982. His research has focused on regional housing dynamics, population movement and the effects of government policies on cities. His study of home seller movement and the capital gain provision led to its change in 1997. He holds a Ph.D. from Case Western Reserve University.

John P. Blair is Professor of Economics at Wright State University. Prior to joining Wright State, he taught at the University of Wisconsin and served on the Urban Policy Staff at the Department of Housing and Urban Development. He has published widely in the field of economic development in such journals as *Economic Development Quarterly, Growth and Change, Urban Affairs Review,* and *Review of Regional Studies.* His latest book is *Local Economic Development: Analysis and Practice* (1995). In addition to teaching, he has served as a consultant to governments and private business on economic development issues. He is a Distinguished Alumni Scholar from West Virginia University, Regional Research Institute and received the Bloomberg Award for Future Studies. He may be reached by e-mail (JBlair@Wright.edu).

Bridget Brown received her master's degree in urban planning and policy from the University of Illinois at Chicago.

Susan E. Clarke is Professor of Political Science at the University of Colorado at Boulder. She has published widely on topics of intergovernmental relationships and urban economics and is coeditor (with Ed Goetz) of *The New Localism* (Sage, 1993) and coauthor (with Gary L. Gaile) of *The Work of Cities.*

Peter B. Doeringer is Professor of Economics at Boston University and has taught at Harvard University, the London School of Economics, and the University of Paris. He publishes widely on topics related to labor markets, regional development, and the economics of industry. He is also Director of the European Program at the Harvard University Center for Textile and Apparel Research where he is completing a book comparing industry performance in the clothing industries of France, the United Kingdom, and the United States.

William J. Drummond is Associate Professor at Georgia Institute of Technology Graduate Program and Associate Director of the Georgia Tech Center for Geographic Information Systems. He teaches in the areas of computer methods and geographic information systems and conducts research in the economic development applications of geographic information systems. He is a graduate of Duke University, Gordon-Conwell Theological Seminary, Union Theological Seminary, and the University of North Carolina at Chapel Hill.

Peter Eisinger is Professor of Urban Affairs and Director of the State Policy Center in the College of Urban, Labor and Metropolitan Affairs at Wayne State University. Prior to this appointment, he taught at the University of Wisconsin for 28 years, where he was Director of the La Follette Institute of Public Affairs from 1991 to 1996. He is author of *The Rise of the Entrepreneurial State* and *Toward an End to Hunger in the United States,* among other books.

Carole R. Endres is Lecturer in economics at Wright State University. Prior to joining Wright State, she was Compensation and Benefits Manager for the city of Dayton, Ohio and an Urban Fellow with the University of Dayton. Her current interests include globalization and alternative economic systems.

David Fasenfest is Associate Professor in the College of Urban, Labor and Metropolitan Affairs, and Director of the Center for Urban Studies, Wayne State University. An economist and sociologist, his research focuses on the nature of economic transformations as they impact on local economic development, the formation and implementation of local development policies, the structure of income inequality, and the impact of recent regulatory changes on the social economy. He is editor of *Community Economic Development: Policy Formation in the U.S. and U.K.*, and has been published in *Economic Development Quarterly, Policy Studies Journal,* and *International Journal of Urban and Regional Research.*

Gary L. Gaile is Professor of Geography at the University of Colorado at Boulder. He has published numerous articles on food security and development in Africa, is coeditor of *Geography in America,* and is coauthor (with Susan E. Clarke) of *The Work of Cities.* His research interests include urban and regional planning, Third World development, spatial statistics, and market-based planning approaches.

Edward G. Goetz is Associate Professor in the Housing Program at the University of Minnesota. He has written on local economic development and housing policy and is author of *Shelter Burden: Progressive Politics and Local Housing Policy.* He received his Ph.D. in political science from Northwestern University.

Edward W. Hill is Senior Research Scholar in the Urban Center and Professor of Urban Studies and Public Administration at Cleveland State University's Maxine Goodman Levin College of Urban Affairs. He has published widely in the field of economic development and urban change and currently is editor of *Economic Development Quarterly.*

Keith R. Ihlanfeldt is Professor of Economics at and Senior Associate of the Policy Research Center at Georgia State University. His recent research has focused on the minority hiring practices of employers, state economic development and tax incentives, and the relationship between transportation infrastructure and employment densification. His book on the relationship between job suburbanization and minority youth unemployment, *Job Accessibility and the Employment and School Enrollment of Teenagers,* was published in 1993.

John D. Kasarda is Kenan Professor of Business Administration and Director of the Frank Hawkins Kenan Institute of Private Enterprise at the University of North Carolina at Chapel Hill. He has produced more than 60 scholarly articles and 9 books on urban development, demographics, and employment issues, has served as a consultant on national urban policy to the Carter, Reagan, Bush, and Clinton administrations and has testified before U.S. congressional committees on urban policy and employment issues. He received his MBA (with Distinction) from Cornell University and his Ph.D. from the University of North Carolina. He has been the recipient of many grants and awards from such organizations as the National Academy of Sciences, U.S. Department of State, and Agency for International Development (AID), has been designated as a Fellow of the American Association for the Advancement of Science, and is a Senior Fellow of the Urban Land Institute.

Terrence Kayser is employed by Metropolitan Council of the Twin Cities, a unique governmental body that provides services for the metropolitan area. He works on policy and economic issues relating to orderly growth of the region and has authored studies on affordable housing issues in the Twin Cities. He received his Ph.D. from the University of Minnesota in 1983.

Larry C. Ledebur is Director of the Urban Center, Levin College of Urban Affairs at Cleveland State University. He previously served as Director of the Economic Development Program and Principal Research Associate at the Urban Institute and as Senior Economist and Associate Director of Research for the White House Conference on Balanced National Growth and Economic Development. His books include *The New Regional Economies: The U.S. Common Market and the Global Economy, Economic Disparities: Problems and Strategy for Black America, Urban Economics: Processes and Problems, Revitalizing the American Economy* (editor), and *Industrial Incentives: Public Promotion of Private Enterprise.*

Mary K. Marvel is Associate Professor in the School of Public Policy and Management and Political Science at Ohio State University. Her research interests include the evaluation of the implementation and impacts of public policies.

Marya Morris is Senior Research Associate at the American Planning Association in Chicago. She has a B.A. in economics from the University of Wisconsin and a master's degree in urban planning and policy from the University of Illinois at Chicago. Her research interests are state reform of planning statutes, urban design, and economic development. She is editor of APA's Planning Advisory Service memo.

Arthur C. Nelson is Professor of City Planning and Public Policy at Georgia Institute of Technology, where he holds appointments in the Colleges of Architecture, Management, and Engineering. He has earned teacher of the year and professional educator of the year honors, and his students have earned national awards. For the past 20 years, he has conducted work in development patterns, economic development, growth management, and public facility finance. His work has been supported by such organizations as the National Science Foundation, National Academy of Sciences, U.S. Departments of Housing and Urban Development, Commerce, and Transportation, the Federal National Mortgage Association, and the Economic Development Administration. He is author of six books and more than 100 scholarly and professional publications. His Ph.D. in regional science and planning is from Portland State University.

Jeremy Nowak is Founding Director of the Delaware Valley Community Reinvestment Fund, the mission of which—poverty alleviation and the revitalization of low- and moderate-income neighborhoods—is met through the provision of capital and technical assistance to community organizations and businesses. He has been Lecturer in community and economic development for the Urban Studies Department at the University of Pennsylvania and authored many articles and papers on neighborhood issues, community development, and the role of religious institutions in urban revitalization. He holds a Ph.D. from the New School for Social Research, and in 1997 Villanova University awarded him an honorary doctorate in recognition of his achievements.

Michael E. Porter is C. Roland Christensen Professor of Business Administration at Harvard Business School, Founder, Chairman, and CEO of the Initiative for a Competitive Inner City, and a leading authority on competitive strategy and international competitiveness. He is author of 14 books, among them *The Competitive Advantage of Nations,* and over 50 articles and has served as a counselor on competitive strategy to many leading U.S. and international companies and governments. He has applied his expertise in competitive strategy to the inner city and published his first article on the topic, "The Competitive Advantage of the Inner City," in the May/June 1995 issue of *Harvard Business Review.*

David C. Ranney is Associate Professor in the College of Urban Planning and Public Affairs at the University of Illinois at Chicago and previously held faculty positions at Southern Illinois University, University of Wisconsin, and University of Iowa. Prior to coming to UIC, he spent five years as a partner in a labor/community service organization on Chicago's Southeast side. He has worked extensively with community and labor organizations in the Chicago area that are concerned with job creation, retention, and assistance to dislocated workers and has wide experience in planning practice as well as academic teaching and research. He has worked in Calcutta, India on the Ford Foundation-sponsored Calcutta Plan and has also served as consultant to the City of Cleveland. He is author of three books and numerous articles and monographs on issues of employment, labor and community organization, city planning, and politics.

Laura A. Reese, a political scientist, is Professor of Urban Planning in the College of Urban and Metropolitan Affairs at Wayne State University. She has published articles on urban politics, local economic development, and public personnel management in *Urban Affairs Review, Economic Development Quarterly, Journal of Politics,* and *Review of Public Personnel Administration.* She is author of *Local Economic Development Policy: The U.S. and Canada* and and most recently coauthor (with Karen E. Lindenberg) of *Implementing Sexual Harassment Policy* (1998).

Herbert J. Rubin is Professor of Sociology at Northern Illinois University. Most of his work is on critical aspects of economic and community development. He has published articles on these topics in *Administration and Society, Public Administration Review, Social Problems,* and *Economic Development Quarterly,* among other journals, and is coauthor (with Irene Rubin) of *Community Organizing and Development* (now being prepared for a third edition) and *Qualitative Interviewing: The Art of Hearing Data.* In a forthcoming book, *Renewing Hope: The Community-Based Develop-*

ment Model, he examines how ideologies and interorganizational relations influence community renewal.

David S. Sawicki is Professor of City Planning and Public Policy at Georgia Institute of Technology and Senior Advisor for Data and Policy Analysis at the Carter Center's Atlanta Project. His current research focuses on the use of data in state and local policy settings, especially local labor markets. Recent articles include "Déjà Vu All Over Again: Porter's Model of Inner-City Redevelopment" (with Mitch Moody) in *The Inner City: Urban Poverty and Economic Development in the Next Century,* "The Democratization of Data: Bridging the Gap for Community Groups" (with Will Craig) and "Neighborhood Indicators; A Review of the Literature and an Assessment of Conceptual and Methodological Issues" (with Patrick Flynn), both in *Journal of the American Planning Association.* Now in its second edition, his *Basic Methods of Policy Analysis and Planning* (with Dr. Carl V. Patton) is the best-selling textbook in public policy.

William J. Shkurti is Vice President of Finance at Ohio State University and Adjunct Professor in the School of Public Policy and Management at OSU. He served as Budget Director for the state of Ohio from 1984 through 1987. He holds a master's degree in public administration and a bachelor's degree in economics from OSU.

David Spitzley received his MPA from Wayne State University in 1996. He is currently completing an internship with the state of Michigan. His research interests are the macroeconomic impacts of income and wealth inequalities and development policies to promote democratized capital ownership.

David G. Terkla is Professor and Chair of the Economics Department at the University of Massachusetts, Boston and is also a faculty member of the Environmental Coastal and Ocean Sciences Program. He has written a book and several articles on the importance of nontraditional cost factors to local economic development. He has been involved in several projects related to environmental management and economic development issues, including the economics of coastal dredging and implications of ocean disposal, valuation of uses of resources in Massachusetts and Cape Cod Bay, analysis of protection policies for water-dependent uses on urban waterfronts, analysis of potential conflicts between tourism and fishing industries in Gloucester, Massachusetts, and economic development and land-use planning in the Buzzards Bay area. In addition to his many writings on the New England fishing industry, he has written on the importance of industry clusters and the location decisions of new Japanese plants in the United States.

Wilbur R. Thompson is Professor Emeritus at Wayne State University and was Albert A. Levin Scholar in Urban Studies at Cleveland State University. He is author of *A Preface to Urban Economics,* one of the foundation books in the field, as well as numerous other books and articles. He received his Ph.D. in economics from the University of Michigan.

Wim Wiewel, Professor of Urban Planning and Policy, is Dean of the College of Urban Planning and Public Affairs at the University of Illinois at Chicago. He was appointed to this post in September 1996. He is coeditor of *Harold Washington and the Neighborhoods: Progressive City Government in Chicago, 1983-1987* and *Challenging Uneven Development: An Urban Agenda for the 1990s.* His articles have appeared in *Economic Development Quarterly, Economic Geography, Journal of the American Planning Association, Administrative Quarterly,* and other professional, academic, and popular journals. He holds degrees in sociology and urban planning from the University of Amsterdam in the Netherlands and a Ph.D. in sociology from Northwestern University.

Harold Wolman is Director of the Policy Sciences Graduate Program at the University of Maryland, Baltimore County. Recent publications include *National Urban Policy* and *The President's National Urban Policy Report* and "Changes in Central City Representation and Influence in Congress Since the 1940's" in *Urban Affairs Review.* He holds a Ph.D. in political science from the University of Michigan and a master's degree in urban planning from Massachusetts Institute of Technology.

Douglas P. Woodward is Director of Division of Research and Associate Professor of Economics at Darla Moore School of Business at the University of South Carolina. His primary research interests are foreign direct investments and regional economic development. He has published articles on these topics and has testified before the U.S. Congress on many occasions. He is coauthor of a book on foreign direct investment in the United States, *The New Competitors,* ranked as one of the top 10 business and economics books of 1989 by *Business Week.*